T0382975

Introduction to LiDAR Remote Sensing

Light detection and ranging, or LiDAR, is an advanced active remote sensing technology developed in the last 30 years to measure variable distances to the Earth. This book explains the fundamental concepts of LiDAR technology and its extended spaceborne, airborne, terrestrial, mobile, and unmanned aerial vehicle (UAV) platforms. It addresses the challenges of massive LiDAR data intelligent processing, LiDAR software engineering, and in-depth applications. The theory and algorithms are integrated with multiple applications in a systematic way and with step-by-step instructions. Written for undergraduate and graduate students and practitioners in the field of LiDAR remote sensing, this book is a much-needed comprehensive resource.

FEATURES

- Explains the fundamentals of LiDAR remote sensing, including theory, techniques, methods, and applications
- Highlights the dissemination and popularization of LiDAR remote sensing technology in the last decade
- Includes new advances in LiDAR data processing and applications
- Introduces new technologies such as spaceborne LiDAR and photon-counting LiDAR
- Provides multiple LiDAR application cases regarding topography mapping, forest investigation, power line inspection, building modeling, automatic driving, crop monitoring, indoor navigation, cultural heritage conservation, and underwater mapping

This book is written for graduate and upper-level undergraduate students taking courses in remote sensing, geography, photogrammetric engineering, laser techniques, surveying and mapping, geographic information system (GIS), forestry, and resources and environmental protection. It is also a comprehensive resource for researchers and scientists interested in learning techniques for collecting LiDAR remote sensing data and processing, analyzing, and managing LiDAR data for applications in forestry, surveying and mapping, cultural relic protection, and digital products.

Introduction to LiDAR Remote Sensing

Cheng Wang, Xuebo Yang, Xiaohuan Xi,
Sheng Nie, and Pinliang Dong

Designed cover image: © Cheng Wang

First edition published 2024
by CRC Press
2385 NW Executive Center Drive, Suite 320, Boca Raton FL 33431

and by CRC Press
4 Park Square, Milton Park, Abingdon, Oxon, OX14 4RN

CRC Press is an imprint of Taylor & Francis Group, LLC

ISBN: 978-1-032-67150-5 (hbk)
ISBN: 978-1-032-67152-9 (pbk)
ISBN: 978-1-032-67151-2 (ebk)

DOI: 10.1201/9781032671512

Typeset in Times
by KnowledgeWorks Global Ltd.

Contents

Foreword

There is a consensus in the science and technology community on advancing Earth-related big data in support of sustainable development goals. Integrated analysis of the land, ocean, and atmosphere systems and human activities can provide multi-scale and multi-temporal information on Earth's environment and resources. LiDAR has become an indispensable tool for data acquisition due to its rapid, direct, and high-precision capabilities in acquiring three-dimensional geospatial data, and has become an important component of Earth-related big data. In the last decade, LiDAR systems have been launched into space, while lightweight airborne and mobile LiDAR systems have experienced rapid development, leading to explosive growth of three-dimensional geospatial data and extensive applications in topographic mapping, carbon sequestration in terrestrial ecosystems, digital city construction, transmission line corridor inspection, cultural heritage protection, and automatic driving, among many others. However, compared with imaging spectrometry and imaging radar technologies, LiDAR remote sensing still has many challenging issues that need to be addressed, such as mechanisms of laser-surface interaction, data processing methods, and quantitative applications. A major objective of LiDAR remote sensing is to utilize massive, multi-source (multi-platform, discrete return, waveform, and photon-counting) LiDAR data to support sustainable development goals.

Professor Cheng Wang's team at the Aerospace Information Research Institute, Chinese Academy of Sciences (CAS), has been conducting research on LiDAR remote sensing for over 20 years. Over the past two decades, they have developed models for quantitative analysis of LiDAR data, constructed workflows for information extraction from multi-source LiDAR data, solved many problems in LiDAR practical applications, and developed free Point Cloud Magic (PCM) software for LiDAR data processing and analysis that has been downloaded by tens of thousands of users worldwide. Their original ideas have been published in major international journals, applied to real-world problems, and awarded science and technology prizes at the provincial and ministry levels.

The book *Introduction to LiDAR Remote Sensing* represents the research results of Professor Cheng Wang's team and the course contents of "LiDAR Remote Sensing" he taught at the University of Chinese Academy of Sciences. It can be used by research professionals and undergraduate/graduate students. Based on real datasets, the book introduces the physical principles of LiDAR remote sensing, multi-platform and multi-mode LiDAR data acquisition and processing, typical applications of LiDAR remote sensing, and trends and advances of LiDAR remote sensing. I highly recommend this comprehensive book to readers and believe that the book can help readers better understand LiDAR remote sensing and its applications.

Huadong Guo
Academician, Chinese Academy of Sciences (CAS)
Fellow, The World Academy of Sciences (TWAS)
Fellow, International Science Council (ISC)
October 31, 2023

Preface

I came to know LiDAR when I started my doctoral research in France in March 2002, with the topic of vegetation mapping in the Venice wetlands using airborne LiDAR and hyperspectral data. Later I conducted post-doctoral research on extracting forest structural parameters using LiDAR data in the United States. In late 2009, I joined the Center for Earth Observation and Digital Earth (CEODE), Chinese Academy of Sciences, and built a research team on LiDAR remote sensing, focusing on extracting surface parameters from LiDAR data and quantitative applications of LiDAR data. Since then, I have supervised over 50 graduate students and delivered presentations on LiDAR remote sensing at universities, research institutions, and industries. Since 2015, I have been teaching LiDAR remote sensing at the University of Chinese Academy of Sciences (formerly the Graduate School of Chinese Academy of Sciences). My colleagues and students often ask me for recommendations of a textbook on LiDAR remote sensing, and I would tell them that the choices are limited.

LiDAR remote sensing is an advanced Earth observation technology that has been extensively used in many fields. An increasing number of universities have introduced LiDAR remote sensing or laser scanning into their curriculum, especially for majors such as surveying engineering, remote sensing science and technology, and geographic information science. From my conversations with instructors of LiDAR remote sensing, it appears that many of them design their own instructional materials with limited access to LiDAR data and software and that a book on LiDAR remote sensing is needed for a comprehensive understanding of LiDAR remote sensing principles, data processing methods, and industry applications. With the support of many colleagues, especially my colleagues in the LiDAR sub-committee of the Chinese National Committee of International Society for Digital Earth (CNISDE-LiDAR), I arranged several co-authors to write this book: *Introduction to LiDAR Remote Sensing*. Most of the LiDAR datasets, algorithms, and case studies in this book are from our research results, including the free software Point Cloud Magic (PCM) for LiDAR data processing and analysis.

I would like to thank my co-authors at the Aerospace Information Research Institute, Chinese Academy of Sciences, Dr. Xuebo Yang (Assistant Research Fellow), Ms. Xiaohuan Xi (Associate Professor), Dr. Sheng Nie (Associate Professor), and Dr. Pinliang Dong (Professor, University of North Texas, USA). I also thank the following graduate students who participated in the research projects: Dong Li, Xiaoxiao Zhu, Di Cao, Shuwen Peng, Jieying Lao, Pu Wang, Lei Liang, Meng Du, Yantian Wang, Jingru Wang, Yiya Qiao, Shichao Chen, Baokun Feng, Yicheng Liu, Shezhou Luo, Haiming Qin, Shaobo Xia, Yong Xiao, Yiping Wan, Yang Liu, Zhao Lei, Pinghua Wang, Haipeng Zhao, Fangjian Wang, Panpan Chen, Haiqing Zhang, Zijia Wang, Lijuan Liu, Dajin Lu, Hongfei Li, and Yu Song. I really appreciate the following colleagues for their comments and suggestions: Prof. Zhizhong Kang (China University of Geosciences, Beijing), Associate Professor Juntao Yang (Shandong University of Science and Technology), Prof. Wei Su (China Agricultural University),

Prof. Haiyan Guan (Nanjing University of Information Science & Technology), Prof. Feifei Pan (University of North Texas, USA), Prof. Jinliang Wang (Yunnan Normal University), Prof. Hebing Zhang and Associate Professor Hongtao Wang (Henan Polytechnic University), and Prof. Wuming Zhang (Sun Yat-sen University). I appreciate the support from the Aerospace Information Research Institute and the University of Chinese Academy of Sciences. Last but not least, I would like to thank Prof. Huadong Guo, Academician of the Chinese Academy of Sciences, for his support and for writing the foreword of this book.

Cheng Wang
Aerospace Information Research Institute
Chinese Academy of Sciences
Beijing
October 31, 2023

Authors

Cheng Wang is a Professor at the Aerospace Information Research Institute, Chinese Academy of Sciences (CAS), Beijing, China. He earned a PhD in remote sensing at the University of Strasbourg, France, in 2005. He presently serves as Director of the LiDAR sub-committee of the Chinese National Committee of International Society for Digital Earth (CNISDE). He has over 20 years of research experience in LiDAR remote sensing mechanism, data processing and applications. He has been the Principal Investigator for over 15 national projects related to LiDAR in China. He has been teaching a LiDAR remote sensing course at the University of Chinese Academy of Sciences for about 10 years. Professor Wang has published more than 200 papers and four books and is the principal award recipient of ten domestic science and technology prizes.

Xuebo Yang is an Assistant Research Fellow at the Aerospace Information Research Institute, Chinese Academy of Sciences, Beijing, China. She earned a PhD at the Chinese Academy of Sciences (CAS), Beijing, China, in 2021. Her research interests include LiDAR remote sensing mechanisms, LiDAR sensor development, and LiDAR applications. She has presided over and participated in scientific research programs such as the National Natural Science Foundation of China and the National Key Research and Development Program of China. She was awarded the Xiaowen Li Remote Sensing Science Youth Award and outstanding PhD thesis of Chinese Academy of Sciences.

Xiaohuan Xi is an Associate Professor and a Master Supervisor at the Aerospace Information Research Institute, Chinese Academy of Sciences, Beijing, China. She earned an MS in environmental geography at Peking University, Beijing, China, in 2000. Her research interests include LiDAR data acquisition, LiDAR urban application, and agricultural application. She is the Secretary-General of the LiDAR sub-committee of the Chinese National Committee of the International Society for Digital Earth. She is the Principal Investigator of eight national scientific research projects of China and the principal award recipient of five domestic science and technology prizes.

Sheng Nie is an Associate Professor at the Aerospace Information Research Institute, Chinese Academy of Sciences, Beijing, China. He earned a PhD at the Chinese Academy of Sciences (CAS), Beijing, China, in 2017. His research interests include multi-mode LiDAR data processing, LiDAR for topography surveying and mapping, and LiDAR vegetation applications. He presides over the National Natural Science Foundation of China and the National Key Research and Development Program of China. He was awarded the Chinese LiDAR Young Scientist award and elected to the Youth Innovation Promotion Association of the Chinese Academy of Sciences.

Pinliang Dong is a Professor and a Doctoral Supervisor in the Department of Geography and the Environment, University of North Texas (UNT), Denton, Texas. He earned a BSc at Peking University, China, an MSc at the Institute of Remote Sensing Applications (IRSA), Chinese Academy of Sciences, and a PhD at the University of New Brunswick, Canada. His research interests include LiDAR data analysis and applications in forestry and vegetation mapping, urban environments, disaster damage assessment, and geosciences.

1 Introduction

1.1 INTRODUCTION TO LiDAR

1.1.1 LiDAR

The word "laser" comes from "light amplification by stimulated emission of radiation." In 1916, Albert Einstein discovered that a single photon of light passing through a substance could stimulate the emission of further photons. Specifically, the electron in the atom absorbs energy and jumps from the low energy level to the high energy level. Then when it goes back from the high energy level to the low energy level, the energy is released in the form of photons.

When mentioning lasers, many people think of laser weapons, laser cutting, laser welding, laser surgery, etc. It seems that laser is a kind of light that is extremely lethal to the human body. In fact, lasers are categorized into Classes I to IV based on how safe they are for the user (Table 1.1).

The maximum permissible exposure (MPE) is used to determine the laser safety class, which is defined as the maximum level of exposure to the human eye or skin immediately or over a long period of time without damage occurring under normal circumstances. Note that the MPE value is not suitable for medical treatment of patients or for cosmetic purposes. The MPE value is related to many factors, such as the safety level of the laser used in the instrument, the laser wavelength, the output power, the pulse duration, the repetition frequency, and the time of exposure to the laser radiation.[1]

Light detection and ranging (LiDAR) is a product of a combination of laser technology and photoelectric detection technology. It uses a laser as a light source to transmit high-frequency laser pulses to the target and a photodetector as a device to receive the returned signal from the target. In terms of the working principle, LiDAR is similar to a traditional microwave radar. The difference is that the former uses laser (e.g., 532-nm or 1064-nm wavelengths) as the carrier to measure the distance and orientation and to identify the target through the position, radial velocity, target scattering, and other characteristics. LiDAR has a very large number of functions and a wide range of applications. If not specified, this book only deals with the range measurement function of the LiDAR system and its land applications.

1.1.2 Characteristics of LiDAR

1.1.2.1 Advantages

a. Active remote sensing. LiDAR systems actively emit high-frequency laser pulses to the object surface being measured and receive the laser signal back from the object surface.
b. Obtain three-dimensional (3D) spatial information of surface objects quickly and directly. This is the most important advantage that distinguishes LiDAR from other traditional remote sensing technologies.

DOI: 10.1201/9781032671512-1

TABLE 1.1
Laser Safety Classes

Laser Class	Safety	Power (mW)	Specificities
Class I	Safe	<0.4	Non-hazardous to human eyes; used for laser presentations, displays, mapping, collimation, leveling, etc.
Class II	Minor hazard	0.4–1	Causes dizziness if looked at directly; need to protect by blinking eyes; avoid observation with telescopic equipment; also used for laser displays.
Class III	a Hazard	1–5	Cannot look at directly into or shine in the human eye; avoid pedestrians when working.
	b Hazard	5–500	Dangerous to aim directly at someone's eyes.
Class IV	Severe hazard	>500	High-energy continuous laser; fire hazard; used for surgery, laser cutting, welding, and other mechanical processes.

c. Good directionality, high angular resolution, high distance resolution, and high velocity resolution. The direction of laser emission is usually limited to a solid angle with a few milliradians, which greatly improves the laser intensity in the emitted direction of illumination. It is also the important basis for laser collimation, guiding, and ranging.
d. Insensitive to electromagnetic interference and strong anti-interference ability. The laser propagates in a straight line. The beam is very narrow and well concealed. The divergence angle is small and the energy is concentrated, which ensure extremely high detection sensitivity and high resolution.
e. Good low-altitude detection performance. Microwave radar is susceptible to the echoes of various objects, and there is a dead zone when detecting at low altitude. In contrast, the laser signal is reflected only when the target is irradiated by LiDAR and is not affected by other objects.
f. Strong penetrability. High-frequency laser pulses can penetrate the small gaps in a vegetation canopy to reach the ground. Blue or green lasers can penetrate a certain depth of water to obtain underwater topography and water quality information, which plays a role in underwater topographic mapping and water quality monitoring in offshore and inland rivers and lakes.

1.1.2.2 Differences with Other Ranging Means

a. Comparison with close-up photogrammetry or aerial photogrammetry.

The form of acquired data is different. LiDAR directly obtains a 3D point cloud, while photogrammetry obtains images and photographs. The means of data acquisition is different. The LiDAR system uses the position and orientation system (POS) to directly calculate the 3D coordinates of the target, and the ground control and field work have less of a workload.

Photogrammetry needs to match the homonymous points. The data solving method, measurement accuracy, and requirements for the measurement environment are also different. The photogrammetry has high requirements for environmental light, temperature, etc., while the LiDAR is not sensitive to these factors.

b. Comparison with microwave radar.

Compared with microwave radar with similar functions, the LiDAR system is smaller and lighter. For example, many unmanned aerial vehicles (UAVs) carry light and small LiDAR systems less than 1 kg. In addition, the LiDAR system has high pulse frequency, long measurement distance, high measurement precision, good directionality, strong anti-interference ability, and a certain degree of covertness.

c. Comparison with total station.

Both LiDAR and a total station can achieve accurate measurement of irregular objects. However, the LiDAR uses non-contact measurement, without setting up reflective prisms, and can carry out the measurement of complex geometric objects and hazardous areas that are difficult for personnel to reach. LiDAR collects high-density, high-resolution data with high efficiency, while a total station measures the objects through the coordinate points connected in a line. When measuring irregular objects such as arcs, it is necessary for a total station to measure a large number of points through multiple operations, while the LiDAR can complete the measurement in only a single scan.

1.1.2.3 Disadvantages

LiDAR technology has obvious advantages, but also has limitations. First, the laser is affected by weather and atmosphere. For example, a laser pulse is sharply attenuated in bad weather such as heavy rain, thick smoke, and dense fog. Atmospheric turbulence may reduce the measurement accuracy of LiDAR. Second, the narrow beam of LiDAR makes it difficult to search for targets, which affects the target interception probability and the detection efficiency. Third, the discrete 3D point cloud acquired by LiDAR is slightly inferior to the traditional 2D remote sensing images for land cover classification.

1.1.3 Classification of LiDAR

After more than half a century of development, there are many types of LiDAR nowadays, which are usually categorized according to the mounting platform, detection mode, and use.

1.1.3.1 Classification by Mounting Platform

Three categories—space-based, air-based and ground-based LiDAR—are available. The space-based LiDAR is also referred to as spaceborne LiDAR, using satellites, space shuttles, or space stations as a platform (most are satellite platforms). The spaceborne LiDAR has a wide range of observations, which meets the large-scale applications. The air-based LiDAR is usually called airborne LiDAR, mainly using

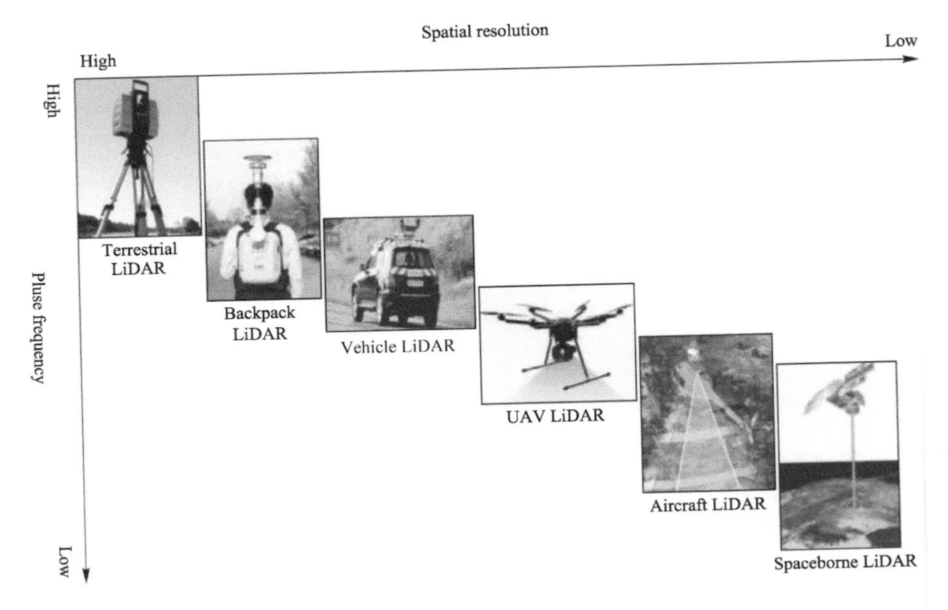

FIGURE 1.1 Characteristics of LiDAR systems with different platforms.

fixed-wing aircraft, helicopters, drones, and other aircraft as platforms, characterized by high efficiency and high point density and is especially suitable for 3D information acquisition of long-distance linear-shaped objects. The ground-based LiDAR mainly includes terrestrial (tripod fixed), shipborne, vehicle-mounted, backpack, and handheld laser scanners, characterized by comprehensive target information acquisition (including indoor space) and flexible access. With the continuous expansion of the visible range of the mounted platform (or the increased elevation of the mounted platform), the laser pulse sampling frequency transitions from high frequency to low frequency, the spatial resolution changes from high to low, and the observation range goes from small scale to regional scale, up to the global scale (Figure 1.1).

1.1.3.2 Classification by Ranging Mode

There are two categories: pulsed LiDAR and phased LiDAR. The former measures by using the round-trip propagation time difference between the transmitted and received laser pulses, characterized by simple and straightforward, long measurement distances. The latter uses the radio frequency to modulate the laser pulse amplitude and measures the phase delay produced in the round-trip process by which the modulated laser detects the target. Then the distance between the LiDAR and target is calculated by the modulated laser wavelength. The phased LiDAR is characterized by a short measurement distance and high pulse frequency.

1.1.3.3 Classification by Laser Medium

There are two categories: gas LiDAR and solid-state LiDAR. Gas LiDAR uses gas or vapor as a working substance to generate a laser, usually represented by a CO_2 laser.

It is characterized by good coherence, a narrow beam, small field of view, strong anti-interference ability, good atmospheric transmission performance, compatibility, and safety in use. The solid-state LiDAR is further divided into semiconductor LiDAR and diode-pumped solid-state LiDAR. The former considers a laser bar as the basic unit, with high output power, operating current, and loss of heat. The latter uses the YAG laser as a representative, which integrates the diode-pumped solid-state laser with high repetition frequency and high peak power and the high sensitivity avalanche diode detector. It is characterized by small size, light weight, and low price.

1.1.3.4 Classification by Use

Uses include ranging LiDAR, fire control LiDAR, shooting range LiDAR, tracking and identification LiDAR, poison detection LiDAR, multifunctional tactical LiDAR, meteorological LiDAR, and navigation LiDAR, among others. The LiDAR in this book refers to ranging LiDAR if not otherwise specified.

1.1.3.5 Classification by Footprint Size

There are categories: large-footprint LiDAR and small-footprint LiDAR. Large-footprint LiDAR usually refers to LiDAR with a ground footprint diameter of more than 10 m, such as the NASA's spaceborne LiDAR—GLAS (Geoscience Laser Altimeter System) with a 70-m footprint diameter and China's Gaofen-7 satellite laser altimeter with a 17-m footprint diameter. Generally, the large-footprint LiDAR has a low footprint density and cannot image the target. However, it can get the data on a regional or global scale, which shows a significant advantage in large-scale geoscientific applications. Small-footprint LiDAR has a footprint diameter of centimeters or even millimeters. The system transmits pulses at a high frequency (currently up to 2000 kHz), resulting in high point density and high data accuracy.

1.1.3.6 Classification by Detection and Recording Method

There are three categories: discrete return LiDAR, full-waveform LiDAR, and photon counting LiDAR. Discrete return LiDAR is the most common and most widely used commercially. For example, building 3D reconstruction for digital cities, high-precision map production for autonomous driving, and cultural heritage digitization 3D reconstruction are all based on point cloud data acquired by discrete return LiDAR systems. Full-waveform LiDAR samples the returned signal continuously, records more detailed information, and acquires the entire object's vertical profile. Photon counting LiDAR is different from the first two. It uses a micro-pulse laser with high repetition frequency and highly sensitive single-photon detectors to record the returned signals (at a single-digit photon level) as the photon points. It has the advantage of being able to detect space targets at long distances with lower laser energy.

1.1.3.7 Classification by Detection Objects

This is categorized into atmospheric LiDAR, oceanic LiDAR, and land LiDAR. The emitted laser pulses of atmospheric LiDAR interact with the aerosols in the atmosphere and various components. Then the backward scattering signals are received by the detector and then processed and analyzed, to provide the information of atmospheric physical elements. The main applications of atmospheric LiDAR include clouds, aerosols and boundary layer detection, detection of atmospheric composition, temperature detection, and so on. The working principle of oceanic LiDAR is that the emitted laser beam penetrates the seawater and produces various scattering and fluorescence, and the received signals are used to retrieve the sound speed; temperature; salinity distribution parameters; and oil, gas, and hydrocarbon indicators of the marine boundary layer. The laser pulse of the land LiDAR detects surface objects such as trees, roads, bridges, and buildings. Sometimes, a portion of the laser is reflected and recorded by the LiDAR receiver, which then calculates the distance from the LiDAR to the object and provides accurate 3D coordinates of the detected object by combining the attitude and position information. If not specifically indicated, this book deals with land LiDAR; atmospheric LiDAR and oceanic LiDAR are not discussed.

1.1.4 Development History of LiDAR

In the 1960s, American scientist Theodore Harold (Ted) Maiman[2] first introduced laser into the practical field and built the world's first laser equipment. In 1989, the University of Stuttgart, Germany, developed the world's first airborne LiDAR prototype. Since the 21st century, LiDAR has entered a period of booming development. Particularly after 2015, LiDAR development efforts have boomed over the world, and various mature commercialized LiDAR products have been introduced. This section summarizes the development of LiDAR in three stages: emergence, development, and explosion.

1.1.4.1 Emergence Period (1960–1990)

The origin of the laser is traced back to 1916. When Albert Einstein first proposed the concept, he theoretically predicted the possibility of laser generation. In 1960, the world's first laser was introduced, followed by the emergence of various types of lasers, such as semiconductor lasers, helium-neon gas lasers, and CO_2 lasers. Their application fields are increasingly wide-ranging, including laser printing, phototypesetting, display, distance measurement, barcode scanning, industrial detection, fiber optic communications, etc. Lasers are known as one of the major scientific and technological discoveries in the 20th century.

After the emergence of laser, scientists applied it to target ranging, bathymetry, tracking, etc. In 1964, the U.S. National Aeronautics and Space Administration (NASA) launched the "Explorer-22" satellite, which first achieved laser ranging by using the carrier's corner reflector. In 1968, Syracuse University in the United States constructed the world's first laser bathymetric measurement system to achieve ocean near-shore bathymetry. In 1969, *Apollo* researchers in the United States used the laser reflector placed on the Moon to accurately measure the distance between the Earth and the Moon. In 1975, Riegl started to produce solid-state diode lasers and laser rangefinders. In the 1980s, the global positioning system (GPS), chronometers, and high-precision inertial measurement units (IMUs) came out successively, which made it possible to obtain precise, real-time positioning and orientation in the process of laser measurement and directly promoted the emergence of the LiDAR system. In 1989, Professor Ackermann[3] of the University of Stuttgart, Germany, developed the prototype of the LiDAR system. This was the world's first airborne laser scanner that combines laser scanning technology with real-time positioning and orientation technology, which is a milestone in the development of LiDAR.

1.1.4.2 Development Period (1990–2000)

Since the 1990s, LiDAR system development has entered a period of rapid development. In 1993, TopScan in Germany and Optech in Canada launched the first commercial airborne LiDAR: ALTM1020 (Airborne Laser Topographic/Terrain Mapping) system, marking the official entrance of the LiDAR system to the commercial stage. Azimuth in the United States began the development of a LiDAR system in 1997. Leica acquired Azimuth in 2001 and launched the ALS40, ALS50, and ALS60 systems (Airborne Laser System). In 1995, Saab in Sweden developed a bathymetric LiDAR system: HAWK Eye. In 1996, Riegl in Austria launched a series of laser scanners that were used on board aircraft, vehicles, and ships.

China started LiDAR-related research in the same period. Professor Li Shukai's team in the Institute of Remote Sensing Applications of the Chinese Academy of Sciences developed the principal prototype of an airborne 3D LiDAR imaging system. After that, China's Zhejiang University, Harbin Institute of Technology, Shanghai Institute of Optics and Fine Mechanics of the Chinese Academy of Sciences, Shanghai Institute of Technical Physics of the Chinese Academy of Sciences, and other institutions all carried out LiDAR system developments.

1.1.4.3 Explosion Period (2000–2020)

Upon entering the 21st century, the global demand for LiDAR technology in various application fields is growing rapidly at a rate of 30% per year. The international LiDAR markets show a blossoming state.

Leica entered the LiDAR field in 2001. It renamed the AeroSensor as ALS40 and produced the ALS50 in 2003, which was upgraded to ALS50-II two years later. In October 2006, the company introduced a new laser transmitter/receiver technology, multiple pulses in air (MPiA), which greatly improved the laser point cloud density and was applied to the ALS70 and ALS80 systems. In 2005, the Blom company introduced the HAWK Eye II airborne LiDAR bathymetry system, which employed two lasers: a 532-nm laser (with a receiver frequency of 4 kHz) for underwater detection and a 1064-nm laser (with a receiver frequency of 6.4 kHz) for shoreline surveys. The German IGI company developed the LiteMapper 2800 and LiteMapper 5600 systems. Starting from 1996, Riegl successively launched a series of laser scanners that can be used for airborne, vehicle-mounted, and ship-mounted measurements, such as the LMS-Q140, LMS-Q560, and a series of terrestrial 3D laser scanning systems, such as VZ400, VZ1000, and ultra-long-range VZ4000 and VZ6000, as well as the dual-channel, dual-band airborne LiDAR system VQ-1560i.

With the popularization of UAVs, airborne LiDAR systems have developed toward lightweight and small-sized configurations. The UAV LiDAR systems that are small, lightweight, and inexpensive have risen rapidly. In 2014, Riegl in Austria released the

VUX-1 scanner, which is the world's first lightweight and small-sized UAV LiDAR, weighing just 3.6 kg. In 2014, Hokuyo in Japan launched the UXM-30LXH-EWA system, weighing 0.8 kg, followed by UST-10LX/20LX, UXM-30LAH-EWA, UST-05LA, etc., most of which are less than 1 kg.

China's commercialized LiDAR systems are rising rapidly and catching up with international standards. Under the support of the Special Program for the Development of Major Scientific Instruments and Equipment of Ministry of Science and Technology of the People's Republic of China, the Shanghai Institute of Optics and Fine Mechanics of the Chinese Academy of Sciences has developed an airborne dual-frequency LiDAR system (He et al., 2018). The Institute of Microelectronics of the Chinese Academy of Sciences has successively developed airborne, vehicle-borne, and ground-based LiDAR systems as well as a mid- and long-range airborne LiDAR: Mars-LiDAR (Li et al., 2013). Several of China's enterprises have achieved remarkable results in the industrialization of LiDAR. Since 2005, Beijing Beike Tianhui Technology Co., Ltd. has launched airborne (A-Polit), vehicle-mounted (R-Angle), and point-station (U-Arm) LiDAR, as well as lightweight and small-scale LiDAR systems: Clouds series and Genius series. Since 2012, Wuhan Haida Digital Cloud Technology Co., Ltd. has launched the self-developed terrestrial laser scanner HS, vehicle-mounted mobile measurement system HISCAN, airborne laser measurement system ARS, and "Zhihui" series of airborne products. Aolunda Technology Co., Ltd. in Chengdu, China, launched the CBI series of LiDAR measurement systems, as well as lightweight CBI-120P and CBI-200P series products in 2019. The Shenzhen DJI Company released the DJI L1 system in October 2020, which integrated the Livox LiDAR and others. For more commercialized LiDAR systems, see Section 1.2.2.

At the beginning of the 21st century, countries around the world aimed at the development and application of spaceborne LiDAR. In general, most of spaceborne LiDAR systems are still led by the national space sectors and have not yet been commercialized. NASA launched the world's first spaceborne laser altimeter ICESat-1 (Ice, Cloud and Land Elevation Satellite)/GLAS (Geoscience Laser Altimeter System) in 2003, the first spaceborne photon-counting LiDAR ICESat-2/ATLAS (Advanced Topographic Laser Altimeter System) in 2018, and GEDI (Global Ecosystem Dynamics Investigation) on board the International Space Station (ISS) in 2018. These spaceborne LiDAR systems provide a reliable data source for research on the production of global control points, changes in polar ice caps, and monitoring of lake levels, as well as the estimation of global forest heights, biomass, and carbon stocks. Some satellite missions of China also carried LiDAR systems. The laser altimeter developed by the Shanghai Institute of Technical Physics and the Shanghai Institute of Optics and Fine Mechanics was carried on Chang'e-1 and Chang'e-2, effectively acquiring the elevation data of the north and south poles of the Moon. The Chang'e-4 carried a laser 3D imager and a laser ranging sensor, which were instrumental in the soft landing on the back of the Moon, providing not only high-precision terrain information but also achieving autonomous obstacle avoidance. The main payloads of China's Gaofen-7 satellite and carbon monitoring satellite for terrestrial ecosystems also include laser altimetry. Other spaceborne LiDAR programs are being planned or deployed around the world.

1.2 LiDAR SYSTEM

1.2.1 LiDAR System Composition

The composition of a LiDAR system with various platforms is slightly different, but their cores are inseparable from the laser scanning system. Figure 1.2 shows the typical spaceborne, airborne, and terrestrial LiDAR and their system composition.

This section introduces the basic components of an airborne LiDAR system as an example (Figure 1.3): the laser scanner, the global navigation satellite system (GNSS), the inertial navigation system (INS), and the monitoring and control system.

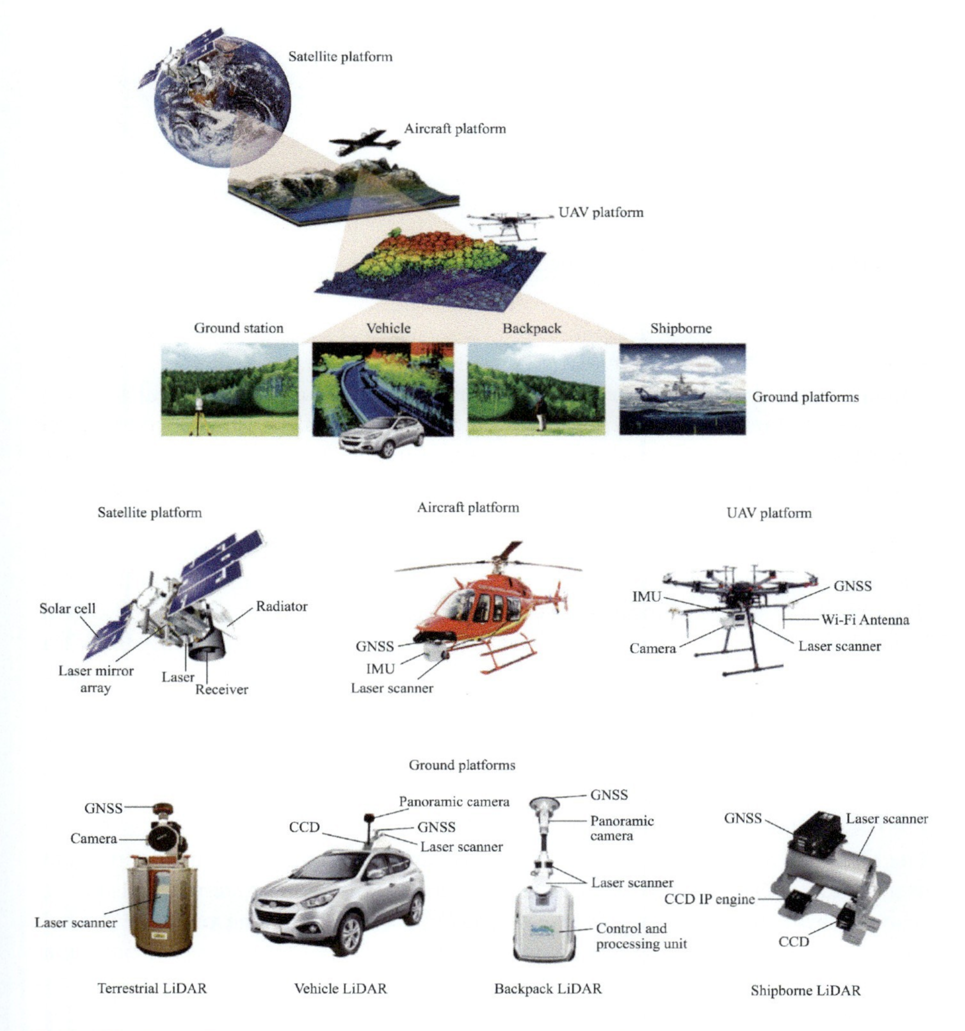

FIGURE 1.2 LiDAR system composition for different platforms.

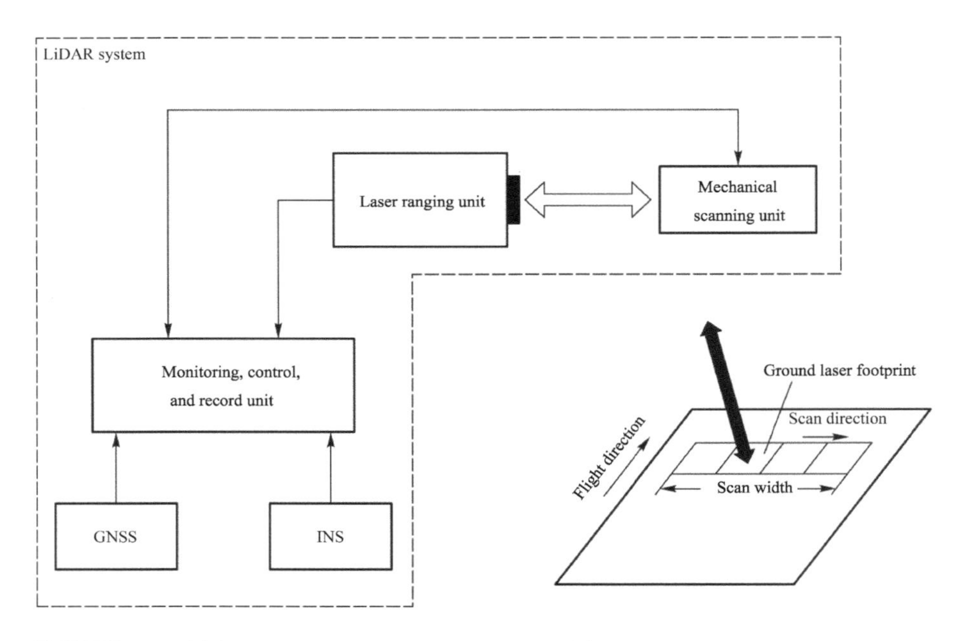

FIGURE 1.3 Airborne laser scanning system composition.

1.2.1.1 Laser Scanning System

The laser scanning system includes a laser ranging unit and a mechanical scanning device. The laser ranging unit consists of a laser transmitter and receiver, which are used for transmitting and receiving laser signals and determining the distance from the target to the LiDAR, the number of returns, and the laser return intensity. The transmitted and received laser beams share the same optical aperture to ensure that the transmitted and received optical paths are the same. The laser beam is usually very narrow with a very small divergence angle. The range of irradiation to the ground is the laser footprint. When the laser is reflected by the ground object, part of the signal returns to the receiver and is recorded by the recording unit. The mechanical scanning device launches out the laser beam from different directions by mechanical rapid rotation. The scanning modes mainly include swing mirror scanning, rotating prism scanning, elliptical scanning, fiber scanning, etc. For details, see Section 2.3.2.

1.2.1.2 Global Navigation Satellite System

The GNSS is a space-based radio navigation and positioning system that provides users at any location on the Earth's surface or in near-Earth space with all-day 3D space coordinates and time. The GNSS usually consists of one or more constellations of satellites and their augmentation systems required to support particular functions. The major GNSS systems around the world include China's Beidou, the United States' GPS, Russia's GLONASS (Global Navigation Satellite System), and the European Galileo system. The main roles of GNSS in LiDAR systems are threefold: first, to achieve time synchronization with the IMU and the laser; second, to combine with the IMU data for navigation and resolving trajectories to improve position and orientation accuracy; and third, to provide navigation data to the flight platform.

1.2.1.3 Inertial Navigation System

The INS consists of the IMU and the navigation processor. IMU is a general term for inertial units such as gyroscopes and accelerometers used to measure attitude. It usually consists of three accelerometers and three gyroscopes, digital circuits, central processing unit (CPU), and so on. Its role is to measure the attitude information of the scanner at the moment of laser emission, including pitch angle, roll angle, and heading angle. The INS and GNSS constitute the POS, providing position and attitude information, the accuracy of which directly determines the accuracy of the point cloud data acquired by the LiDAR system.

1.2.1.4 Monitoring and Control System

The monitoring and control system controls the working condition of the laser scanner, GNSS, and IMU. Its core is designed to keep the coordinated and synchronized work of the LiDAR system and store the acquired data. The acquired data includes (1) distance and intensity data, (2) position and attitude data from the GNSS system and INS system, and (3) auxiliary data.

1.2.2 Major Commercial LiDAR Systems

The spaceborne LiDAR systems usually have a high development cost and maintenance cost. Coupled with their limitations of applications (most for scientific research), the spaceborne LiDAR systems have not been commercialized. Hence, this section introduces commercial airborne and terrestrial LiDAR systems.

Tables 1.2–1.5 list some commercial airborne and terrestrial LiDAR systems. It is seen that, with more than a decade of development, the main performance parameters of the devices have been greatly improved, and most of the systems are lightweight, small, and easy to use.

1.2.3 Data Format and Processing Software

1.2.3.1 Data Format

As mentioned before, there are many types of LiDAR and various data forms, such as full waveform data, photon counting data, and discrete point cloud data, among which discrete point cloud data are the most widely used. This subsection focuses on several common discrete point cloud data formats, such as formats specifically designed for storing point cloud data (LAS/LAZ, PTS/PTX, PCD, etc.) and file

TABLE 1.2
Some Airborne LiDAR Systems

Manufacturer	Optech	IGI	Leica Geosystems	Riegl	TopoSYS	TopEye	Fugro	Velodyne
Systems	ALTM Gemini/3100	LiteMapper/5600	ALS70	LMSQ1560	Harrier 56/24, Falcom III	TopEye Mk II	FLI-MAP 400	Puck LITE
Laser wavelength (nm)	1060	1550	1064	1550	1550	1064	1500	903
Pulse frequency (kHz)	167	40–200	500	800	25–200	1–50	250	300
Maximum scanning angle (°)	50	60	75	60	60	+/− 20, +/− 14 BWD/FWD	60	30
Measuring range (m)	200–4000	30–1800	200–5000	50–5600	30–1000	60–750	50–400	10–500
POS post-processing software	POS PAC	AEROoffice	GrafNav/IPAS Pro	AEROoffice/ POS PAC	Applanix Pos /Pac	POS GPS TopEye PP	GrafNav/IPAS Pro	POS PAC
Post-processing software	DASHMap	AEROoffice, GeocodeWF, TerraScan, TerraModeler	Leica ALS Post Processor, TerraScan/ TerraModeler	RiPROCESS560	TopPIT	TopEye PP&TASQ	FLIP7	Vella

TABLE 1.3
Some Airborne LiDAR Systems Developed in China

Manufacturer	Beike Tianhui	Haida Digital Cloud	Shenzhen DJI	Shanghai Institute of Optics and Fine Mechanics, Chinese Academy of Sciences	Shenzhen University	Alundar
Systems	ALTM Gemini/3100	Zhihui PM-1500	DJI L1	Dual-Band Mapper 5000	Full waveform bathymetry	CBI-200P
Laser wavelength (nm)	1060	1550	905	Land mode: 1550 Ocean mode: 532/1064	532	905
Pulse frequency (kHz)	167	40–200	240	Land mode: 100–400 Ocean mode: 5	10–100	1280
Maximum scanning angle (°)	50	60	95	Land mode: 30 Ocean mode: 15	20	360
Measuring range (m)	200–4000	30–1800	100	Land mode: 300–1600 Ocean mode: 300–1100	200–5000	200
POS post-processing software	POSPac trajectory data processing software	Zhihui airborne LiDAR post-processing software	DJI POS post-processing module	/	/	ALiDAR data processing software

TABLE 1.4
Some Terrestrial 3D Laser Scanner Systems

Manufacturer	Leica	Riegl	Optech	Faro	Z + F	Topcon	Trimble
Country	Switzerland	Austria	Canada	United States of Ameria	Germany	Japan	United States of Ameria
Product	ScanStation P30/P40	VZ-1000	ILRIS-3D	Focus S350	Imager 5010	GLS-2000	Trimble TX8
Scan type	Pulsed	Pulsed	Pulsed	Phased	Phased	Pulsed	Pulsed
Maximum pulse frequency (kHz)	1000	300	3.5	976	1016.7	120	1000
Wavelength (nm)	1550	1550	1535	1550	1350	1064	1500
Maximum measuring range (m)	270	1400	1700	0.6–350	0.3–187	350	120
Field of view (horizontal ×vertical) (°)	360 × 270	360 × 100	360 × 110	360 × 300	360 × 320	360 × 270	360 × 317
Ranging accuracy (mm)	1.2 (+10 ppm)	5 @100 m	7 @100 m	1 @25 m	1 @50 m	3.5 @150 m	2 @100 m
Digital camera	Internally	Externally	Internally	Internally	Externally	Internally	Internally
Operating temperature (°C)	–20–50	0–40	0–40	5–40	–10–45	0–40	0–40
Scanning control and data processing Software	Cyclone, Cloudworks	RiScan Pro	ILRIS-3D Polyworks	SCENE, Geomagic, Rhinoceros	Laser Control, Light Form Modeller	ScanMaster	Rrimble FX Controller, 3Dipsos RealWorks

TABLE 1.5
Some Terrestrial 3D Laser Scanner Systems Developed in China

Manufacturer	Haida Digital Cloud	Beike Tianhui	Stonex	Zhongke Tianwei	Hualang 3D	Xunneng Optoelectronic
Product	HS1200	U-Arm1500	X300 Plus	TW-A1000	HL1000	SC500
Scan type	Pulsed	Pulsed	Pulsed	Pulsed	Pulsed	Pulsed
Maximum pulse frequency (kHz)	500	300	40	500	36	36
Wavelength (nm)	1545	1550	915	1064	905	905
Laser safe class	Level 1	Level 1	Level 1	Level 1	Level 1	Level 1
Maximum measuring range (m)	1200	1500	300 @ 80% reflectance	1000	200 @ 90% reflectance	3000 @ 90% reflectance
Field of view (horizontal × vertical) (°)	360 × 100	360 × 300	360 × 180	360 × 300	360 × 100	360 × 300
Ranging accuracy (mm)	5 @100 m	5–8 @100 m	4 @50 m	7 @100 m	1.2 @50 m	1.2 @50 m
Digital camera	Externally	Externally	Internally	Externally	Externally	Internally
Operating temperature (°C)	–20 to 65	–20 to 55	–10 to 50	0–50	0–40	0–40
Scanning control and data processing software	HD 3LS Scene	UIUA, JRC 3D Reconstructor	Si-Scan	/	/	3D Cloud Processor

formats with the capability to represent and store point cloud data, such as model files in the field of computer graphics (e.g., PLY, OFF).

1. LAS/LAZ file format. Laser file format (LAS) is a file format specially designed for 3D point cloud data, managed and maintained by the American Society for Photogrammetry and Remote Sensing (ASPRS). The LAZ file format is a lossless compressed version of LAS. The LAS file format adopts binary storage, which saves the 3D coordinates and intensity of laser points, returns, red-green-blue (RGB), scanning angle, and other information. The LAS is the most widely used point cloud data format.

   In the latest version of LAS 1.4 released by the ASPRS in 2019, the LAS file consists of a public file header area, a variable length record (VLR) area, a point data record area, and an optional extended VLR area. The public file header area contains some records describing the overall situation of the data, such as the number of point records and coordinate boundaries. The VLR area is used to store some variable length data, such as projection information, metadata, waveform packet, and user application data.

   Point data records the 3D coordinates and attribute information of each laser point. The LAS 1.4 file format supports Point Data Record Formats (PDRFs) 0–10, for a total of 11 point types, of which PDRFs 6–10 are the recommended point types for use by the ASPRS, and PDRFs 0–5 are used mainly for compatibility with older versions. Each LAS file records only one type of point, identified by the "Point Data Format" field in the public file header area.

   The point coordinates in the LAS file are stored in long integer type (4 bytes), which saves half of the space compared with the double precision floating point type (8 bytes). When the file is read or written, the scaling factor (X_{scale}, Y_{scale}, Z_{scale}) and offset (X_{offset}, Y_{offset}, Z_{offset}) in the public file header are used to convert the long integer values (X_{record}, Y_{record}, Z_{record}) in the point data record area to derive the real coordinate information ($X_{\text{coordinate}}$, $Y_{\text{coordinate}}$, $Z_{\text{coordinate}}$), which is expressed by Equation (1.1).

$$\begin{cases} X_{\text{coordinate}} = X_{\text{record}} \cdot X_{\text{scale}} + X_{\text{offset}} \\ Y_{\text{coordinate}} = Y_{\text{record}} \cdot Y_{\text{scale}} + Y_{\text{offset}} \\ Z_{\text{coordinate}} = Z_{\text{record}} \cdot Z_{\text{scale}} + Z_{\text{offset}} \end{cases} \tag{1.1}$$

   The waveform packets in the LAS file can be stored as extended variable length records (EVLRs) at the end of all point data records to facilitate separation or materialization. The EVLRs are stored in an unsigned, extra-long-integer (8-byte) format that allows more information to be stored than VLRs.

   The LAZ format uses the chunked compression method to reduce the file size. However, the compression reduces file reading and writing efficiency. Hence, the LAZ format is mainly used for situations with a high requirement for storage space and low requirement for read-write efficiency.

2. PCD format. PCD (Point Cloud Data) is the file format of PCL (Point Cloud Library), an open-source programming library for point cloud processing. PCD has two formats: text and binary. It can be used to store and process ordered/disordered point cloud datasets and support *n*-dimensional point type extensions. Compared with other point cloud file formats, PCD can adapt PCL to the greatest extent possible and maximize the performance of PCL applications.

   The PCD format consists of a file header and point data. The file header is saved in ASCII code, which declares and stores information such as the number, attributes, and type of point cloud data. The point data record the coordinates and attributes of the points. Each dimension and attribute of points can be obtained from the "FIELDS" in the file header.
3. PTS/PTX format. PTX and PTS are file formats used by Leica scanners and supporting software, both of which are stored in text format. The PTX file format employs the concept of individual scanning. Each file records one or more groups of point clouds. Generally, each scanning site corresponds to one group of point clouds, and each group of point clouds provides separate header information, including the number of rows and columns, the scanner position, the scanner spindle and the transformation matrix, etc. Based on the header information and stored point coordinates, the users can not only calculate the coordinates of the laser points in the unified coordinate system but also possibly recover the scan line information of each laser point. The PTS file format does not store the original scanning site information and is simpler than the PTX format. Its first line contains the number of point clouds, and subsequent lines contain information of each laser point, including coordinates, intensity, RGB values, etc.
4. Model file format. Model files are generally well standardized and generalized, supported by many software or open-source libraries, and partly applied to the saving of point cloud files, commonly including PLY, OFF, etc. The PLY (Polygon File Format) is a file format used in the field of computer graphics to save a collection of graphical objects with text or binary storage. A typical PLY file consists of a list of *XYZ* coordinate triples of vertices and elements described by vertex list indices. The file includes a file header, a vertex list, a face list, and a list of other elements. The OFF (Object File Format) is a file format that uses polygons to represent the geometry of a model and is stored in text format as well. The OFF file consists of a file header, a vertex list, and a polygon list. Each polygon has any number of vertices, and the number of vertices, facets, and edges is recorded in the header. Compared with the formats specially designed for point clouds such as LAS and PCD, the model file format records the topological relationship between vertices in addition to the coordinates, as well as some other attribute information. For example, the PLY file format can also record the RGB values, normal vectors, and so on.
5. Text format. In addition to the previously mentioned file formats with clear standards, text files are often used to save point cloud data by virtue of their wide compatibility. The commonly used file suffixes include xyz.,

asc., neu., txt., csv., and others. These non-standardized files are more flexible. They use ASCII code to store point cloud data by line. However, when reading them, it is generally necessary to know the file recording rules in advance; otherwise, they cannot be parsed correctly.

1.2.3.2 LiDAR Data Processing Software

In recent years, a variety of advanced LiDAR systems have been emerging, and data acquisition has become more and more convenient. The increasing demand for LiDAR applications has posed a great challenge in terms of the efficient processing of massive point cloud data. There are many software packages for LiDAR data processing, such as TerraSolid and open-source CloudCompare. In addition, some commercial software packages, such as ENVI, ERDAS, ArcGIS, and ArcGIS Pro, have LiDAR data processing and analysis functionality. Also, many research institutes, enterprises, and universities have successively developed LiDAR data processing software with independent property rights, such as Point Cloud Magic and LiDAR360. Table 1.6 lists several popular LiDAR data processing software programs and their functions.

1. TerraSolid. The TerraSolid software is developed by the University of Helsinki, Finland. It is the first commercialized airborne LiDAR data processing software in the world, which is developed based on Bently's Microstation CAD software and covers most of the functions of point cloud data processing. It includes Terra Scan, Terra Modeler, Terra Photo, Terra Match, and other modules. The Terra Scan is used for processing point cloud data; the Terra Modeler is used for building surface models; the Terra Photo is used for producing orthophotos; the Terra Match is used for point cloud aerial strip splicing.

   TerraSolid supports reading multiple point cloud formats, images, digital elevation model (DEM) data, and vector data (dgn, dwg) and supports the display of point cloud and image data, route management, point cloud amplitude and batch processing, profile interaction, contour production, DEM production, terrain analysis, calibration, and coordinate conversion. The integrated point cloud filter algorithm uses the progressive triangulated irregular network (TIN) filtering. Professional applications include power line extraction, forestry analysis, and hydrological analysis. The disadvantage of TerraSolid is that it is based on the secondary development of MicroStation, and users have to install MicroStation before using TerraSolid. Otherwise, some of the software's functions and application extensions are limited, e.g., visualization and human-computer interaction.
2. CloudCompare. The CloudCompare is developed using the C++ programming language and can be compiled on Windows, Linux, and Mac operating systems. It is a 3D point cloud (and triangle mesh) processing software, originally designed to compare two dense 3D point clouds (e.g., those acquired by laser scanner) or to compare point clouds with triangle meshes. The software relies on a specific octree structure and has high computational performance. For example, a dual-core processor laptop can compute

TABLE 1.6
Major LiDAR Data Processing Softwares and Their Functions

Software	3D Display	Classification	Interactive Editing	Batch Processing	Topographic Application	Forest Application	Powerline Application	Building Application
ArcGIS	√	√	√	×	√	×	×	×
CloudCompare	√	×	√	×	×	×	×	×
ENVI LiDAR	√	√	√	×	×	×	√	×
FUSION	×	×	×	×	×	√	×	×
Global Mapper LiDAR Module	√	√	√	×	×	×	×	×
Quick Terrain Modeler	√	×	×	×	×	×	×	×
LiDAR360	√	√	√	√	√	√	√	×
LP360	√	√	√	√	√	×	×	×
PCC (Point Cloud Catalyst)	√	√	√	×	√	√	√	×
PCM (Point Cloud Magic)	√	√	√	√	√	√	√	√
RiALITY	√	×	×	×	×	×	×	×
TerraSolid	√	√	√	√	√	×	×	×
ALiDAR	√	√	√	√	√	×	√	×

3 million points to 14,000 triangle meshes in about 10 s. Subsequently, the software was expanded into a general-purpose point cloud processing software including many advanced algorithms (alignment, resampling, color/normal vectors/scales, statistical computation, sensor management, interactive or automatic segmentation, display enhancement, etc.). Also, some functions, such as normal vector optimization, Poisson mesh construction, and point cloud filtering, can be easily used.

3. ENVI LiDAR. The predecessor of ENVI LiDAR is Environment for 3D Exploitation (E3De) developed by the EXELIS VIS company (original manufacturer of ENVI/IDL). ENVI LiDAR allows users to write a program to implement a specified function by creating an Interactive Data Language (IDL) project file according to the users' needs. This indicates its excellent secondary development capabilities. As highly integrated software, ENVI LiDAR is easy to operate and has low requirements for users. It supports LAS, NITF LAS, ASCII, and LAZ file formats, among others. It has functions for interactive cross-section visualization, visual domain analysis, 3D visualization flight browsing and editing, point cloud feature extraction and classification, and some professional applications of forest resources survey, urban expansion mapping, terrain visualization, and power line survey decision-making.
4. Point Cloud Magic (PCM). PCM is LiDAR data processing and application software developed by the Institute of Aerospace Information Research Institute, Chinese Academy of Sciences. It was first released in October 2015, and version 2.0 was released in November 2020. It has a flat theme style and data management platform. The software functions include point cloud basic tools, point cloud filtering, land classification, mine mapping, forest application, building 3D models, 3D modeling of cultural heritage sites, safety analysis of power lines, 3D reconstruction of transmission corridors, crop monitoring, etc. It also provides customizable workflow settings to further enhance the user experience. Its basic functions include (1) basic platform: support opening point cloud data, model data, image data, and vector data and support rendering point cloud data by elevation, category, RGB, intensity, GPS time, etc.; support profile operation, single-point and multipoint selection, distance measurement; support interactive operations such as point cloud data cropping and attribute change; (2) basic tools: support basic operations such as partitioning, merging, cropping, filtering, format conversion, attribute statistics of point cloud data; (3) machine learning classification: three classifiers with customizable parameters, i.e., random forest, neural network, and Light Gradient Boosting Machine (LightGBM); and (4) other functions: provide up to 20 interface styles to meet the user's visual experience and support the setup of operations according to the user's personal operating habits.
5. LiDAR360. LiDAR360 is a LiDAR point cloud data processing and analysis software independently developed by Beijing Digital Green Earth Technology Co., Ltd. It supports the visualization, classification, analysis, extraction, editing, modeling, and multivariate data exporting of massive

point clouds. It supports multiple data formats of point cloud, images, DEM data, vector data (shp/dxf), and other customized data formats (LiData, LiModel). It can automatically match aerial strips with different flight paths to generate a highly precise point cloud. It provides automatic or semi-automatic classification and quickly classifies ground, vegetation, buildings, and powerlines. It supports the interactive editing of point clouds through the profiling tool. Its terrain application module includes generation of a high-precision digital terrain model; interactive editing of DEM; generation of maps of slope, aspect, contour lines, and surface roughness; and generation of orthophoto models. Its powerline application module includes classification, fitting of power lines, and monitoring of hazardous points. It also supports 3D reconstruction of buildings and forest statistical variable extraction.

1.3 LiDAR REMOTE SENSING APPLICATIONS

LiDAR technology was mainly used in the military field in the early stage, and then it was popularized in the civil field. Its application scope not only includes the land and ocean on the Earth but also covers the atmosphere and the surface of the Moon, Mercury, and Mars. This book focuses on LiDAR land applications, including most of the land features and many aspects of the national economy and social development. The following briefly describes several representative applications. More detailed introductions are provided in Chapter 5.

1.3.1 Topographic Mapping

The major advantage of LiDAR is its ability to directly obtain high-precision 3D spatial information. For example, spaceborne LiDAR can carry out sub-meter-level elevation measurements on a global scale, which provides support for the production of global high-precision control points. The high-density and high-precision point cloud acquired by airborne LiDAR can be classified into the ground point cloud by filtering. Then DEM, digital line graphic (DLG), and contour lines are generated by constructing a TIN from ground point clouds. These topographic products provide the surveying and mapping data for many applications.

1.3.2 Forest Resource Investigation

High-frequency laser pulses can penetrate forest canopy gaps to reach the ground, which acquire data not only on the fine canopy vertical structure but also the understory topography. The LiDAR signal can be used to accurately retrieve forest structural parameters, such as tree height, biomass, crown size, and Leaf Area Index (LAI). LiDAR overcomes the problem of vegetation index saturation of optical imagery in forest LAI inversion and significantly improves the accuracy of forest LAI inversion. In addition, terrestrial LiDAR can acquire the diameter at breast height (DBH) and height under branches of trees, which provides efficient and accurate data support for forest resource investigation.

1.3.3 Digital Cities

Building 3D models is an important part of digital city construction; especially in a real 3D world, the demand for 3D spatial information is unprecedented. Airborne and terrestrial LiDAR systems can perform multiangle and all-around rapid scanning of urban buildings and the surrounding environment. The acquired 3D point cloud data of buildings can be reconstructed to provide high-precision and measurable true 3D digital models required in digital city construction through data processing and 3D modeling. In addition, these 3D models can be published on the Internet, achieving real-time and interactive presentation of the urban scene, as well as an immersive user experience.

1.3.4 Digital Power Grid

The application of LiDAR in the digital power grid covers the whole process of power grid construction, such as power line design and planning, 3D digitization of the power infrastructure, hazard detection, and early warning analysis. For the power line design, 3D point cloud data can intuitively show the topography and surface coverage of the entire power transmission area, providing a scientific basis for power line selection and design, survey and positioning, 3D simulation, and construction volume estimation. For the power line safety inspection, LiDAR can accurately detect the 3D position of power lines and power towers; intuitively display the 3D spatial position relationship between the power lines and the other objects in the power transmission channel; and help analyze the safety distance between the power lines and the ground, the power line span distance, and the safety distance between the power lines and the vegetation. In addition, combined with other parameters (temperature, wind speed, etc.) acquired by monitoring equipment installed on the power tower, the power line arc sag changes under different working conditions can be simulated to conduct hazard analysis and provide an early warning. A comparative analysis of multiperiod LiDAR point cloud data can also analyze the changes (e.g., tree growth, illegal building construction) in the power transmission corridors and the possible dangers to the safe operation of power lines.

1.3.5 Crop Monitoring

LiDAR returns can accurately characterize light penetration into the canopy and provide fine canopy structure information in the vertical direction. This makes its application in low vegetation such as crops possible. For example, the airborne and terrestrial LiDAR data can achieve accurate estimation of crop parameters, such as crop plant height, LAI, leaf angle distribution, fraction of absorbed photosynthetically active radiation (FPAR), and aboveground biomass. In addition, the LiDAR and hyperspectral data can be integrated for classifying crops and mapping various crop parameters.

1.3.6 Cultural Heritage Digitization and Preservation

Stone carvings, grottoes, ancient buildings, and other cultural heritage sites are valuable treasures left to us by the ancients. Their digitized 3D models are not only formally displayed in an all-round way but also of great significance for their protection, 3D digital archiving, and permanent preservation and dissemination. LiDAR technology can directly and quickly acquire high-precision, high-density 3D spatial information on the surface of these heritage sites. Also, the high-resolution digital camera carried by LiDAR systems can acquire the fine feature information of objects. By combining the LiDAR point cloud and camera photos, the real 3D digital model of cultural heritage sites can be constructed. In addition, the terrestrial 3D laser scanning technology can digitally record and preserve the archaeological site, such as the 3D measurement of cultural relics and analysis of the erosion of the site surface, to achieve the dynamic display of the archaeological process, digital record, and analysis of the changes. The airborne LiDAR can obtain the understory topographic information, providing basic data support for understory archaeology.

1.3.7 Unmanned Driving

LiDAR calculates the distance to the obstacle based on the round-trip time between the laser to the obstacle. In the driverless field, LiDAR helps cars autonomously sense the road environment and automatically plan driving routes to reach the intended destination. LiDAR can also accurately measure the relative distance between the object's contour edge in the field of view and the LiDAR. By using the contour point cloud, a 3D environment map can be drawn through 3D modeling, with an accuracy of up to centimeters.

1.3.8 Transportation Route Planning

Fine road modeling is very important in road engineering design, pavement inspection, and 3D visualization. Traditional methods, such as single-point measurement by GNSS or total station, only obtain discrete data and are affected by road vehicles, which makes it difficult to know the road situation completely and accurately. Different from the single-point measurement method, LiDAR can obtain high-precision 3D coordinates of the road through scanning and does not require a control point network to be established. After data processing, a high-precision 3D digital model of the road panorama is obtained. In addition, LiDAR technology is used as an important detection means in railroad tunnel construction, such as monitoring and sampling in the tunnel construction process, generation of high-precision cross-section maps, blasting area and volume analysis, excavation earthwork volume calculation, flatness analysis of the excavated tunnel wall, over- or under-digging analysis, and check and calibration of the tunnel boring direction. In recent years, high-speed railroads have been developing rapidly over the world, and LiDAR plays an important role in the rapid detection of track micro-changes.

1.3.9 Mine Monitoring

There are more and more large and super-large mines. In particular, the open-pit mines are facing the problem of expanding mining scale and deepening mining depth gradually, which brings threats to the stability of mine slope rocks and soil bodies. LiDAR technology extends the point measurement to the surface measurement to a large extent. It can be used for mapping open-pit mine slopes and other complex fields and reconstructing 3D models of the mine scene. Through multitemporal LiDAR data, the users can extract the deformation information of the slopes and ore body to provide data support for the mine safety monitoring.

1.3.10 Other Applications

In addition to the previous applications, LiDAR is used in near-coastal topographic mapping, river surveying, flood assessment, dam deformation monitoring, and other fields. This section does not describe such applications in detail. Interested readers can refer to the relevant literature for more information.

1.4 SUMMARY

This chapter first briefly introduces the origin, characteristics, classification, and development history of LiDAR, as well as the LiDAR system composition. Then, the current mainstream commercial LiDAR equipment and data processing software products are summarized. Finally, several representative applications of LiDAR remote sensing are briefly introduced.

EXERCISES

1. Briefly describe the characteristics of LiDAR.
2. What are the components of a LiDAR system? Explain the role of each component.
3. List five ways to categorize LiDAR.
4. What are the LiDAR point cloud data formats? Please introduce one of the data formats in detail.

NOTES

1. IEC 60825-1:2007 Safety of laser products — Part 1: Equipment classification and requirements.
2. Theodore Harold (Ted) Maiman, an American physicist produced a collimated, monochromatic, coherent beam of light by amplifying a line of optical radiation at a narrow-amplitude frequency through stimulated radiation amplification and necessary feedback resonance on July 8, 1960. Since then, the world's first ruby laser has been introduced.
3. Friedrich Ackermann, a very prominent leader in the field of photogrammetry worldwide, is known for his outstanding achievements in photogrammetric research.

REFERENCES

He, Y., Hu, S., Chen, W., Zhu, X., Wang, Y., Yang, Z., Zhu, X., Lv, D., Huang, T., Xi, X., Qu, S., & Yao, B. (2018). Research progress of domestic airborne dual-frequency LiDAR detection technology (in Chinese). *Laser & Optoelectronics Progress*, *55*(8), 082801.

Li, M., Zuo, J., Zhu, J., & Meng, T. (2013). Research on dual-channel 3D imaging LiDAR technology (in Chinese). *Science of Surveying and Mapping*, *38*(3), 49–51.

2 LiDAR Remote Sensing Principles

2.1 LiDAR RANGING PRINCIPLE

Ranging is a very important application of LiDAR systems. There are two main LiDAR ranging modes: pulse-based ranging and phase-based ranging. The ranging principles of the two modes are described in the following sections.

2.1.1 RANGING PRINCIPLE OF PULSED LiDAR

At present, most LiDAR systems work in the pulsed-based ranging mode. That is, the laser emits a very narrow pulse (the pulse width is usually less than 50 ns) toward the target, and then part of the pulse is reflected by the target and returns to the receiver. The LiDAR system records the time of flight (TOF) between the emission and reception of a laser pulse and then calculates the distance between the senor and the target according to the speed of light, as in Equation (2.1). Figure 2.1 shows the ranging principle of pulsed LiDAR.

$$R = \frac{1}{2} \cdot c \cdot t \tag{2.1}$$

where R is the distance between the sensor and target, c is the speed of light in air, and t is the TOF of the laser between emission and reception.

Differentiating Equation (2.1), we have Equation (2.2):

$$\Delta R = \frac{1}{2} \cdot c \cdot \Delta t \tag{2.2}$$

where ΔR is the ranging resolution, representing the minimum distance at which two objects can be distinguished, depending on the time sampling interval.

Additionally, in order to ensure that the returns of multiple laser beams can be distinguished, the pulsed LiDAR system must transmit the next laser pulse after receiving the return of the previous laser pulse. In this case, the maximum measurement distance (R_{max}) needs to be considered (Lai, 2010). It is calculated by the longest time measurement (t_{max}), as in Equation (2.3):

$$R_{max} = \frac{1}{2} \cdot c \cdot t_{max} \tag{2.3}$$

In practical applications, the maximum measurement distance is affected by many factors, such as laser power, beam divergence, atmospheric transmission,

DOI: 10.1201/9781032671512-2

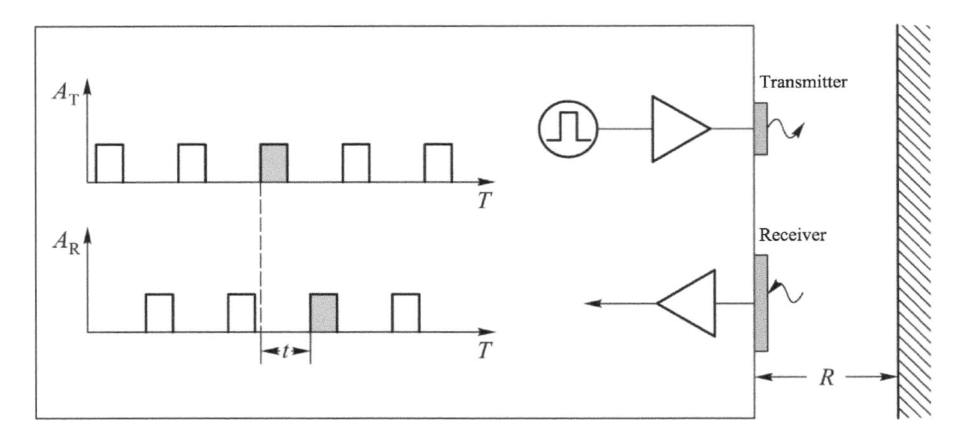

FIGURE 2.1 Ranging principle of pulsed LiDAR.

target reflection characteristics, detector sensitivity, flight altitude, and deviation of flight attitude record.

2.1.2 Ranging Principle of Phased LiDAR

The phase-based LiDAR is also called continuous wave LiDAR. It transmits an intensity-modulated continuous laser which is reflected by the target and then returns to the receiver. The distance between the target and sensor is calculated by the phase variation of the laser during the round trip, as in Figure 2.2. Generally, the ranging precision of a phased LiDAR system is higher than that of a pulsed LiDAR.

Assuming that the modulated continuous laser wave is a sine waveform with a period of T, the ratio of the time interval of the measured phase difference to the

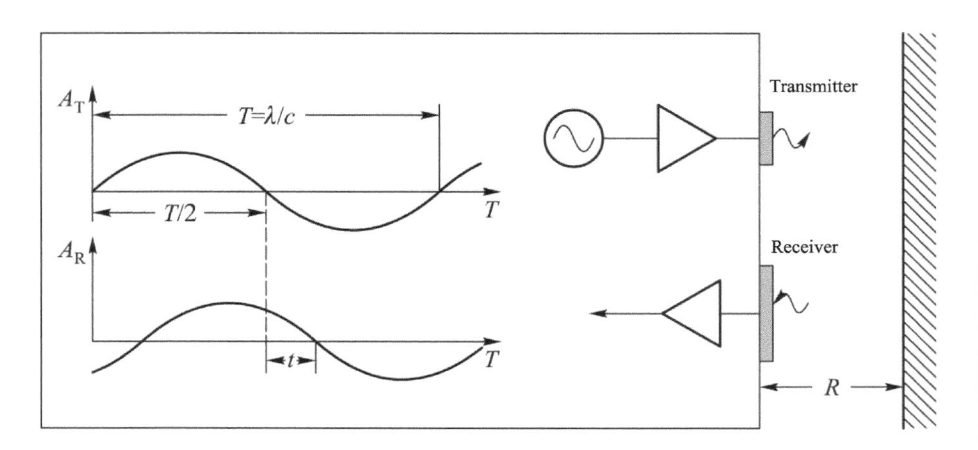

FIGURE 2.2 Ranging principle of phased LiDAR.

period T is exactly the ratio of the phase difference of the transmitted wave and the received wave to the phase 2π, as in Equation (2.4).

$$\frac{t}{T} = \frac{\phi}{2\pi} \tag{2.4}$$

where ϕ is the phase difference of the transmitted wave and the received wave. According to Equation (2.4), the distance between the sensor and target (R) is expressed as in Equation (2.5).

$$R = \frac{c}{2} \cdot \frac{T}{2\pi} \cdot \phi \tag{2.5}$$

where the period T is the reciprocal of the modulated frequency f of the sine wave. In this case, R can be expressed as in Equation (2.6).

$$R = \frac{1}{4\pi} \cdot \frac{c}{f} \cdot \phi \tag{2.6}$$

where the ratio of the speed of light (c) to the modulated frequency (f) is exactly the wavelength (λ). Hence, the distance R is as in Equation (2.7).

$$R = \frac{\lambda}{4\pi} \cdot \phi \tag{2.7}$$

Differentiating Equation (2.7), we get the ranging resolution (ΔR) of phased LiDAR, as in Equation (2.8).

$$\Delta R = \frac{\lambda}{4\pi} \cdot \Delta\phi \tag{2.8}$$

Equation (2.8) indicates that the ranging resolution of the phased LiDAR is related to not only the phase resolution but also the laser wavelength. In contrast, the ranging resolution of the pulsed LiDAR is only related to the time resolution [Equation (2.2)]. With the same phase (time) resolution, the ranging resolution of the phased LiDAR system could be improved by using a shorter wavelength. Hence, the phased LiDAR is usually used in applications with high ranging precision. For example, for a continuous laser wave with the period of 1 second, assuming that the phase resolution is 0.4°, its ranging resolution is 0.1 mm. However, to achieve the same ranging resolution, the pulsed LiDAR system must have a time resolution of 1 picosecond, which requires higher standards for the time interval counting hardware system.

Although the phased LiDAR system is more suitable for high-precision ranging applications, it is usually used for short-range detection. This is because it requires high power to emit a continuous laser. Also, it may cause ranging ambiguity. The maximum measured distance (R_{max}) of the phase-based LiDAR system which can uniquely determine the target is the distance measured when the continuous wave travels exactly one period, as in Equation (2.9).

$$R_{\max} = \frac{\lambda}{4\pi} \cdot \phi_{\max} = \frac{2\pi}{4\pi} \cdot \lambda = \frac{\lambda}{2} \tag{2.9}$$

Equation (2.8) indicates that the ranging resolution (ΔR) of the phased LiDAR system is related to the phase resolution ($\Delta\phi$) and the laser wavelength (λ). The shorter the laser wavelength used, the higher the ranging resolution. Hence, phased LiDAR makes it relatively easy to achieve high ranging resolution. However, Equation (2.9) shows that the maximum measured distance ($R_{\max}$) of phased LiDAR is determined by the wavelength (λ). The longer the wavelength, the longer the maximum measured distance. Long-distance measurement and high ranging resolution both cannot be achieved in a continuous wave LiDAR system. Therefore, in order to meet the requirements of practical applications, the phased LiDAR system is usually equipped with a signal frequency modulation device that can adjust multiple frequencies. Additionally, high-power phased lasers are difficult to produce. Currently, most LiDAR systems use a pulsed ranging system. The LiDAR mentioned in the rest of this book refers to the pulsed LiDAR.

2.1.3 Ranging Precision of LiDAR

Ranging precision is a pivotal parameter of a LiDAR system and is different from ranging resolution. Specifically, the ranging resolution refers to the limit of the ability to measure the distance between two objects along the same line of sight, while the ranging accuracy is the standard deviation of the target distance estimation in the presence of noise. The ranging precision is related to the signal-to-noise ratio (SNR) of the returned laser signal and the emitted laser pulse (Wehr & Lohr, 1999).

Usually, the ranging precisions of pulsed LiDAR and phased LiDAR are expressed as in Equations (2.10) and (2.11), respectively.

$$\sigma_{R_{\text{pulse}}} \propto \frac{c}{2} \cdot t_{\text{rise}} \cdot \frac{\sqrt{B_{\text{pulse}}}}{P_{R_{\text{peak}}}} \tag{2.10}$$

$$\sigma_{R_{\text{cw}}} \propto \frac{\lambda}{4\pi} \cdot \frac{\sqrt{B_{\text{cw}}}}{P_{R_{\text{av}}}} \tag{2.11}$$

where $\sigma_{R_{\text{pulse}}}$ is the ranging precision of pulsed LiDAR, $\propto$ indicates a proportional relationship, t_{rise} is the pulse rise time, B_{pulse} is the noise width of pulsed LiDAR, $P_{R_{\text{peak}}}$ is the peak power of the received signal of pulsed LiDAR, $\sigma_{R_{\text{cw}}}$ is the ranging precision of phased LiDAR, λ is the continuous laser wavelength, B_{cw} is the noise width of phased LiDAR, and $P_{R_{\text{av}}}$ is the average power of the received signal of phased LiDAR.

Here we compare the performance of the two ranging modes with the assumption of measuring the same target. In this case, the received power can be replaced by the transmitted power. That is, the peak power $P_{R_{\text{peak}}}$ of the received laser of pulsed LiDAR is replaced by the peak power of the transmitted pulse $P_{T_{\text{peak}}}$, and the average power of the received laser of phased LiDAR $P_{R_{\text{av}}}$ is replaced by that of the transmitted laser $P_{T_{\text{av}}}$. The ratio of ranging precisions of pulsed LiDAR to phased LiDAR is expressed as in Equation (2.12).

$$\frac{\sigma_{R_{\text{pulse}}}}{\sigma_{R_{\text{cw}}}} \propto 2\pi \cdot \frac{c}{\lambda} \cdot t_{\text{rise}} \cdot \frac{P_{T_{\text{av}}}}{P_{T_{\text{peak}}}} \cdot \sqrt{\frac{B_{\text{pulse}}}{B_{\text{cw}}}} \tag{2.12}$$

Assuming that:

$$t_{\text{rise}} \propto \frac{1}{B_{\text{pulse}}} \tag{2.13}$$

Equation (2.12) can be rearranged as:

$$\frac{\sigma_{R_{\text{pulse}}}}{\sigma_{R_{\text{cw}}}} \propto 2\pi \cdot f \cdot \sqrt{\frac{t_{\text{rise}}}{B_{\text{cw}}}} \cdot \frac{P_{T_{\text{cw}}}}{P_{T_{\text{peak}}}} \tag{2.14}$$

With the assumption of the pulse rise time of pulsed LiDAR of 1 ns, the peak power of transmitted pulse of 2000 W, the frequency of phased LiDAR of 10 MHz, the noise width of phased LiDAR of 7 kHz, and the average transmit power of 1 W, the ratio of ranging precisions of pulsed LiDAR to phased LiDAR can be calculated by Equation (2.14), that is about 0.0012. This indicates that although the transmit power of the pulsed LiDAR is 2000 times that of the phased LiDAR, the ranging precision of the pulsed LiDAR can theoretically only reach 85 times that of the phased LiDAR.

2.2 LiDAR RADIATION PRINCIPLE

The LiDAR system performs non-contact detection of an object's surface by transmitting laser pulses and receiving the returned energy. Its remote sensing (RS) process includes four steps (Figure 2.3): (1) transmission of the emitted laser toward the target, that is, the pulse downward; (2) interaction of pulse and target surface, including interception, scattering, absorption, etc.; (3) transmission of the scattered laser toward the sensor, that is, pulse upward; and (4) reception of the scattered laser by the receiver. Through this process, the received LiDAR signal is a comprehensive expression of the sensor configuration, the target surface characteristics, and the surrounding environment. In order to interpret the LiDAR signal, revert the LiDAR RS process, and retrieve the target information, it is necessary to understand the radiation principle of LiDAR RS. This section presents the radiation principle of the LiDAR system, which is also the core of LiDAR's three-dimensional (3D) imaging mechanism.

2.2.1 LiDAR EQUATION

The physical process of the laser beam penetrating the atmosphere, interacting with scatterers, and returning to the receiver is very complex. Usually, beginners would

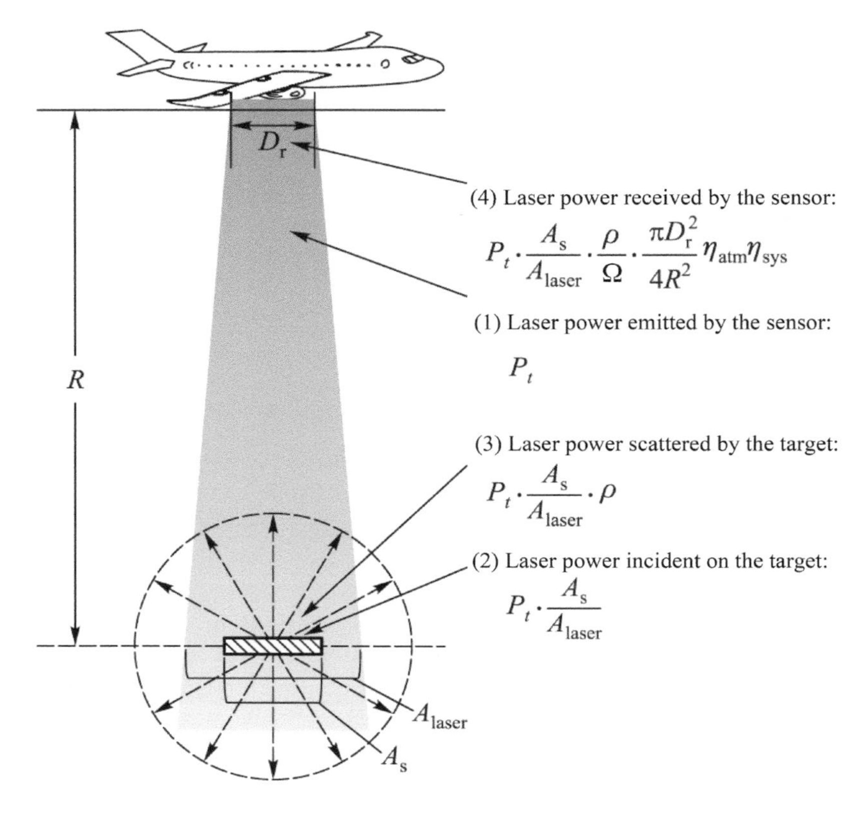

FIGURE 2.3 LiDAR RS process. (P_t: Laser power emitted by the sensor; A_s: effective area of target [scatterer]; A_{laser}: laser footprint area; ρ: reflectance of target [scatterer]; Ω: scattered solid angle of target toward the receiver; R: distance between sensor and target; D_r: diameter of LiDAR receiver; η_{atm}: effect of atmospheric transfer; η_{sys}: efficiency of sensor transmitting, receiving, and processing signals.)

simplify this complex process of LiDAR RS. The LiDAR equation is a physically based formula to compute the power returned to a receiver for given transmitted laser power, optical properties of the medium through which the LiDAR beam passes, and target properties. It is deducted based on two strong assumptions: (1) the power emitted by the laser is uniformly distributed within the footprint and (2) the scatterer uniformly disperses the incident power into a conical solid angle of Ω (Wagner et al., 2006). Based on the assumptions, the derivation process of the LiDAR equation is as follows.

The footprint area of the laser beam at the distance of the scatterer is:

$$A_{laser} = \frac{\pi R^2 \beta_t^2}{4} \tag{2.15}$$

where R is the distance between the sensor and scatterer and β_t is the tangent of the laser beam divergence angle. Based on assumption (1) earlier, the incident laser power density of the scatterer surface S_S is expressed as in Equation (2.16):

$$S_S = \frac{4P_t}{\pi R^2 \beta_t^2} \tag{2.16}$$

where P_t is the total emitted laser power. Equation (2.16) describes the property by which the laser power density gradually decays with the transmitted distance R. Considering the reflectance ρ and the effective area A_s of the scatterer, the laser power dispersed by the scatterer P_s is expressed as in Equation (2.17).

$$P_s = \frac{4P_t}{\pi R^2 \beta_t^2} \rho A_s \tag{2.17}$$

Based on assumption (2) earlier, if the scattered conical solid angle of the scatterer overlaps with the field of view of the LiDAR receiver, the returned laser power density is expressed as in Equation (2.18).

$$S_r = \frac{4P_t}{\pi R^2 \beta_t^2} \rho A_s \frac{1}{\Omega R^2} \tag{2.18}$$

Finally, considering the receiver aperture and the influence of the atmosphere and hardware systems, the LiDAR equation of a single scatterer can be expressed by Equation (2.19) as:

$$P_{R,i}(t) = \frac{4P_t}{\pi R^2 \beta_t^2} \rho A_s \frac{1}{\Omega R^2} \cdot \frac{\pi D_r^2}{4} \eta_{\text{atm}} \eta_{\text{sys}} \tag{2.19}$$

where $P_{R,i}(t)$ represents the returned laser power from the i-th scatterer at the time t, D_r is the diameter of the LiDAR receiver, Ω is the scattered solid angle of the scatterer, η_{atm} is the effect of atmosphere transfer on laser power, and η_{sys} is the effect of the sensor transmitting, receiving, and processing signals.

All the parameters related to the scatterer can be combined into one, which is called the laser radar cross section σ (LRCS) (Dai, 2002):

$$\sigma = \frac{4\pi}{\Omega} \rho A_s \tag{2.20}$$

Equation (2.20) shows that LRCS is determined by the effective area A_s, reflectance ρ, and scattered solid angle Ω of the scatterer.

In the case where multiple scatterers are distributed in the propagated path of the laser, the received signal can be expressed as a general expression of the LiDAR equation, as in Equation (2.21).

$$P_{\text{sum}}(t) = \sum_{i=1}^{N} \eta_{\text{atm},i} \eta_{\text{sys},i} \frac{D_r^2}{4\pi R_i^4 \beta_t^2} P_t * \sigma_i(R_i) \tag{2.21}$$

where $P_{\text{sum}}(t)$ is the accumulation of the return signals of N scatterers at time t and the relationship of the time t and the distance R_i is $t = \frac{2R_i}{c}$, * represents the convolution operation.

2.2.2 LiDAR Waveform Model

Once there are multiple scatterers at different distances from the sensor in the laser propagation direction, the emitted laser might be reflected multiple times, and the reflected laser intensity and the traveled distance are both different for each time. The LiDAR waveform records the laser intensity according to the returned time delay, that is, laser intensity as a function of time. As in Figure 2.4, the LiDAR waveform can be quantified as a one-dimensional waveform with time as the horizontal axis and intensity as the vertical axis. It is theoretically feasible to inverse the characteristics of scatterers by accurately interpreting the LiDAR waveform. However, many parameters in the LiDAR equation are difficult to obtain accurately, such as η_{atm} and η_{sys}. Also, the parameters such as Ω and ρ are different for different objects and under different observation conditions. This makes it very difficult to directly inverse the surface characteristics from the LiDAR waveform.

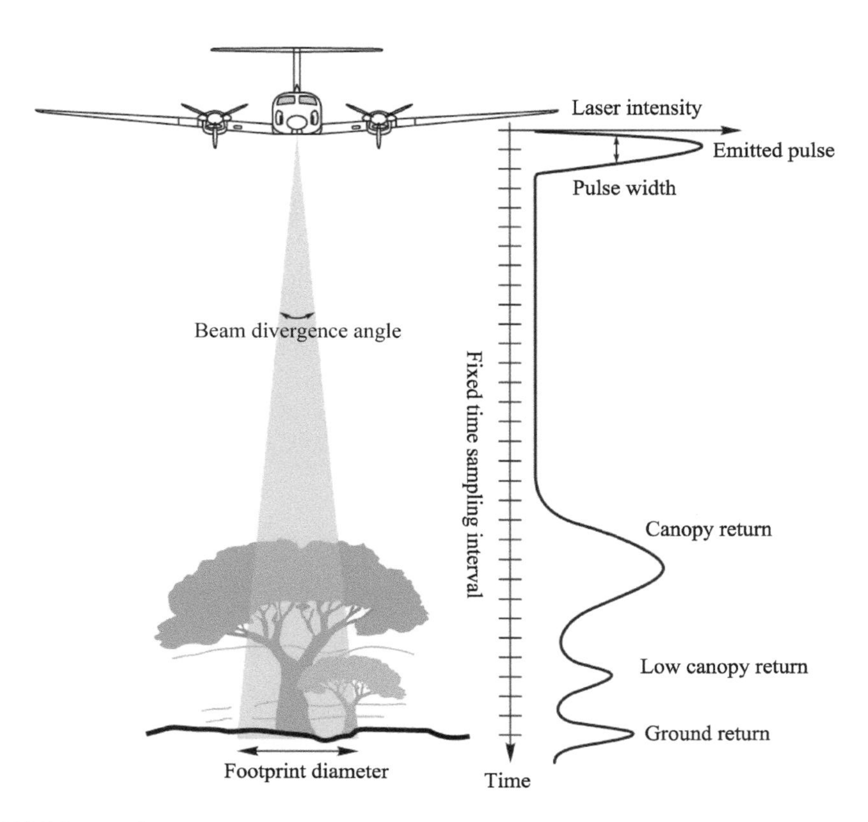

FIGURE 2.4 Schematic diagram of LiDAR waveform.

Therefore, in practical applications, the LiDAR equation need to be further simplified based on some specific assumptions.

The LiDAR waveform is a product of the interaction between the laser pulse and Earth's surface. The laser-surface interaction process can be expressed as the transfer function. Since the emitted laser pulse is close to a one-dimensional Gaussian distribution in the time domain, the LiDAR waveform is just the convolution of the emitted pulse and the transfer function in the time domain, which is usually approximated by a mixed Gaussian model. That is, the LiDAR waveform is regarded as a superposition of multiple Gaussian functions, as in Equations (2.22) and (2.23).

$$P_{\text{sum}}(t) = \sum_{i=1}^{N} f_i(t) \tag{2.22}$$

$$f_i(t) = A_i e^{-\frac{(t-\mu_i)^2}{2\sigma_i^2}} \tag{2.23}$$

where $f_i(t)$ is the i-th Gaussian component and A_i, μ_i, and σ_i are the amplitude, mean, and standard deviation of the i-th Gaussian component, respectively. Based on this principle, the mixed Gaussian function can be used to decompose the waveform, and the characteristic parameters of the ground object (such as reflectivity and geometric shape) can be obtained by inversion. For more details on waveform decomposition, see Section 4.2 of this book.

2.2.3 LiDAR Radiative Transfer Model

In addition to the LiDAR equation and the LiDAR waveform model based on very strong assumptions, some researchers have proposed relatively complex LiDAR radiative transfer models (RTMs). The RS radiative transfer models are considered celebrated tools for modeling, validating, and interpreting the RS signals, as they can simulate the interaction mechanisms of electromagnetic waves and Earth's surface that give rise to the signals.

To date, many researchers have developed a wide range of LiDAR RTMs with diverse complexity and efficiency. According to the degree of introduction of physical mechanism, the LiDAR RTMs can be divided into semi-empirical models, analytical models, and computer simulation models (Table 2.1). The semi-empirical models use a digital surface model (DSM) or similar methods to represent the elevation distribution of the Earth's surface and simulates the LiDAR signal as a temporal summation of Gaussian profiles that are computed by convolving the target reflectance and pulse energy distribution (Blair & Hofton, 1999). However, they only consider the single scattering of the laser on the target surface. Physically based analytical models use strong simplification of landscapes and ray propagations. For example, Sun and Ranson (2000) established a 3D LiDAR waveform simulation model, which simplifies the vegetation landscape into a series of 3D turbid voxels and only considers the single scattering of laser rays. Ni-Meister et al. (2001) created a LiDAR waveform simulation model based on 3D geometric optics radiative transfer (GORT),

TABLE 2.1
LiDAR Radiative Transfer Model

Category	Model	Proposer	Features
Semi-empirical model	DSM-based model	(Blair & Hofton, 1999)	DSM is needed as prior knowledge; only single scattering is considered
Analytical model	3D LiDAR model	(Sun & Ranson, 2000)	Canopy is simplified as turbid mediums; only single scattering is considered
	3D GORT model	(Ni-Meister et al., 2001)	Canopy is simplified as single-layer and double-layer forest; the ground is a flat or a terrain with fixed slope
Computer simulation model	RAPID	(Huang & Wynne, 2013)	Radiosity method
	DART-RC	(Gastellu-Etchegorry et al., 2016)	Forward ray tracing
	FLIGHT	(North et al., 2010)	Backward path tracing
	DIRSIG	(Goodenough & Brown, 2017)	Forward ray tracing
	LIBRAT	(Disney et al., 2009)	Backward path tracing
	RAYTRAN	(Govaerts & Verstraete, 1998)	Forward ray tracing
	FILES	(Kobayashi & Iwabuchi, 2008)	Forward ray tracing
	DART-Lux	(Yang et al., 2022)	Bidirectional path tracing, efficient and accurate

which simplifies the ground as a terrain with a fixed slope or a flat surface without undulations and assumes that the ground has the same reflectance and the vegetation is a single-layer or double-layer forest. The semi-empirical and analytical models all make strong simplifications and assumptions on Earth's scenery and laser propagation process. Although computationally efficient, these models have an increasingly unsuitable accuracy for most potential LiDAR applications, such as signal modeling, parameter inversions of multiple targets, and LiDAR sensor design.

With the development of computer graphics, some LiDAR computer simulation models based on Monte Carlo ray tracing (MCRT) (Pharr et al., 2016) were proposed, such as DART (Gastellu-Etchegorry et al., 2016; Yang et al., 2022) and FLIGHT (North et al., 2010). These MCRT-based LiDAR 3D RTMs are the most commonly used and accurate models since they are more adaptive to complex scene structures and usually do not make simplifications of ray interactions. According to the direction of ray tracing, they are usually categorized as forward models if rays are traced from the laser source, backward models if rays are traced from the receiver, and bidirectional path tracing (BDPT) models if rays are traced from the source and receiver simultaneously. Specifically, the forward ray tracing models use the laser source as the starting point for ray tracing and record the signal that finally enters the receiver. The backward tracing models trace the path from the receiver to ensure that the traced signal is received. The bidirectional tracing model traces the rays from the source and receiver synchronously. The purpose of all these ray tracing models is to generate a series of random paths that connect the source and receiver. For LiDAR systems with the received field of view larger than the illumination area (footprint), the bidirectional tracing model usually has the highest accuracy and efficiency, followed by the forward tracing models and the backward tracing model with the lowest.

Here we introduce a BDPT-based LiDAR modeling method called DART-Lux (Yang et al., 2022). The DART-Lux LiDAR model transforms the laser transport problem into an integration over all possible paths in the 3D landscape that connect the laser source and the receiver. It relies on the light transport equation, as in Equation (2.24).

$$L(r',\Omega_o) = L_e(r',\Omega_o) + \int_{\Omega} L(r',\Omega_i) \cdot f(r',-\Omega_i,\Omega_o) \cdot |\cos\theta_i| \cdot d\Omega_i \quad (2.24)$$

where the exitant radiance $L(r',\Omega_o)$ at vertex r' along direction Ω_o is the sum of the emitted radiance $L_e(r',\Omega_o)$ and the scattered radiance due to incident radiance $L(r',-\Omega_i)$ along direction Ω_i, $f(r',-\Omega_i,\Omega_o)$ is the bidirectional scattering distribution function (BSDF) of the surface at vertex r', and θ_i is the incident angle between the incident direction Ω_i and the surface normal vector. Equation (2.24) can be transformed into an area integration form instead of a solid angle integration form, as in Equation (2.25).

$$L(r' \to r) = L_e(r' \to r) + \int_{A} L(r'' \to r') \cdot f(r'' \to r' \to r) \cdot G(r'' \to r') \cdot dA(r'') \quad (2.25)$$

where r'' and r are the previous vertex and next vertex of r' in the light transport process, $dA(r'')$ is the area at vertex r'', and the connection function $G(r'' \leftrightarrow r')$ between vertices r'' and r' is the product of $|\cos\theta_i|$ and the Jacobian term to transfer solid angle integration over area integration.

Since laser rays might be scattered multiple times in the 3D landscape, the received laser power can be represented by incrementally expanding Equation (2.25) to an infinite sum of a multidimensional integral, which can be expressed as a Lebesgue integral, as in Equation (2.26).

$$\Phi_{\text{LiDAR}}(\tau(\bar{\tau})) = \int_{\mathcal{D}} f(\bar{r}) \cdot d\mu(\bar{r}) \quad (2.26)$$

where $\tau(\bar{r})$ is the laser travel time along path $\bar{r}$ and $\Phi_{\text{LiDAR}}(\tau(\bar{r}))$ is the returned laser power at the time $\tau(\bar{r})$. $\mathcal{D}$ is the set of all paths. $\mathcal{D} = \bigcup_{n=1}^{\infty} \mathcal{D}_n$ with $\mathcal{D}_n$ the set of paths with n edges (n is also called the path depth). $\bar{r}$ is a path of the set of paths that connect the laser source and the receiver; it passes through a series of vertices in the scene: $\bar{r} \in \mathcal{D} = \{\bar{r}_n | \bar{r}_n = r_0, r_1, \ldots, r_n;\ r_{k=0,1,\ldots,n} \in A,\ n \in \mathbb{Z}^+\}$. $f(\bar{r})$ is the power contribution of path $\bar{r}$. $d\mu(\bar{r})$ is the area-product measure of path $\bar{r}$, $d\mu(\bar{r}_n) = dA(r_n) \cdot dA(r_{n-1}) \cdots dA(r_0)$. The solution of Equation (2.26) is the simulated LiDAR signal. The DART-Lux LiDAR model uses a Monte Carlo–based BDPT

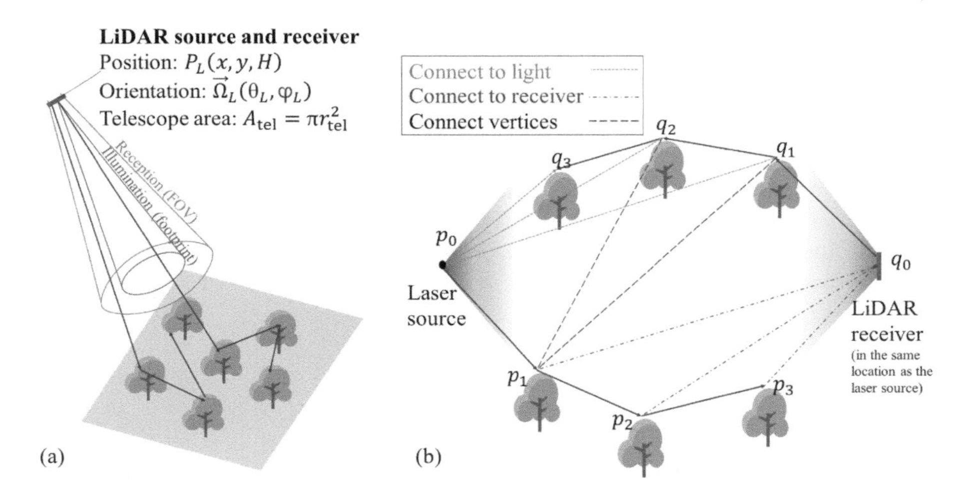

FIGURE 2.5 Schematic diagram of the DART-Lux LiDAR model. (a) A sub-path starts from the LiDAR source and a sub-path starts from the receiver (at the same location as the source). (b) A series of paths that connect the LiDAR source and receiver are generated by the BDPT method.

algorithm to solve Equation (2.26). Generally, the source and receiver of the LiDAR sensor are considered to have the same central position $P_L(x,y,H)$ and orientation $\vec{\Omega}_L(\theta_L,\varphi_L)$. The paths from laser source to receiver are constructed based on the combination of two sub-paths that start from the source and the receiver, respectively, as in Figure 2.5.

In the DART-Lux LiDAR model, the first random walk starts from the laser source and samples a light sub-path $\overline{p}_{N_s} = p_0, p_1, \ldots, p_{N_s\text{-}1}$ with N_s vertices, of which vertex p_0 is on the laser source. Each vertex is sampled with the probability distribution function (PDF) $p\left(\overrightarrow{p_k}\right)$ $(k = 0,1,\ldots,N_s\text{-}1)$. The second random walk starts from the receiver and samples a receiver sub-path $\overline{q}_{N_t} = q_0, q_1, \ldots, q_{N_t\text{-}1}$ with N_t vertices, of which vertex q_0 is on the receiver. Each vertex is sampled with the PDF $p\left(\overleftarrow{q_k}\right)$ $(k = 0,1,\ldots,N_t\text{-}1)$. The combination of the two sub-paths leads to the construction of a series of complete paths from source to receiver by using three sampling techniques: (1) "connect to light": the vertex on the receiver sub-path is connected to a sampled vertex on the laser source; (2) "connect to receiver": the vertex on the light sub-path is connected to a sampled vertex on the receiver; and (3) "connect vertices": a vertex on the light sub-path is connected to a vertex on the receiver sub-path. Also, a light sub-path can randomly hit the LiDAR receiver, and a receiver sub-path can randomly hit the laser source. The length $l\left(\overline{r}_{s,t}\right)$ of any path $\overline{r}_{s,t}$ is the sum of the lengths of the light sub-path, the receiver sub-path, and the distance between the connected vertices p_{s-1} and q_{t-1}. The path length is further converted to the laser travel time $\tau(\overline{r})$. The power contribution divided by the path PDF $\left(\frac{f(\overline{r}_{s,t})}{p(\overline{r}_{s,t})}\right)$ of a random path $\overline{r}_{s,t}$ can unbiasedly estimate the laser returned power Φ_{LiDAR}. The simulated temporal power profile is the so-called LiDAR waveform. Furthermore, the DART-Lux LiDAR model extends the single-pulse waveform LiDAR modeling component to simulate multiplatform (satellite, airborne, terrestrial), multitype (waveform, discrete return, photon counting), and multipulse LiDAR signals. The multipulse simulation of laser scanning systems is an iterative loop of single-pulse modeling with specific geometry configurations and methods of detection and digitization. The point cloud of discrete return LiDAR is converted from the simulated waveform using Gaussian decomposition or peak detection methods. The points of photon counting LiDAR are derived per waveform using a statistical method based on the instrumental parameters of the single-photon detector.

2.3 PRINCIPLES OF LiDAR ON DIFFERENT PLATFORMS

According to different application requirements, LiDAR systems can be equipped on different platforms, including spaceborne (satellites and the International Space Station [ISS]), airborne (manned and unmanned aerial vehicles), ground (car, backpack, terrestrial tripod), etc. This section introduces the principles of LiDAR on different platforms.

2.3.1 Spaceborne LiDAR

Until 2022, the existing spaceborne LiDAR systems for Earth observation include Ice, Cloud, and Land Elevation Satellite/Geoscience Laser Altimeter System (ICESat/GLAS), Advanced Topographic Laser Altimeter System (ICESat-2/ATLAS), Global Ecosystem Dynamics Investigation (GEDI), and Gaofen (GF)-7. Among them, the ICESat-2/ATLAS implements the micro-pulse photon counting LiDAR technique, which is introduced in Section 2.4.3. The other systems use the full-waveform LiDAR. Here we mainly introduce the principles of spaceborne full-waveform LiDAR.

The spaceborne LiDAR, with high platform altitude and repeated, large-area observations, can provide reliable and multitemporal height measurement data, which have been widely used in mapping the surface vertical structure of the Earth and the surface topography of other planets. For example, the first LiDAR satellite ICESat/GLAS system detects the Earth's surface by emitting the laser beam, and each beam forms a nearly round footprint with a diameter of about 70 m on the ground (Schutz et al., 2005). The GEDI system, whose main mission is to monitor global forest biomass, carbon cycling, and biodiversity, has a footprint diameter of about 25 m on the ground (Dubayah et al., 2020). The GF-7 laser altimeter system has a footprint diameter of about 17 m.

The footprint diameter of spaceborne LiDAR is usually very large (>10 m) due to the high observation platform. The large footprint diameter might cause the ground vertical extent to overlap with the object vertical extent. The degree of overlapping is affected by the terrain slope, sensor observation orientation, and footprint diameter. Usually, to avoid signal aliasing as much as possible, the spaceborne LiDAR is designed with an on-nadir viewing mode, which means that the laser pulse is emitted vertically to the Earth's surface. Even so, sometimes, in order to collect more altimeter data, some spaceborne LiDAR systems are designed with multiple laser ground tracks by beam splitting and dithering, e.g., GEDI and

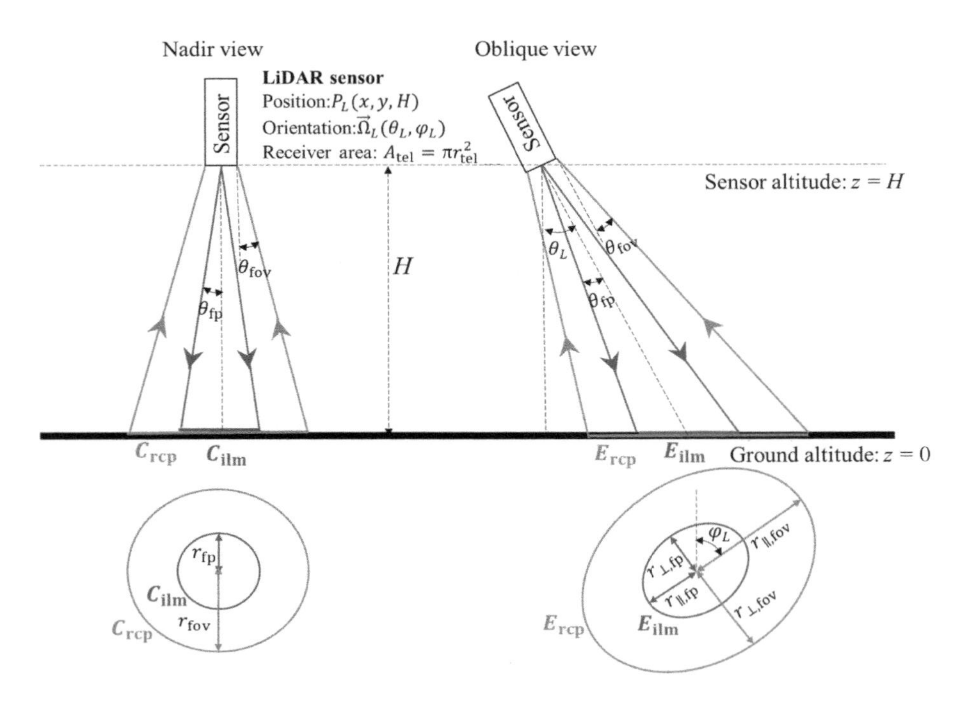

FIGURE 2.6 Pulse geometry of spaceborne LiDAR for nadir and oblique orientations, with the illumination and reception areas at the horizontal plane.

GF-7 systems. The modulation of multibeam splitting and dithering would cause the emitted laser pulse to be slightly incline with the milliradian level, which is usually negligible.

Figure 2.6 shows the geometry configuration of the laser pulses for nadir and oblique views. Generally, the LiDAR transmitter (laser) and receiver are considered to have the same location $P_L(x, y, H)$ and the same center orientation $\vec{\Omega}_L(\theta_L, \varphi_L)$. θ_L and φ_L are the zenith angle and azimuth angle of the LiDAR orientation, respectively; H is the sensor altitude. The laser can be regarded as a point source that emits laser power into a fixed conical solid angle with the divergence half-angle of θ_{fp}. The LiDAR receiver is usually a circular aperture with a radius of r_{tel} and the field of view (FOV) half-angle of θ_{fov}.

For the nadir view, the zenith and azimuth angles of the laser pulse are both 0°. A circular illumination area (i.e., C_{ilm} in Figure 2.6, also called the footprint) would be generated on the flat ground with an elevation of 0. The FOV area of the LiDAR receiver is also circular (i.e., C_{rcp} in Figure 2.6). The radii of the illumination area (r_{fp}) and FOV area (r_{fov}) are calculated as in Equation (2.27).

$$r_{fp} = H \cdot \tan\theta_{fp}, \quad r_{fov} = H \cdot \tan\theta_{fov} + r_{tel} \tag{2.27}$$

For the oblique view, the zenith angle and azimuth angle of the laser pulse are θ_L and φ_L, respectively. The illumination area (i.e., E_{ilm} in Figure 2.6) and the FOV area (i.e., E_{rcp} in Figure 2.6) of the LiDAR system on flat ground are both elliptical. The long axis of the ellipse is in the same direction as the laser azimuth. In this case, the semi-major long axes (parallel to the laser azimuth) and the semi-minor axes (perpendicular to the laser azimuth) of the illumination area and the FOV area are calculated as in Equation (2.28).

$$r_{\perp,\mathrm{fp}} = \frac{H \cdot \tan\theta_{\mathrm{fp}}}{\cos\theta_L}, \quad r_{\parallel,\mathrm{fp}} = \frac{H \cdot \tan\theta_{\mathrm{fp}}}{\cos^2\theta_L}$$
$$r_{\perp,\mathrm{fov}} = \frac{H \cdot \tan\theta_{\mathrm{fov}}}{\cos\theta_L} + r_{\mathrm{tel}}, \quad r_{\parallel,\mathrm{fov}} = \frac{\dfrac{H \cdot \tan\theta_{\mathrm{fov}}}{\cos\theta_L} + r_{\mathrm{tel}}}{\cos\theta_L} \tag{2.28}$$

where $r_{\perp,\mathrm{fp}}$ and $r_{\parallel,\mathrm{fp}}$ are the semi-major and semi-minor axes of the illumination area, respectively; $r_{\perp,\mathrm{fov}}$ and $r_{\parallel,\mathrm{fov}}$ are the semi-major and semi-minor axes of the FOV area, respectively.

2.3.2 Airborne LiDAR

Airborne LiDAR is defined as a LiDAR system equipped on the flying platforms such as drones and helicopters. It is also called airborne laser scanning (ALS), airborne laser altimetry (ALA), airborne laser topographic/terrain mapping (ALTM), and airborne laser mapping (ALM), among others. Although the names are slightly different, the principles are similar. That is, the LiDAR senor generates laser pulses and receives returned signals, and the scanning device controls the orientation of the LiDAR sensor. The main features of an airborne LiDAR system are as follows:

1. High precision. The vast majority of airborne LiDAR systems on the market use the pulsed laser ranging system with the consideration of its long measurement distance. However, it should be noted that the pulsed LiDAR requires advanced hardware and processing methods to achieve high ranging precision. Additionally, the ranging precision is influenced by the working environment to some extent. The LiDAR usually works best in dry, cold, and clear atmospheric conditions. In contrast, the LiDAR performs poor in the daytime with strong sunlight.
2. High power. The airborne LiDAR system scans the Earth's surface in the air. It requires a relatively high laser power, so that the returned laser energy through long-distance atmospheric loss and target absorption is strong enough to be recorded by the detector.
3. Small size. The aircraft's carrying capacity and volume are limited. It is required to load LiDAR equipment and carry operators in a limited space. Hence, airborne LiDAR systems are commonly designed to be small and lightweight.
4. Suitable wavelength. It would be better that the selected laser wavelength satisfies the following requirements: (1) the wavelength is located in the atmospheric window, so that the energy is less absorbed by the atmosphere;

(2) the wavelength is positioned where target has strong reflectance, so that the target can return a strong laser signal; (3) the wavelength is positioned where the detector sensitivity is high, so as to reduce the entry of light in other bands; and (4) it is safe for human eyes. Most existing airborne LiDAR systems usually use a 1064-nm, 1550-nm, 905-nm, or 532-nm wavelength. Some use other special wavelengths, but relatively few.

The airborne LiDAR systems scan the target in specific modes (Wehr & Lohr, 1999). The four main scanning modes are as follows:

1. Swing mirror scanning mode (Figure 2.7). This scanning mode drives the mirror with a motor to repeatedly swing within a certain angle, to achieve a series of scanning lines of laser beams on the ground. In one swing mirror cycle, the laser forms a periodic movement track on the ground, usually Z-shaped or sinusoidal. The advantage of the swing mirror mode is the simple principle, high scanning efficiency, and large scanning angle range. One of its significant disadvantages is that the laser spot density is not even, being denser at both ends and sparse in the middle, caused by the mechanical device continuously undergoing acceleration and deceleration processes

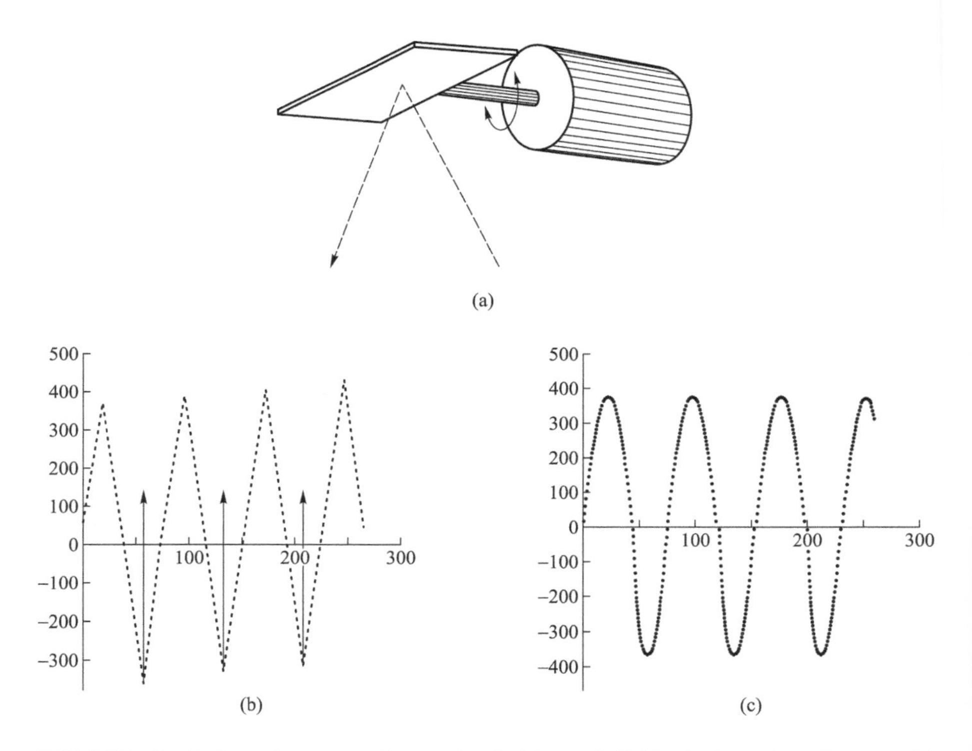

FIGURE 2.7 Swing mirror scanning mode of airborne LiDAR. (a) Scanning diagram; (b) Z-shaped ground scan line; (c) sinusoidal ground scan line.

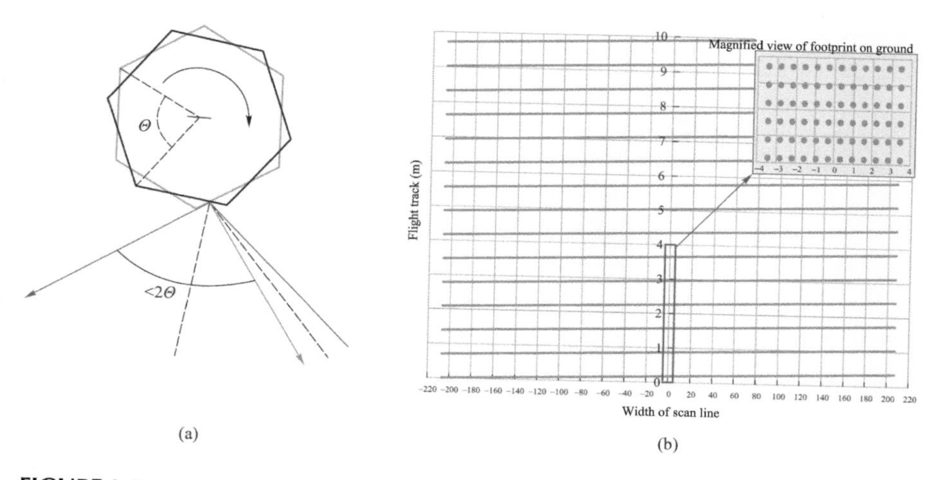

FIGURE 2.8 Rotating prism scanning mode of airborne LiDAR. (a) Scanning diagram; (b) parallel ground scan lines.

during the swing periods. Also, this mode has the disadvantages of poor seismic performance and high requirements for motor performance.

2. Rotate prism scanning mode (Figure 2.8). This mode causes the polygonal prism with the motor to rotate, causing the direction of the reflected laser beam to reciprocate within a certain range and scan on the ground. The scanning lines formed are usually parallel lines. Its advantage is that the obtained pulse density is uniform, as the angular velocity of the prism rotation is constant. The disadvantage is its relatively low scanning efficiency.
3. Elliptical scanning mode (Figure 2.9). This mode forms an elliptical scan line on the ground. With the movement of the sensor, a series of overlapping elliptical scan lines is presented on the ground. Its advantages include the simple scanning principle, good seismic performance, high ranging accuracy, and similar influence of inclination angle and atmospheric transfer on all laser points. However, the sampling points of the elliptical trajectory are

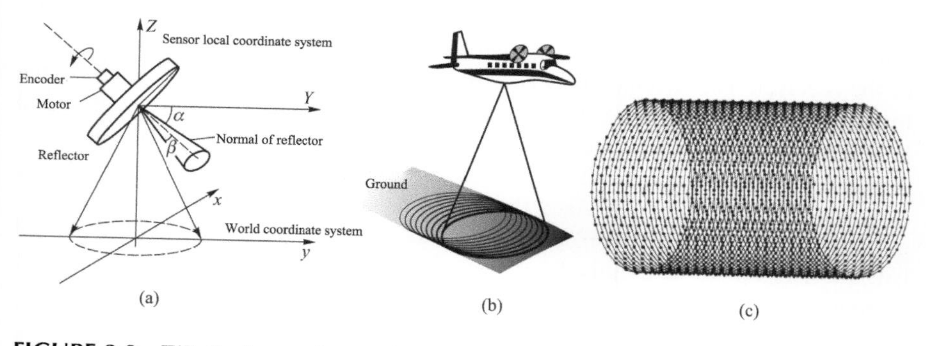

FIGURE 2.9 Elliptical scanning mode of airborne LiDAR. (a) Scanning diagram; (b) and (c) elliptical ground scan lines.

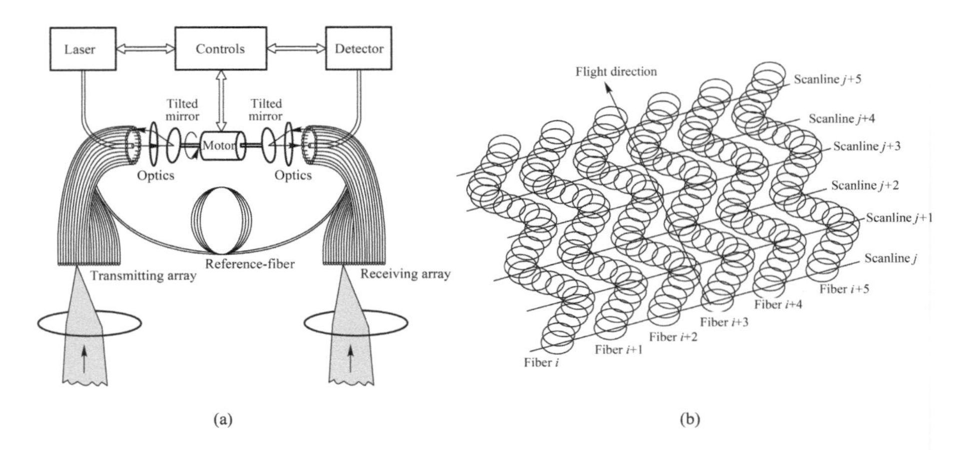

FIGURE 2.10 Fiber scanning mode of airborne LiDAR. (a) Scanning system; (b) ground scan lines.

unevenly distributed, which brings difficulties in terms of data processing. Also, the high repetition rate of the spots on the ground cause relatively low scanning efficiency.

4. Fiber scanning mode (Figure 2.10). Its principle is that the laser is transmitted through the fiber and enters the fiber group in turn; the terminals of the fiber group are arranged in a row; the fibers in different positions correspond to different fields of view; the lasers are emitted from the fiber arrays, in turn, to scan the target. The advantage is fast scanning speed and evenly distributed scanning points. The even points greatly shorten the data processing time.

2.3.3 Terrestrial LiDAR

The terrestrial LiDAR systems include fixed LiDAR and mobile LiDAR such as vehicle-mounted and backpack-based scanning systems. This section introduces the working principle of the terrestrial LiDAR system based on fixed tripod.

The terrestrial LiDAR fixes the 3D laser scanner on a tripod or other base station. The laser pulse is transmitted, then reflected by the object, and finally returns to the receiver. The distance between the scanner and object is calculated by the time delay of emitting and receiving the pulse. Also, the transverse scan angle θ and longitudinal scan angle φ of each pulse are recorded. Generally, the operator sets the transverse scan angle range $\Delta\theta$ and longitudinal scan angle range $\Delta\varphi$ to scan the scene in different directions. Also, the transverse scan angle resolution $\delta\theta$ and longitudinal scan angle resolution $\delta\varphi$ can be set to adjust the pulse density of terrestrial LiDAR. The number of emitted laser pulses (N_{tls}) can be calculated according to the scanning angle range and resolution, as in Equation (2.29).

$$N_{\text{tls}} = \frac{\Delta\theta}{\delta\theta} \cdot \frac{\Delta\varphi}{\delta\varphi} \tag{2.29}$$

Note the point cloud acquired by the terrestrial LiDAR is usually not uniform. Specifically, the objects close to the scanner have large point densities; otherwise, the point density is relatively small. Additionally, due to the occlusion between the objects, the single-station scanning cannot acquire comprehensive 3D information of all the objects in the scene. Therefore, in practice, the operators usually conduct multistation scanning with the terrestrial LiDAR and then splice the acquired data. More information about terrestrial laser scanning is provided in Section 3.2.

2.4 PRINCIPLES OF LiDAR WITH DIFFERENT METHODS OF DETECTION AND DIGITIZATION

Full waveform, discrete return, and photon counting are three common methods of detection and digitization for LiDAR systems (Mandlburger et al., 2019). Figure 2.11 shows how the three methods record the emitted and received signals when the laser penetrates a complex forest canopy. Among them, the full waveform LiDAR (fwLiDAR) and discrete return LiDAR (drLiDAR) record the returned signal through an analog-to-digital convertor, while the photon counting LiDAR (pcLiDAR) directly records the returned laser photons. For the two modes of analog electronic detection and direct detection, the received signal is composed of a valid laser signal and invalid noise signal. The noise signal is mainly caused by the sensor system and radiation from the sun. The difference is that the analog electronics detector converts the received laser power into the voltage signal, producing the returned intensity that varies over time. Also, the LiDAR with analog-to-digital record mode generally emits relatively high-power laser beams. The received valid laser signal has thousands of photons, which greatly exceeds the detector's internal noise and solar noise, i.e., high SNR. It helps to accurately identify a valid signal and

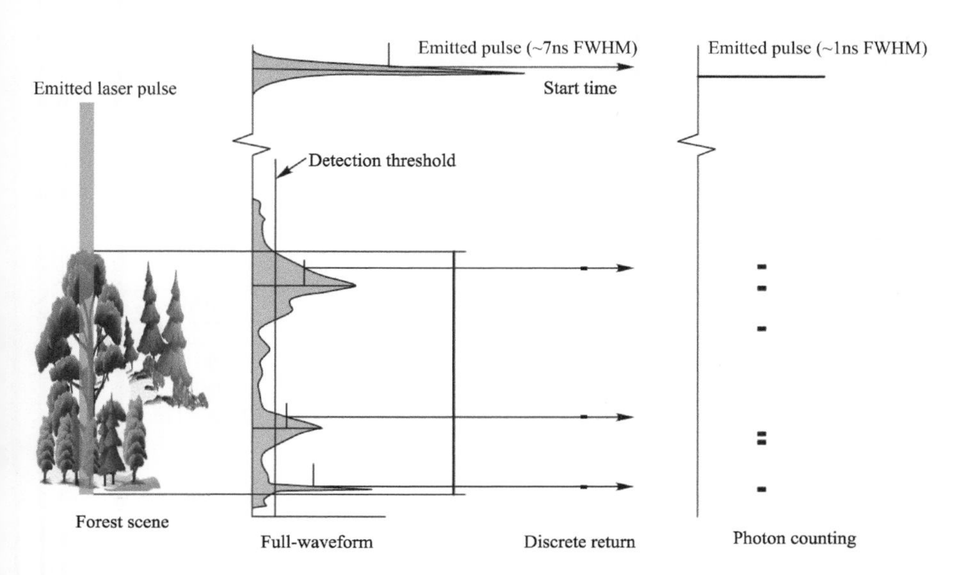

FIGURE 2.11 Principles of LiDAR with different methods of detection and digitization.

then calculate the distance between the sensor and target. In contrast, the photon counting mode employs a micro-pulse laser that emits a very weak laser beam. A single-photon detector is used to detect the returned photons and record their return time. However, the solar noise is usually obvious compared with the weak laser energy. This causes the low SNR of pcLiDAR.

We present the detailed principles of the fwLiDAR, drLiDAR, and pcLiDAR systems next.

2.4.1 Full-Waveform LiDAR

Full-waveform LiDAR uses an analog-to-digital converter (ADC) to digitize the returned laser power time series into a waveform. The waveform can characterize the complete vertical structure of the target, such as the gray shaded signal in Figure 2.11 (Harding et al., 2001). The pulse width of the returned waveform is influenced by the width of the emitted pulse and the vertical extent of the target within the footprint. For a target with a simple structure, such as a flat surface or building roofs, we would observe one returned laser peak. For a target with a complex structure, multiple return peaks are formed. For example, the laser hitting a building edge would be reflected by the building roof and the adjacent ground, thus forming two returned peaks. For objects with gaps, e.g., vegetation canopy, the laser pulse would irradiate the multiple objects in the transmitted direction, including leaves, stems, branches, and understory, forming a waveform signal with complex shapes. This process demonstrates the capability of LiDAR to capture vegetation's vertical structure.

Generally, the fwLiDAR system provides the digital recordings of emitted and received waveforms. The valid returned signal is calculated by detecting the first waveform location and the last waveform location exceeding the detected threshold. Then the ranging information is derived from the time delay between the center of the emitted pulse and the center of a valid received signal. The commonly used methods of calculating the pulse center include the centroid method, mean value method, and Gaussian fitting method (Tang et al., 2016).

Figure 2.12 displays the development of fwLiDAR systems. The earliest waveform sampling experiment for land terrain detection was the Airborne Oceanographic

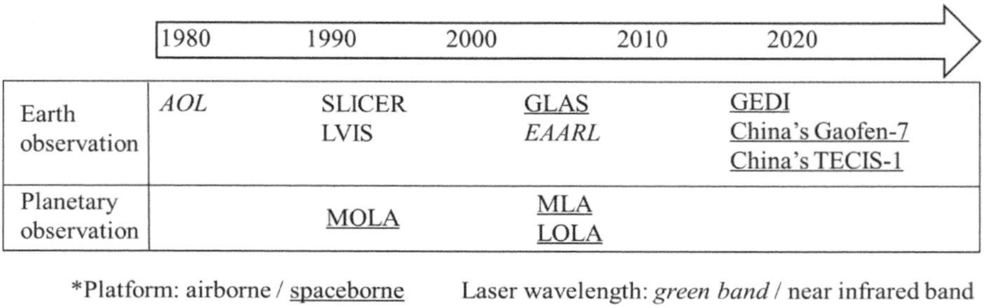

	1980	1990	2000	2010	2020
Earth observation	*AOL*	SLICER LVIS	GLAS *EAARL*		GEDI China's Gaofen-7 China's TECIS-1
Planetary observation		MOLA	MLA LOLA		

*Platform: airborne / spaceborne Laser wavelength: *green band* / near infrared band

FIGURE 2.12 Existing full-waveform LiDAR systems.

LiDAR (AOL) system developed by the National Aeronautics and Space Administration (NASA) (Krabill et al., 1984). Subsequently, scientists developed multiple types of airborne fwLiDAR systems by combining the full-waveform LiDAR sensor with the airborne scanning device. The typical ones are Airborne Topographic Mapper (ATM) (Krabill et al., 2002), Scanning Lidar Imager of Canopies by Echo Recovery (SLICER) (Harding et al., 2000), Laser Vegetation Imaging Sensor (LVIS) (Blair et al., 1999), and Experimental Advanced Airborne Research LiDAR (EAARL) (Nayegandhi et al., 2005).

The main purpose of the SLICER and LVIS systems is to measure the vegetation canopy structure. They are designed with a large LiDAR footprint with a diameter larger than 10 m, of which obtained waveform sampling can reveal the forest vertical structure at a large scale. In contrast, the purpose of the ATM and EAARL systems is high-resolution topographic mapping with a small footprint (diameter <1 m). In the forest region, the small footprint cannot cover a tree crown completely. Hence, the acquired waveform cannot characterize the large-scale vegetation vertical structure.

In particular, the EAARL system uses a 532-nm laser and transmits the returned signal to channels with different ADC ratios. Its purpose is to adapt the significant variance of laser intensity returned from different scatter surfaces, such as land, vegetation, surface of the water, and bottom of the water. Hence, the EAARL system is capable of acquiring bathymetric data and land topographic data. If the returned intensity is not scaled properly, the recorded waveform signal would be saturated or distorted.

In addition to airborne platforms, full-waveform LiDAR is also equipped on spaceborne platforms to measure the Earth's 3D surface structure on a global scale. However, the existing LiDAR satellites equip the laser altimetry, rather than the laser scanner. The ICESat/GLAS is the world's first satellite-based, full-waveform laser altimeter for Earth observation. It was launched by NASA in 2003 and stopped working in 2009. In 2018, NASA carried the spaceborne full-waveform laser altimetry GEDI onto the ISS for global forest monitoring. China's GF-7, launched in 2019, and Terrestrial Ecosystem Carbon Inventory Satellite-1 (TECIS-1) launched in 2022, are both equipped with full-waveform laser altimeters. Their missions are global topography mapping and forest biomass monitoring, respectively. More details about the spaceborne LiDARs for Earth observation are introduced in Section 3.3 of this book.

Except for Earth observation, full-waveform LiDAR systems were also applied to other planetary surveying, such as the Mars Orbiter Laser Altimeter (MOLA) for Mars surface mapping (Smith et al., 1999), Messenger Laser Altimeter (MLA) for Mercury surface detection (Cavanaugh et al., 2007), and Lunar Orbiter Laser Altimeter (LOLA) for lunar surface detection (Smith et al., 2010).

2.4.2 Discrete Return LiDAR

In contrast to fwLiDAR, discrete return LiDAR records only a few locations in the waveform time series output by the detector. Threshold detection, recording the position exceeding the threshold in the waveform time series, is the simplest method of recording discrete return points. With this method, only one or a limited number of objects irradiated by the laser are detected, which greatly reduces the recorded data volume.

A crucial system parameter of the drLiDAR system is the maximum number of returns that can be recorded per pulse. The earliest drLiDAR can only record one point per pulse. The drLiDAR systems developed subsequently gradually record two, three, and up to five points per pulse. Taking as an example of Figure 2.11, the drLiDAR detector records three discrete points from the waveform time series. In addition to recording the 3D coordinates (X, Y, Z) of each discrete point, the intensity (I) of each discrete point is recorded by using the peak detection method and the Gaussian decomposition method. The peak detection method records the peak center of the waveform series as the point bin location and the peak amplitude as the point intensity. The Gaussian decomposition method performs the Gaussian decomposition algorithm on the waveform series. The center of the decomposed Gaussian component is recorded as the point bin location, and the amplitude is recorded as the point intensity. The time delay between the point bin location and the center location of the emitted pulse is used to calculate the distance between the target corresponding to each point and the sensor. Then combined with the global positioning system (GPS) and inertial measurement unit (IMU) data of the sensor, the 3D coordinates of each point are derived. Through multiple pulse laser scanning, massive discrete points with 3D geographic coordinates are generated, usually called a "point cloud".

The drLiDAR systems are usually designed with a small footprint and high pulse repetition frequency. Figure 2.13 shows an example of a discrete point cloud in a forest area acquired by a small-footprint drLiDAR system, recording up to four returns per pulse. The position, intensity, and corresponding return number of each discrete point depend on the spatial structure and optical properties of the surface objects.

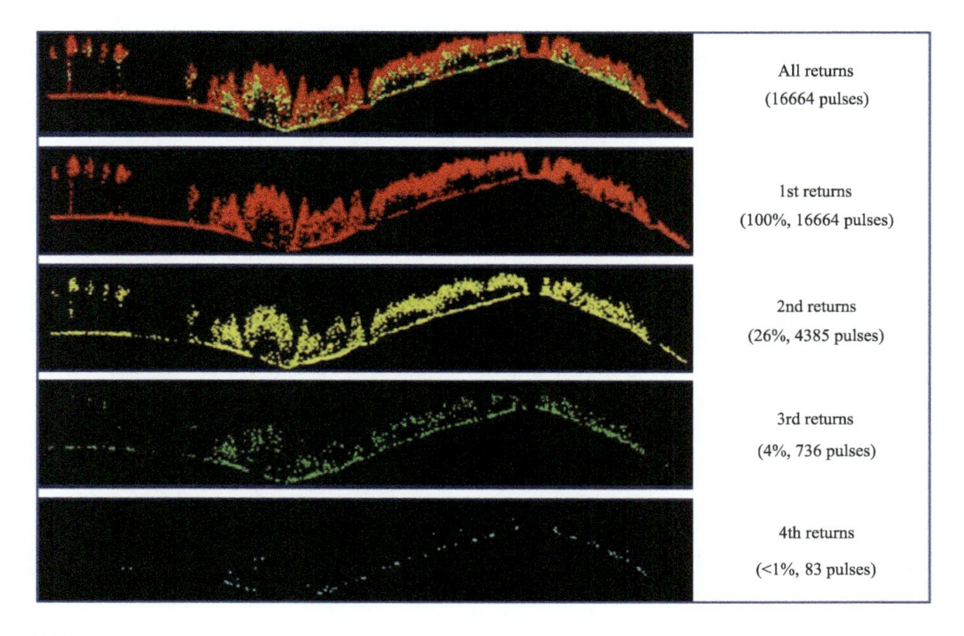

FIGURE 2.13 Multiple returns of discrete return LiDAR.

According to the LiDAR equation (Equation 2.20), the return laser intensity is related to the intercepted area and reflectance of the corresponding object. Once the object property makes it generate a strong enough laser intensity to exceed the detection threshold, the detector records it as one discrete point. The first discrete point corresponds to the location of the waveform signal that exceeds the intensity threshold for the first time, which might come from the top of the vegetation canopy, or the interior of the vegetation canopy, or the ground. The last discrete point corresponds to the location of the detected waveform signal for the last time, which might come from the interior of the vegetation canopy or from the ground. Sometimes, some ground points can be detected in the vegetation-covered region. This is because the laser pulse can penetrate gaps in the vegetation canopy to reach the ground. The frequency of detecting the ground under the canopy depends on many factors, such as canopy gap fraction, laser scanning angle, beam divergence angle, footprint diameter, pulse repetition frequency, and ground reflectance.

2.4.3 Photon Counting LiDAR

Different from the first two detection modes, the photon counting LiDAR records the arrival time of each laser photon to achieve the sampling of the target vertical structure (Figure 2.11). Through detection of multiple pulses, the accumulated photon points potentially construct the spatial structure of the Earth's surface (Howland et al., 2011).

Full-waveform LiDAR and discrete return LiDAR systems usually emit high-intensity laser pulses, thus obtaining the laser signal with a high SNR. In contrast, photon counting systems emit very weak micro-pulses, with only a small number of photons returning per pulse. Hence, a single-photon detector is required to detect the returned photons. The very narrow pulse width of the emitted pulse (<1 ns), low jitter of the detector, and high resolution of the timing electronics make the pcLiDAR reach decimeter-level ranging accuracy. Note that the single photon detector of pcLiDAR usually has some dark counts, i.e., system noise. The dark counts are quantified by the parameter "dark count rate," which is the detector's thermal noise count rate without any light source. The dark count rate is usually very low compared with the frequency of received photon counts.

In addition to good detection sensitivity and low dark count noise, the single photon detector can detect multiple photons for one pulse. That is, a few photon points are recorded for each pulse. This characteristic is determined by a key detector parameter, dead time, which is defined as the period needed by the detector to recover from a detection event, before another photon can be detected. The dead time of a single photon detector is very short, so that the pcLiDAR can distinguish between consecutive photons returned from closely spaced surfaces, e.g., the returns from vegetation canopy. This characteristic of pcLiDAR not only can detect multiple photons for one pulse, improving the photon density, but also record the acquired data in the form of discrete photon points.

When the photon counting LiDAR system works in the daytime, the solar radiations reflected from the Earth's surface and clouds are the main noise source. The solar noise rate is influenced by the surface reflectance, solar irradiance, atmospheric environment, and sensor configuration, among other things. Generally, there is a filter with a very narrow bandwidth located before the pcLiDAR detector. Its purpose is to prevent the electromagnetic wave other than the laser band from entering the detector. Also, pcLiDAR systems are typically designed to have a small FOV, confining the received signal within the FOV region.

To summarize, the pcLiDAR has the characteristics of low laser power, small receiving FOV, and high pulse repetition frequency. Compared with the analog-to-digital LiDAR, it has strong potential in measuring fine 3D structures of the Earth's surface at the large scale. Also, the output of pcLiDAR is the discrete photon points with specific 3D spatial coordinates. This makes the existing point cloud algorithms of data processing, analysis and visualization, and the software potentially suitable for pcLiDAR data.

2.5 INFLUENCE MECHANISM OF THE SKY ON THE LiDAR SIGNAL

The LiDAR is inevitably affected by the surrounding environment during the measurement process, especially spaceborne and airborne LiDAR. Figure 2.14 shows the radiative transfer process of sunlight energy (red arrows) and laser energy (green

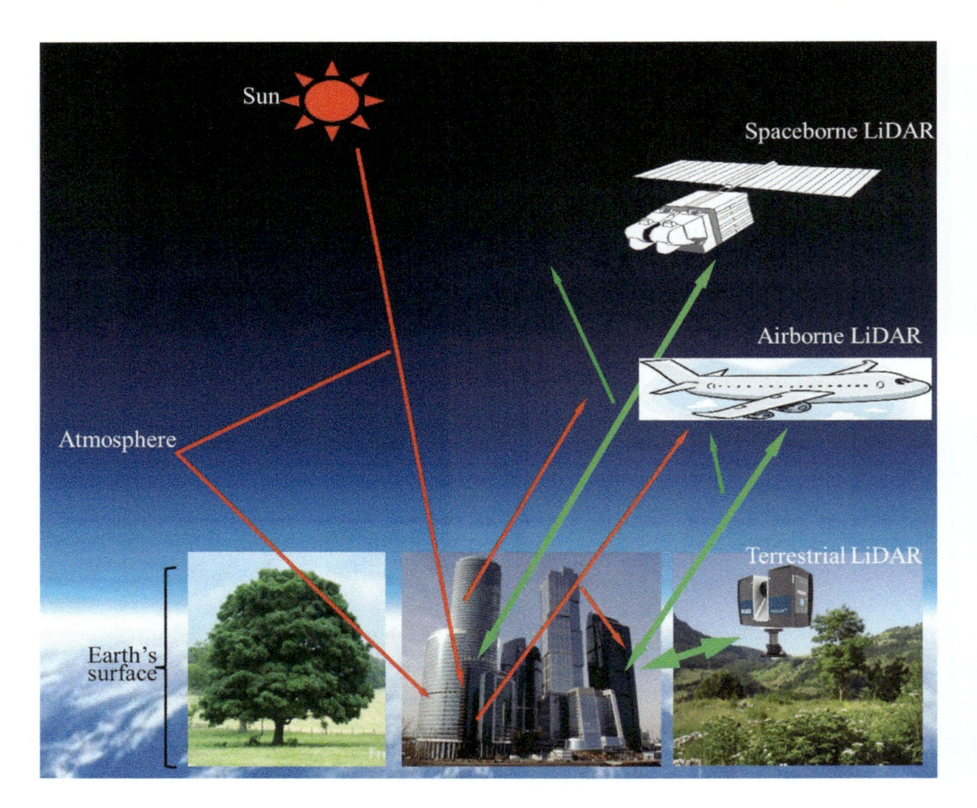

FIGURE 2.14 Radiative transfer process of sunlight and laser among the atmosphere, Earth's surface, and LiDAR sensors.

arrows) among the atmosphere, Earth's surface, and sensors. It is seen that in the transmitted process, the laser pulse interacts with not only the Earth's surface but also the atmosphere layer. Also, the sensor receives some noise signals from the sunlight, including the solar radiation scattered by the atmosphere and surface. This section presents the influential mechanisms of the atmosphere and sunlight on the LiDAR signal.

2.5.1 Effect of Atmosphere

The LiDAR systems usually use a laser with a short wavelength. The atmosphere will absorb and scatter the laser signal, causing the LiDAR sensor to be very sensitive to the atmosphere (Qiang et al., 2000). In particular, rain, dust, fog, and haze in the atmosphere interfere with the laser signal. The main influences of atmosphere on the LiDAR signal include (1) energy attenuation: the atmospheric gas, aerosol, dust, fog, and rain would absorb or scatter the laser signal, causing the attenuation of laser energy, and (2) laser refraction: the uneven spatial distribution of the atmosphere causes the laser to be refracted rather than directly fired. Additionally, atmospheric turbulence affects the fluctuation of laser energy distribution on the beam cross section and the beam expansion and drift. Atmospheric absorption in the laser transmission path causes variations of the density gradient and refractive index of air, i.e., atmospheric thermal halo effect. This results in the nonlinear thermal distortion of the laser beam. The effects of atmospheric attenuation and atmospheric refraction in the LiDAR transmission process are introduced in detail next.

2.5.1.1 Atmospheric Attenuation

A laser has excellent monochromaticity. With the assumption that the atmosphere is uniform or stratified uniform, the laser energy through the atmosphere layer is expressed by the Beer-Lambert law:

$$I(\lambda,z) = I(\lambda,0)e^{-\sigma(\lambda)\cdot z} \tag{2.30}$$

where $I(\lambda,0)$ and $I(\lambda,z)$ are the initial laser intensity and the laser intensity passing through the atmosphere with a thickness of z, respectively, λ is the laser wavelength, and $\sigma(\lambda)$ is the atmospheric attenuation coefficient related to the wavelength.

The atmospheric attenuation includes two parts: atmosphere absorption and atmosphere scattering. Therefore, $\sigma(\lambda)$ can be expressed as Equation (2.31):

$$\sigma(\lambda) = \sigma_{\mathrm{m}} + K_{\mathrm{m}} + \sigma_{\mathrm{a}} + K_{\mathrm{a}} \tag{2.31}$$

where σ_{m} is the gas molecular scattering coefficient, K_{m} is the gas molecular absorption coefficient, σ_{a} is the aerosol scattering coefficient, and K_{a} is the aerosol absorption coefficient.

The absorption coefficient of a gas molecule and aerosol for laser light can be calculated based on the transmission theory of monochromatic light. When the laser pulse is transmitted in the lower atmosphere below 20 km, the width of the atmosphere absorption spectrum is mainly determined by the pressure broadening caused by gas molecular collision. The gas molecular absorption coefficient of the low atmosphere layer (K_{mL}) is expressed as Equation (2.32):

$$K_{\mathrm{mL}} = \frac{S}{\pi} \cdot \frac{\gamma_L}{(\nu - \nu_0)^2 + {\gamma_L}^2} \tag{2.32}$$

where ν is light frequency, ν_0 is the frequency of the center of the laser spectrum, S is the integrated intensity of the laser spectrum, γ_L is a half-width of the Lorentz spectral line, and S and γ_L are both influenced by atmospheric temperature and pressure.

In the upper atmosphere above 60 km, the gas molecular absorption coefficient (K_{mD}) is calculated by the Doppler-broadened line-shaped function, as in Equation (2.33).

$$K_{\mathrm{mD}} = \frac{S}{\gamma_D} \cdot \left(\frac{\ln 2}{\pi}\right)^{1/2} \cdot \mathrm{e}^{-\ln 2 \cdot (\nu - \nu_0)^2 / {\gamma_D}^2} \tag{2.33}$$

where γ_D is the Doppler line half-width. In the middle atmosphere at altitudes of 20–60 km, the collision broadening and Doppler broadening exist simultaneously. The gas molecular absorption coefficient (K_{mV}) is given by the Voigt line shape, as in Equation (2.34).

$$K_{\mathrm{mV}} = \frac{S \cdot {\gamma_L}^2}{\gamma_D \cdot \sqrt{\pi}} \cdot \int_{-\infty}^{\infty} \mathrm{e}^{-t^2} \cdot \left[\left(\frac{\gamma_L}{\gamma_D}\right)^2 + \left(\frac{\nu_0 - \nu}{\gamma_D} - t\right)^2\right] \mathrm{d}t \tag{2.34}$$

The atmospheric scattering effect includes Rayleigh scattering by atmosphere gas molecular and Mie scattering by spherical or quasi-spherical particles such as raindrops, fog droplets, haze, and aerosols.

Rayleigh scattering occurs when the laser wavelength is much larger than the particle size, and the gas molecular scattering coefficient σ_{m} is expressed as in Equation (2.35).

$$\sigma_{\mathrm{m}} = \frac{8\pi^2}{3} \cdot \frac{(n^2 - 1)^2}{N_s^2 \lambda^4} \cdot \frac{6 + 3\delta_p}{6 - 7\delta_p} \tag{2.35}$$

where n is the particle refractive index, N_s is particle density in the air, and δ_p is the depolarization factor. Based on experience, σ_{m} is expressed as:

$$\sigma_{\mathrm{m}} = 2.677 \times 10^{-17} P \nu^4 / T \tag{2.36}$$

where P is pressure and T is temperature.

Mie scattering occurs when the size of atmospheric particles is comparable to the laser wavelength. Generally speaking, the Mie scattering is the aerosol scattering. The total atmosphere attenuation coefficient can be approximately expressed as:

$$\sigma_{\mathrm{T}} = \sigma_{\mathrm{a}} \cdot \sigma_{\mathrm{m}} = \frac{3.912}{V_{\mathrm{m}}} \cdot \left(\frac{0.55}{\lambda}\right)^b \tag{2.37}$$

where V_m is visibility (unit: km), which is defined as the maximum distance at which the human eye can distinguish the target, and the coefficient b is related to the visibility. For normal visibility ($6 \text{ km} \le V_m \le 20 \text{ km}$), $b = 1.3$; For good visibility ($V_m > 20$ km), $b = 1.6$; For poor visibility ($V_m < 6$ km), $b = 0.585 \cdot V_m^{1/3}$.

In general, the scattering coefficients of atmospheric gas molecules and aerosols obey the negative exponential distribution law of altitude. That is, with the increase of atmospheric altitude, the scattering coefficients decrease rapidly. For example, the aerosol scattering coefficient above 5 km decreases more than an order of magnitude compared with that on the ground. Additionally, in actual situations, the aerosol scattering is the main interaction process in the low atmosphere layer near the ground, while the contributions of gas scattering and aerosol scattering are comparable in the middle and upper atmosphere layers.

2.5.1.2 Atmospheric Refraction

The atmospheric refraction effect is defined as the phenomenon that when the laser passes through the atmosphere, the light transmission path is bent and the traveled distance increases, caused by uneven atmosphere density or existence of an atmospheric refractive index gradient. Its effect on LiDAR measurement is to increase the measured error of target bearing and distance.

The atmosphere density varies with the altitude, so the atmosphere refractive index for light at different altitudes is different. The refractive index n is related to the laser wavelength λ, atmosphere temperature T, humidity e, and pressure P, which is generally expressed as Equation (2.38):

$$n = 1 + N(\lambda, T, P, e) \tag{2.38}$$

where N is the refractive index modulus with the unit of order 10^{-6}. For the standard atmosphere ($P = 1$ atm, $T = 288.15$ K, $e = 0$), the atmosphere refractive index from visible to near-infrared band is expressed as Equation (2.39):

$$N_0(\lambda) = 272.5794 + 1.5932\lambda_0^{-2} + 0.015\lambda_0^{-4} \tag{2.39}$$

where $N_0(\lambda)$ is the standard atmospheric refractive index in the wavelength λ.

The atmospheric refractive index $N(\lambda)$ in any atmospheric condition can be calculated from the standard atmospheric refractive index $N_0(\lambda)$, as in Equation (2.40):

$$N(\lambda) = N_0(\lambda) \times \left(2.8434 \times 10^{-3} \times \frac{P}{T} - 0.1127 \times \frac{e}{T} \right) \tag{2.40}$$

The atmosphere attenuation effect in the LiDAR radiative transfer process was introduced earlier. For practical application, other complex effects, such as atmospheric turbulence and thermal halo, should also be considered. The propagation of the LiDAR signal in the atmosphere directly affects the sensor performance. Therefore, it is necessary to fully understand the atmospheric propagation characteristics of the laser, find ways to avoid or reduce the atmospheric effects, and select the suitable working wavelength and working mode for the LiDAR system according to the laser propagation mechanism in the atmosphere to finally achieve the high atmosphere transmittance of the laser beam.

2.5.2 Effect of Sunlight

The existence of sunlight makes the LiDAR RS radiative transfer process a RS configuration of "two light sources (laser and sunlight) + one sensor" (Figure 2.15). The laser source is regarded as a point light source emitting laser power onto a fixed conical solid angle. The emitted laser only lasts for a short time. The sun emits solar energy from an infinite distance to the Earth's surface. Without the consideration of atmospheric scattering, the sunlight reaching the Earth's surface can be regarded as continuous parallel beams with the same direction. Once the sunlight irradiates the surface within the LiDAR receiving FOV, the solar energy might be scattered by the surface and then received by the LiDAR sensor. The influence of sunlight on the LiDAR signal is different for different types of LiDAR systems. Next we introduce its influences separately.

The large-footprint LiDAR systems (footprint diameter >10 m, usually mounted on spaceborne and airborne platforms) have a relatively large laser illumination region and receiving FOV region. The solar energy scattered into the LiDAR sensor

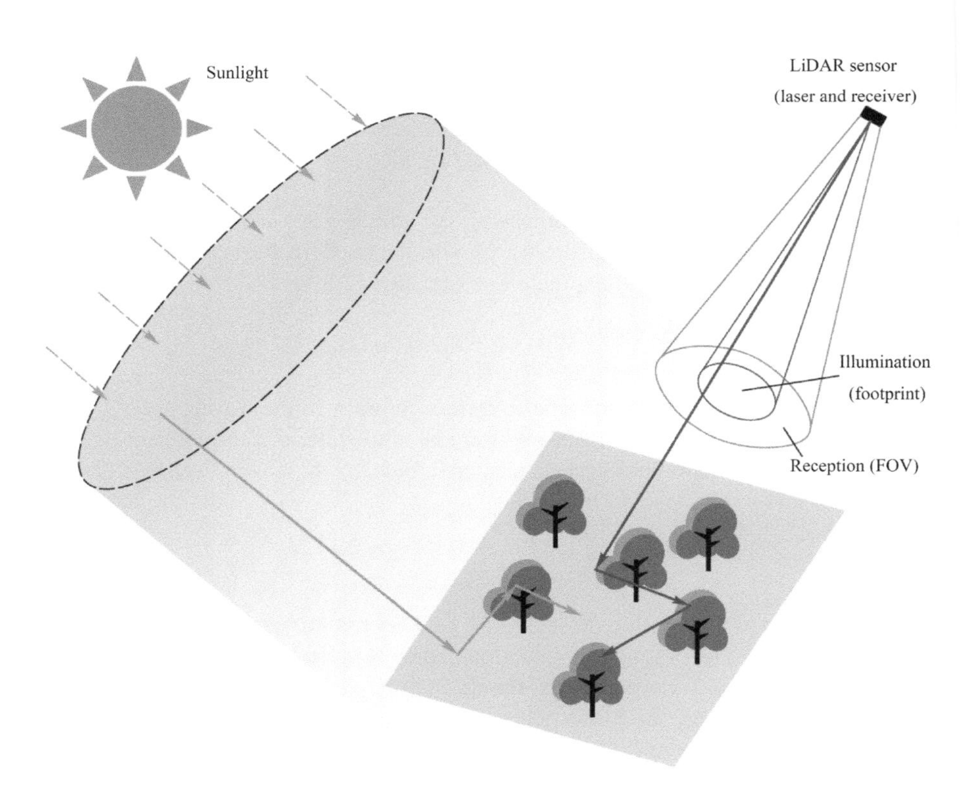

FIGURE 2.15 Radiative transfer process of dual light sources (laser and sunlight).

is relatively high. In contrast, the small-footprint LiDAR systems (footprint diameter <10 m, usually equipped on spaceborne and airborne platforms) have a small illumination region and receiving FOV region. Only very little solar energy enters the LiDAR receiver. Hence, the sunlight has less effect on the small-footprint LiDAR signal.

For fwLiDAR, the influence of sunlight is shown as a relatively constant solar background noise. This is because the persistence of sunlight makes its radiation transfer process to adopt a dynamic equilibrium state, and the solar energy entering the LiDAR receiver almost remains constant. The solar noise of a LiDAR waveform can be easily removed by setting a background noise threshold.

For drLiDAR, the solar noise is extremely weak due to the small footprint. Also, drLiDAR usually records discrete point clouds from waveform data by using the threshold detection method, peak detection method, or Gaussian decomposition method. The constant and very weak solar noise does not affect the calculated 3D coordinates of these points.

For pcLiDAR, the detector treats solar photons in the filter band the same as laser photons. That is, the recorded photons are a mixture of solar photons, laser photons, and dark noise photons. Due to the extremely low laser energy of pcLiDAR, solar noise has a significant effect on the pcLiDAR data acquired in the daytime. Furthermore, due to the persistence of solar irradiation, the solar noise photons exist over the vertical extent of the entire scene, greatly increasing the density of the photon points. Therefore, in order to obtain accurate surface structural information, the solar noise photons must be removed by postprocessing algorithms. The denoising algorithms for photon counting data are introduced in Section 4.3.

2.6 SUMMARY

This chapter first introduces the ranging principle of LiDAR systems, including pulsed LiDAR and phased LiDAR. Then, the LiDAR radiation principle is introduced based on the LiDAR equation. The commonly used LiDAR waveform model and LiDAR radiative transfer model are also presented. After that, we systematically introduce the principles of LiDAR systems with different platforms (spaceborne/airborne/terrestrial) and different methods of detection and digitization (full waveform/discrete return/photon counting). Finally, the influence mechanisms of the sky environment (atmosphere and sunlight) on the LiDAR signal are analyzed.

EXERCISES

1. Why do most LiDAR systems employ the pulse ranging method?
2. What is the ranging principle of the pulsed LiDAR? How is it different from that of the phased LiDAR?
3. Please write the LiDAR equation and explain the meaning of each parameter.
4. Please list three LiDAR 3D radiation transfer models and introduce their core ideas.
5. Please state the commonly used scanning modes of airborne LiDAR systems and their advantages and disadvantages.

6. Please explain the differences of the LiDAR systems with three different methods of detection and digitization, i.e., full waveform LiDAR, discrete return LiDAR, and photon counting LiDAR.
7. How are the sizes of the LiDAR footprint and field of view (FOV) calculated?

REFERENCES

Blair, J. B., & Hofton, M. A. (1999). Modeling laser altimeter return waveforms over complex vegetation using high resolution elevation data. *Geophysical Research Letters*, *26*(16), 2509–2512.

Blair, J. B., Rabine, D. L., & Hofton, M. A. (1999). The laser vegetation imaging sensor: A medium-altitude, digitization-only, airborne laser altimeter for mapping vegetation and topography. *ISRPS Journal of Photogrammetry and Remote Sensing*, *54*(2–3), 115–122.

Cavanaugh, J. F., Smith, J. C., Sun, X. L., Bartels, A. E., Ramos-Izquierdo, L., Krebs, D. J., McGarry, J. F., Trunzo, R., Novo-Gradac, A. M., Britt, J. L., Karsh, J., Katz, R. B., Lukemire, A. T., Szymkiewicz, R., Berry, D. L., Swinski, J. P., Neumann, G. A., Zuber, M. T., & Smith, D. E. (2007). The mercury laser altimeter instrument for the MESSENGER mission. *Space Science Reviews*, *131*(1–4), 451–479.

Dai, Y. (2002). *The principle of LiDAR* (in Chinese). National Defense Industry Press.

Disney, M. I., Lewis, P. E., Bouvet, M., Preto-Blanco, A., & Hancock, S. (2009). Quantifying surface reflectivity for spaceborne LiDAR via two independent methods. *IEEE Transactions on Geoscience and Remote Sensing*, *47*(9), 3262–3271.

Dubayah, R., Blair, J. B., Goetz, S., Fatoyinbo, L., Hansen, M., Healey, S., Hofton, M., Hurtt, G., Kellner, J., Luthcke, S., Armston, J., Tang, H., Duncanson, L., Hancock, S., Jantz, P., Marselis, S., Patterson, P. L., Qi, W. L., & Silva, C. (2020). The global ecosystem dynamics investigation: High-resolution laser ranging of the Earth's forests and topography. *Science of Remote Sensing*, *1*, 100002.

Gastellu-Etchegorry, J.-P., Yin, T., Lauret, N., Grau, E., Rubio, J., Cook, B. D., Morton, D. C., & Sun, G. (2016). Simulation of satellite, airborne and terrestrial LiDAR with DART(i): Waveform simulation with quasi-Monte Carlo ray tracing. *Remote Sensing of Environment*, *184*, 418–435.

Goodenough, A. A., & Brown, S. D. (2017). DIRSIG5: Next-generation remote sensing data and image simulation framework. *IEEE Journal of Selected Topics in Applied Earth Observations and Remote Sensing*, *10*(11), 4818–4833.

Govaerts, Y. M., & Verstraete, M. M. (1998). Raytran: A Monte Carlo ray-tracing model to compute light scattering in three-dimensional heterogeneous media. *IEEE Transactions on Geoscience and Remote Sensing*, *36*(2), 493–505.

Harding, D. J., Blair, J. B., & Rabine, D. L. (2000). *SLICER airborne laser altimeter characterization of canopy structure and sub-canopy topography for the BAREAS northern and southern study regions: Instrument and data product description*. National Aeronautics and Space Administration, Goddard Space Flight Center.

Harding, D. J., Lefsky, M. A., Parker, G. G., & Blair, J. B. (2001). Laser altimeter canopy height profiles: Methods and validation for closed-canopy, broadleaf forests. *Remote Sensing of Environment*, *76*(3), 283–297.

Howland, G. A., Dixon, P. B., & Howell, J. C. (2011). Photon-counting compressive sensing laser radar for 3D imaging. *Applied Optics*, *50*(31), 5917–5920.

Huang, H., & Wynne, R. H. (2013). Simulation of LiDAR waveforms with a time-dependent radiosity algorithm. *Canadian Journal of Remote Sensing*, *39*, S126–S138.

Kobayashi, H., & Iwabuchi, H. (2008). A coupled 1-D atmosphere and 3-d canopy radiative transfer model for canopy reflectance, light environment, and photosynthesis simulation in a heterogeneous landscape. *Remote Sensing of Environment*, *112*(1), 173–185.

Krabill, W. B., Abdalati, W., Frederick, E. B., Manizade, S. S., Martin, C. F., Sonntag, J. G., Swift, R. N., Thomas, R. H., & Yungel, J. G. (2002). Aircraft laser altimetry measurement of elevation changes of the Greenland ice sheet: Technique and accuracy assessment. *Journal of Geodynamics*, *34*(3–4), 357–376.

Krabill, W. B., Collins, J., Link, L., Swift, R., & Butler, M. (1984). Airborne laser topographic mapping results. *Photogrammetric Engineering and Remote Sensing*, *50*(6), 685–694.

Lai, X. (2010). *Basic principles and applications of airborne LiDAR* (in Chinese). Publishing House of Electronics Industry.

Mandlburger, G., Lehner, H., & Pferfer, N. (2019). A comparison of single photon and full-waveform LiDAR. *ISPRS Annals of Photogrammetry, Remote Sensing and Spatial Information Sciences*, *4*, 397–404.

Nayegandhi, A., Brock, J. C., & Wright, C. W. (2005). Classifying vegetation using NASA's Experimental Advanced Airborne Research LiDAR (EAARL) at Assateague island national seashore. American Society for Photogrammetry and Remote Sensing - Annual Conference 2005, 2, 769–777.

Ni-Meister, W., Jupp, D. L. B., & Dubayah, R. (2001). Modeling LiDAR waveforms in heterogeneous and discrete canopies. *IEEE Transactions on Geoscience and Remote Sensing*, *39*(9), 1943–1958.

North, P. R. J., Rosette, J. A. B., Suarez, J. C., & Los, S. O. (2010). A Monte Carlo radiative transfer model of satellite waveform LiDAR. *International Journal of Remote Sensing*, *31*(5), 1343–1358.

Pharr, M., Jakob, W., & Humphreys, G. (2016). *Physically based rendering: From theory to implementation*. Morgan Kaufmann.

Qiang, X., Zhang, H., Tu, Q., Yuan, R., & Li, Z. (2000). Atmospheric attenuation effects of Ladar signals (in Chinese). *Journal of Applied Optics*, *21*(4), 21–25.

Schutz, B. E., Zwally, H. J., Shuman, C. A., Hancock, D., & DiMarzio, J. P. (2005). Overview of the ICESat mission. *Geophysical Research Letters*, *32*(21), 97–116.

Smith, D. E., Zuber, M. T., Neumann, G. A., Lemoine, F. G., Mazarico, E., Torrence, M. H., McGarry, J. F., Rowlands, D. D., Head, J. W., Duxbury, T. H., Aharonson, O., Lucey, P. G., Robinson, M. S., Barnouin, O. S., Cavanaugh, J. F., Sun, X. L., Liiva, P., Mao, D.-D., Smith, J. C., & Bartels, A. E. (2010). Initial observations from the lunar orbiter laser altimeter (LOLA). *Geophysical Research Letters*, *37*(18), L18204.

Smith, D. E., Zuber, M. T., Solomon, S. C., Phillips, R. J., Head, J. W., Garvin, J. B., Banerdt, W. B., Muhleman, D. O., Pettengill, G. H., Neumann, G. A., Lemoine, F. G., Abshire, J. B., Aharonson, O., Brown, C. D., Hauck, S. A., Ivanov, A. B., Mcgovern, P. J., Zwally, H. J., & Duxbury, T. C. (1999). The global topography of Mars and implications for surface evolution. *Science*, *284*(5419), 1495–1503.

Sun, G., & Ranson, K. J. (2000). Modeling LiDAR returns from forest canopies. *IEEE Transactions on Geoscience and Remote Sensing*, *38*(6), 2617–2626.

Tang, X., Li, G., Gao, X., & Chen, J. (2016). The rigorous geometric model of satellite laser altimeter and preliminarily accuracy validation (in Chinese). *Acta Geodaetica et Cartographica Sinica*, *45*(10), 1182–1191.

Wagner, W., Ullrich, A., Ducic, V., Melzer, T., & Studnicka, N. (2006). Gaussian decomposition and calibration of a novel small-footprint digitizing airborne laser scanning. *ISPRS Journal of Photogrammetry and Remote Sensing*, *60*(2), 100–112.

Wehr, A., & Lohr, U. (1999). Airborne laser scanning-an introduction and overview. *ISPRS Journal of Photogrammetry and Remote Sensing*, *54*(2–3), 68–82.

Yang, X., Wang, Y., Yin, T., Wang, C., Lauret, N., Regaieg, O., Xi, X., & Gastellu-Etchegorry, J.-P. (2022). Comprehensive LiDAR simulation with efficient physically-based DART-Lux model (I): Theory, novelty, and consistency validation. *Remote Sensing of Environment*, *272*, 112952.

3 LiDAR Data Acquisition

3.1 AIRBORNE LiDAR DATA ACQUISITION

The LiDAR scanning system mounted on the aircraft platform is collectively referred to as an airborne LiDAR system or airborne laser scanning (ALS) system, which is an active data acquisition technology that integrates a variety of high technologies (Lai, 2010). At present, the light and small unmanned aerial vehicles (UAVs) and manned aircraft are the main flight platforms of airborne LiDAR systems. The airborne LiDAR data acquisition process usually includes the plan preparation stage, flight implementation stage, and data preprocessing stage (Zhang, 2007). The plan preparation stage is the preparatory work before executing the flight mission, which mainly includes three steps: project analysis, flight plan design, and flight preparation. The flight implementation stage is designed to perform flight tasks as planned, including equipment installation, equipment debugging, base station placement, ground coordination, flight operations, data collection. In the data preprocessing stage, high-precision airborne LiDAR data are obtained through point cloud coordinate solution, error calibration of inertial measurement unit (IMU) boresight angle, strip adjustment, radiometric calibration, quality control, and other steps. Figure 3.1 shows the overall workflow for data acquisition of the airborne LiDAR system.

3.1.1 Plan Preparation Stage

Airborne LiDAR data acquisition is a complex process involving many specific works, such as airspace application, aircraft leasing or purchase, and equipment selection and installation. Therefore, an effective plan needs to be formulated in advance in the planning preparation stage, which mainly includes three steps: project analysis, flight plan design, and flight preparation.

3.1.1.1 Project Analysis

First of all, it is necessary to have a comprehensive understanding of the topographic characteristics, climatic characteristics, and airspace characteristics of the survey area, such as the geographical location, longitude and latitude range, geomorphic features, climate zone, solar radiation, temperature, precipitation, main surface feature types, and distribution of surrounding airports.

In addition to considering the status of the survey area, the project task analysis should be carried out, including the task content, objectives, scope, workload, time limit, key points, difficulties of the task, and schedule of each stage of the task. It is necessary to reasonably formulate a project schedule based on the survey area and task conditions. Table 3.1 shows the example schedule of an airborne LiDAR data acquisition project.

DOI: 10.1201/9781032671512-3

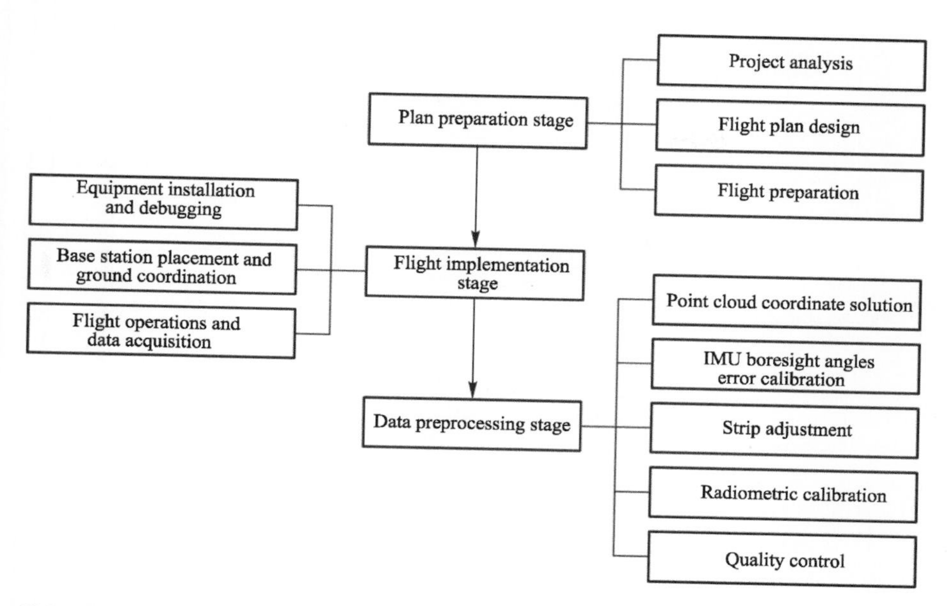

FIGURE 3.1 Airborne LiDAR data acquisition process.

3.1.1.2 Flight Plan Design

The design of the flight plan shall be based on the considerations of safety, cost, thoroughness, and efficiency. According to the project requirements and the actual situation of the survey area, combined with the characteristics of airborne LiDAR equipment, appropriate aerial survey parameters should be selected to carry out the flight operations. The plan design includes ground base station layout, flight route design in the survey area, layout and measurement of the calibration field,

TABLE 3.1
Project Schedule of Airborne LiDAR Data Acquisition

Procedure	Content	Time Schedule
Preparatory stage	Information collection, survey area scouting	mm/dd/yyyy-mm/dd/yyyy
Airspace application	Preparation of application materials, application for airspace	mm/dd/yyyy-mm/dd/yyyy
Aerial survey and submission of the first batch of data	Airspace coordination, data acquisition, data preprocessing, and data inspection	mm/dd/yyyy-mm/dd/yyyy
Aerial survey of other data	Airspace coordination, data acquisition, data preprocessing, and data inspection	mm/dd/yyyy-mm/dd/yyyy
Quality inspection of all data	Data test	mm/dd/yyyy-mm/dd/yyyy
Submission of all data	Data submission	mm/dd/yyyy-mm/dd/yyyy

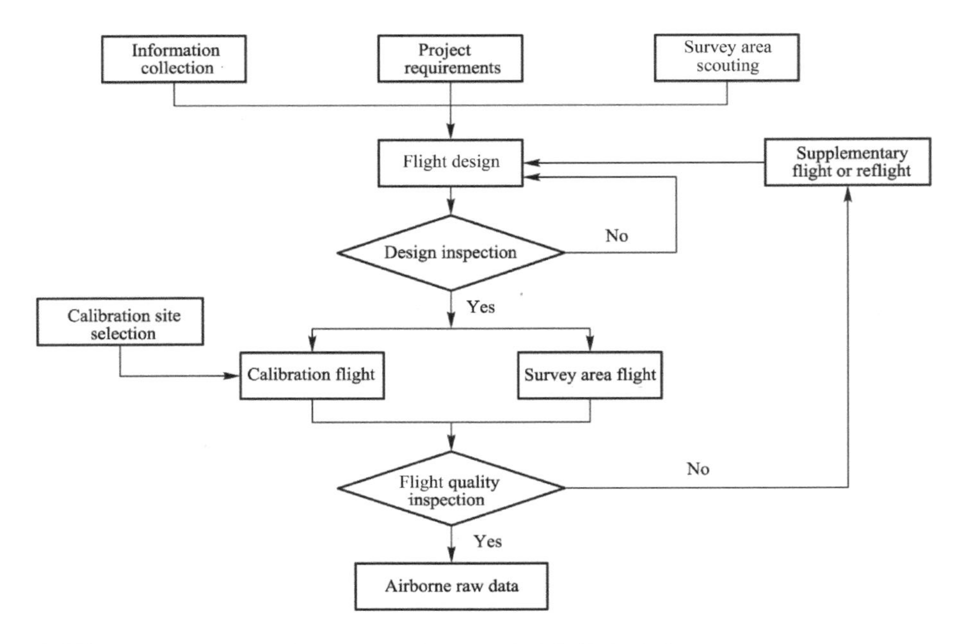

FIGURE 3.2 Technical flowchart of airborne LiDAR flight design.

and supplementary flight or reflight. The specific technical flowchart is shown in Figure 3.2.

1. Ground base station layout. The ground base station layout mainly considers the location and coverage of the base station. Generally, a ground station is placed on a known point provided by the task party. It is required to be in an open and unobstructed place and away from water and high-voltage lines. The coverage of the base station should also be considered. In recent years, some location service providers have provided cloud trajectory calculation services that do not require the installation of base stations. For example, the "FindTrace" service allows users to upload the global navigation satellite system (GNSS) raw observation data from satellite signal–receiving equipment and then generate the matched virtual based station data used for trajectory calculation. This service is widely used in UAV LiDAR flight, and there is no need for a base station.
2. Flight route design in survey area. For the route design of the survey area, refer to the technical standards and specifications for airborne LiDAR data acquisition, such as China's Specifications for data acquisition of airborne LiDAR (CH/T 8024—2011), Specifications for data processing of airborne LiDAR (CH/T 8023—2011), Specifications for IMU/GPS supported aerial photography (GB/T 27919—2011), Specifications for global positioning system (GPS) surveys (GB/T 18314—2009), and Specifications for aerial photogrammetric digital mapping of 1:500 1:1000 1:2000 topographic maps (GB/T 15967—2008), as well as the project professional technical design

TABLE 3.2
Airborne LiDAR Point Density Indicators

Framing Scale	DEM Grid Spacing (m)	Point Cloud Density (points/m²)*
1:500	0.5	≥16
1:1000	1.0	≥4
1:2000	2.0	≥1
1:5000	2.5	≥1
1:10,000	5.0	≥0.25

* Indicates the point cloud density calculated with the grid spacing (no greater than 1/2) of DEM.

book and other relevant technical requirements reviewed and approved by the buyer of a contract. According to the requirements of relevant specifications, the side overlap design of the LiDAR route should reach 20%, but at least 13%. For the demands of different scales, the corresponding point cloud density requirements are shown in Table 3.2.

The flight route design of the survey area includes establishing a route design project, loading digital elevation model (DEM) data, importing the survey area range, and inputting technical parameters (Figure 3.3) (Wang et al., 2010). A qualified flight route design in the survey area should submit flight record forms, flight schematic files, knowledge markup language (KML) files, and other results. Figure 3.4 shows an example of flight route design in a survey area.

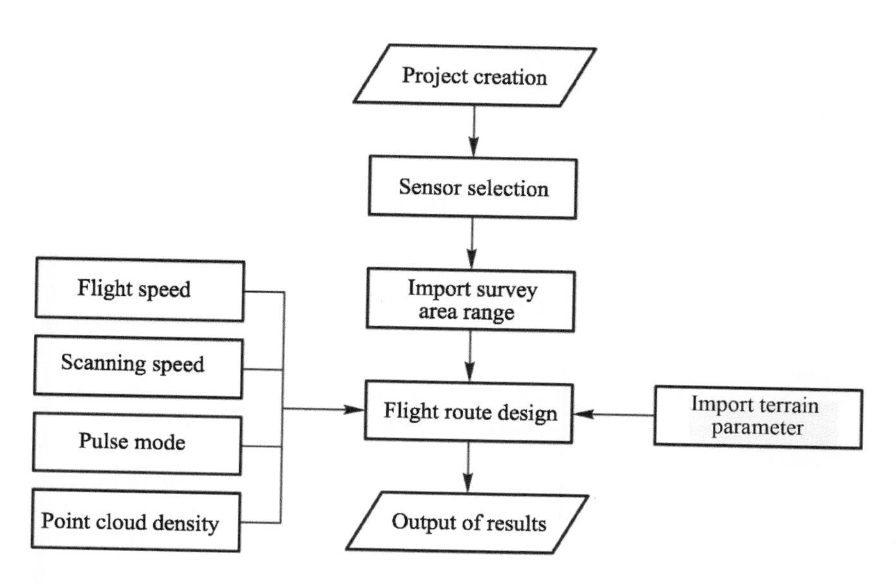

FIGURE 3.3 Design process of airborne LiDAR flight route in survey area.

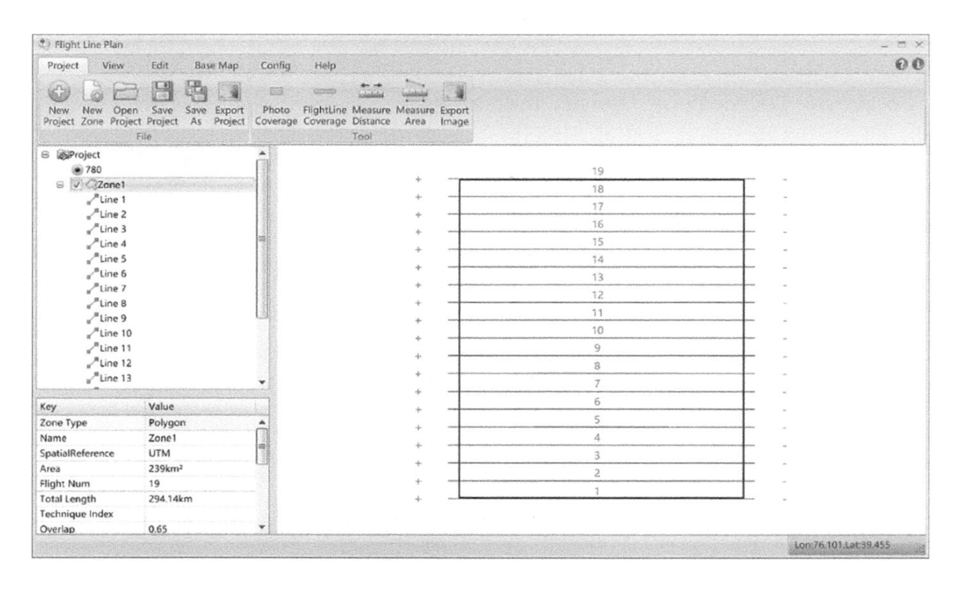

FIGURE 3.4 An example of a flight route design.

3. Layout and measurement of the calibration field. For airborne LiDAR systems, there is a relative position offset and angle deviation between the position and orientation system (POS) and the laser scanner. Specifically, the GNSS records have relative position offset error, which is generally reduced or eliminated through ground static calibration. That is, the relative position offset can be derived from the comparison of the ground measurement and total station measurement. There are certain systematic errors between the angle values recorded by IMU and the angle values of the laser pointing. Generally, a calibration field is set up to eliminate the errors through flight dynamic calibration.

   According to the requirements of laser scanner calibration, the layout scheme of the calibration field is formulated. The relative and absolute elevations of the LiDAR system are calibrated with the control field, while the roll and pitch attitude angles are calibrated by the buildings. The calibration field should be as far away as possible from areas with low reflectivity such as water surfaces (lakes and rivers). The basic requirements are that (1) the calibration field should be flat with a bare ground surface; (2) the calibration field should have buildings or obvious protruding ground objects for calibration; (3) the targets in the field should have high reflectivity; and (4) there are obvious object points (such as road corner points).
4. Supplementary flight or reflight plan. When the following problems occur, supplementary flight or reflight is required: local data records of the POS system are missing; the inspection according to the evaluation indicators of each equipment does not meet the requirements; and the quality of the original data has local defects, further affecting the accuracy and density of the point cloud.

For manned aircraft, the two ends of the supplementary flight or reflight route should generally exceed the half-picture outside the supplementary flight range. The excess part should not be less than 500 m and not more than 2000 m. The requirements of overlap with the side direction and heading direction of the original route shall also be met.

3.1.1.3 Flight Preparation

The acquisition of airborne LiDAR data is related to the aircraft platform. If a manned aircraft platform is used, it is necessary to apply for airspace and lease an aircraft. If a UAV LiDAR system is to fly, it is also necessary to apply for airspace, but UAVs are relatively inexpensive and can be purchased or leased. Flight preparation includes two parts: airspace coordination and flight preparation.

In terms of airspace coordination, it is necessary to comply with relevant national laws and regulations. If a manned aircraft platform is used, an application for airspace should be made to the authority before the mission begins. At the same time, it is essential to obey air traffic rules and flight control. For example, the conditions required for UAVs to fly legally in China include real-name registration, flight license, flight altitude, and other flight requirements. The real-name registration means that all the UAVs, except micro-UAVs, need real-name registration. The flight license refers to the need to obtain an operator's license to operate UAVs over 7 kg. The flight altitude refers to the corresponding flight altitude requirements for different types of UAVs in areas such as flight control departments, important targets, government offices, squares, and bus stations. Other requirements mean that UAVs (except micro-UAVs) are allowed not to be driven at night; cannot carry contraband, dangerous goods, and other unapproved items; cannot throw objects or spray liquids on the ground; and cannot be operated or flown on mobile vehicles or aircraft.

In terms of flight preparation, according to the technical requirements of the mission and the terrain of the survey area, the appropriate LiDAR equipment and flight platform are determined. Tables 3.3 and 3.4 list the commonly used manned aircrafts and UAVs and their performance parameters, respectively. If the survey area

TABLE 3.3
Common Manned Aircrafts and Their Performance Parameters

	Aircrafts			
Performance	**Y-5**	**Y-12**	**Citation Jet**	**Antonov An-30**
Maximum ceiling(m)	4000	8000	13,105	7000
Maximum speed(km/h)	250	320	746	540
Cruise speed(km/h)	180	250	713	430
Maximum range(km)	1376	1440	3167	2630
Endurance time(h)	6	6	4	6
Maximum climb rate(m/s)	3	12	15.1	7.7
Operating height range(m)	500–4000	1000–6000	1500–12,000	1500–6000

TABLE 3.4
Common UAV and Their Performance Parameters

	UAVs			
Performance	**DJI M300**	**FEIMA D20**	**DAPENG CW-100**	**ZEROTECH ZT-30V**
Maximum ceiling(m)	7000	6000	4500	3500
Cruise speed(km/h)	83	65	100	110
Maximum range(km)	76	86	800	660
Endurance time(h)	0.9	1.3	8	6
Maximum climb rate(m/s)	7	5	17	15

is large, it is advisable to choose a manned aircraft platform (commonly used ones include Y-5, Y-12, etc.). If the survey area is small, a UAV platform is better. The UAV is usually divided into fixed-wing UAVs and multi-rotor UAVs.

There are many commercial LiDAR devices (see Section 1.2.2 for details). Each device has different configuration parameters and needs to be selected according to specific conditions. In most cases, the ranging capability of LiDAR is a major factor to be considered. In addition, it is necessary to prepare equipment such as GNSS receivers and off-road vehicles to help professionals formulate flight tasks and to allow technicians to guide the operation of flight equipment. The quality inspectors should be responsible for checking equipment failures and errors before the task and checking the equipment and data quality after the task is completed. It is also necessary to require operators to follow the aircraft throughout the flight mission.

3.1.2 Flight Implementation Stage

The flight implementation is to perform the flight observation mission and obtain the airborne LiDAR data according to the plan. It includes three steps: equipment installation and debugging, base station placement and ground coordination, flight operations and data acquisition.

3.1.2.1 Equipment Installation and Debugging

The implementation of aviation flight involves the use of a variety of equipment. It is very important to manage the equipment and keep its performance stable for the smooth implementation of aviation flight. Equipment management includes equipment storage environment requirements (storage space, dust proof, moisture proof, heat proof, heat exhaustion, and anticorrosion), routine equipment maintenance (daily testing of equipment parameters), equipment transportation conditions and precautions (e.g., whether the use of transport vehicles affects the performance of the equipment), the safety of equipment use (whether there will be accidents such as failures in high-altitude operations), etc. For manned aircraft flight, due to equipment management requirements, mission aircraft changes, and other reasons, equipment usually needs to be transported from the warehouse to the mission aircraft for installation and debugging. Installation and debugging include equipment inventory, transition plate preparation, equipment installation, GNSS eccentric component (the distance from the GNSS geometric center to the laser scanner geometric center after the equipment is installed) measurement, equipment ground energization test, equipment condition evaluation, etc.

3.1.2.2 Base Station Placement and Ground Coordination

The observation time of the ground base station should cover the flight time, and the GNSS static observation is generally used. The battery power and GNSS storage space can maintain at least one full flight time. Multiple base stations and multiple instruments can be set up synchronously to observe at the same time. The FindTrace service mentioned before does not need to set up a base station. This service can directly output high-precision trajectory data by generating virtual base station data and matching the virtual data with the GNSS original observation data and inertial navigation data provided by the user.

3.1.2.3 Flight Operations and Data Acquisition

The manned aircraft should perform the figure-eight-curve flight to activate the inertial navigation before starting to acquire data, so as to avoid error accumulation of the inertial navigation device. Before entering the survey area every time, the aircraft should fly flat for 3–5 minutes, and then do the figure-eight curve flight. When the flight is over, the aircraft should first do the figure-eight curve flight, and then fly flat for 3–5 minutes. The UAV can perform an "M" flight to activate inertial navigation.

The flight operation should also meet the requirements of ground static observation. That is, the surrounding area of the aircraft parking position before flight should have a wide field of view, and the height angle of obstacles in the field of view should be less than 20°. During the flight, the turning slope is generally no more than 15° and the maximum turning slope is no more than 22°, so as to prevent the GNSS satellite signal from being blocked; the pitch angle and roll angle on the route are generally no greater than 2°, the maximum pitch angle and the maximum roll angle are no more than 4°, the curvature of the route is no greater than 3%, and the ascent and descent rates of the aircraft within the same route shall not exceed 10 m/s.

Before acquiring LiDAR data, it is important to set the parameters of the equipment according to the task requirements, such as laser pulse frequency and scanning mode. For a manned aircraft, there is generally a supporting professional flight management software, and the route design file is imported into it before takeoff. During the flight, the software will design the route according to the aircraft position, open or close the data recording, and provide information to the pilot to ensure that the aircraft is flying on the line. For a UAV, the flight control is usually controlled by the flight controller; the easiest way is to record data throughout the whole process or start recording data after the aircraft takes off and reaches a certain relative altitude.

3.1.3 Data Preprocessing Stage

Data preprocessing is designed to solve and calibrate the data collected during the flight. It is the basis of subsequent point cloud operations. First, the coordinates of the point cloud in the calibration field are solved. The IMU boresight angle error

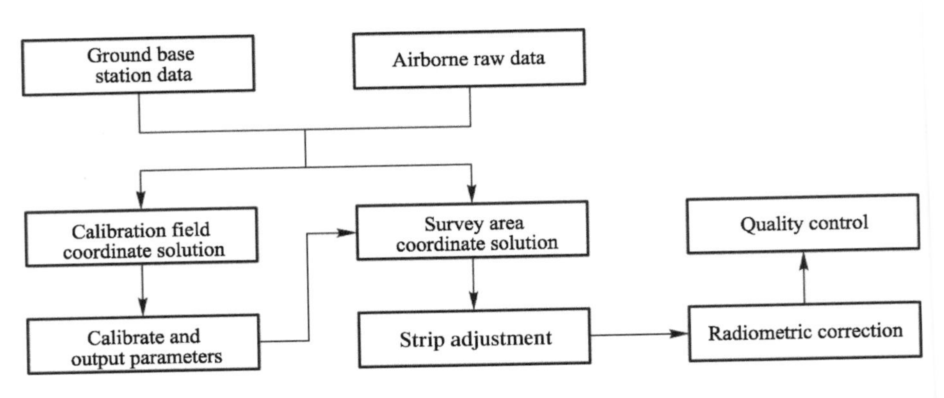

FIGURE 3.5 Airborne LiDAR data preprocessing process.

calibration is carried out to eliminate the system errors. The coordinates of the survey area are further solved to improve the positioning accuracy of the airborne LiDAR for the entire survey area. Second, in order to reduce the three-dimensional (3D) coordinate offsets between flight bands caused by airborne LiDAR system errors and random errors, it is necessary to perform strip adjustment on the acquired data. In order to make the data obtained under different conditions fully indicate the radiation characteristics of ground objects, radiometric correction of the data is required. Third, quality control is carried out in terms of point cloud density and point cloud elevation accuracy. The airborne LiDAR data preprocessing process is shown in Figure 3.5.

3.1.3.1 Point Cloud Coordinate Solution

The airborne LiDAR system uses the direct positioning method to calculate the 3D coordinates of the laser footprints. Specifically, the LiDAR system calculates the distance from the target to the LiDAR system by recording the time interval between the laser transmitting and receiving pulse signals and using the airborne raw data obtained by the flight and the data obtained by the ground base station.

The GNSS and the inertial navigation system (INS) measure the position and attitude at the moment when the laser emits pulses. That is, the spatial vector $\boldsymbol{P}$ between the laser emission position and the target is determined, so that the coordinates of the target point can be solved. The laser scanner obtains the measurement perpendicular to the flight direction through different scanning methods, such as swing mirror scanning and rotating polygon mirror scanning. As the flight platform moves, the measurement of the flight direction is completed and the point cloud data covering the entire survey area is obtained. The accuracy of point cloud coordinates depends on the high-precision measurement values of each component, the precise time synchronization among components, and the correction of various errors. The following describes the coordinate solution equation.

The airborne LiDAR system uses the scanner to record the distance and scanning angle, uses the POS to measure the scanner position and attitude, and solves the geometric coordinates of the point cloud in the geospatial reference system through a series of coordinate transformations. Equation (3.1) shows the calculation of 3D coordinates of a laser point cloud in the World Geodetic System (WGS84):

$$X = R_W R_G R_N \left[R_M R_L \begin{bmatrix} 0 \\ 0 \\ \rho \end{bmatrix} + P \right] + X_{GPS} \tag{3.1}$$

where X is the coordinate of the laser footprint in the WGS84 coordinate system, ρ is the distance from the laser emission center to the target, R_L is the rotation matrix from the instantaneous laser coordinate system to the scanner coordinate system, and R_M is the rotation matrix from the scanner coordinate system to the IMU reference coordinate system. The vector P is the GNSS eccentric component, which consists of the vector from the scanner laser emission center to the IMU reference center and the vector from the IMU reference center to the GNSS antenna phase center (both in the IMU reference coordinate system). R_N is a matrix composed of three attitude angles measured by IMU, i.e., roll angle, pitch angle, and heading angle. It converts the IMU reference coordinate system to the local navigation coordinate system. R_G is used to correct the vertical deviation and transform the local navigation coordinate system to the local ellipsoidal coordinate system. R_W is the transformation matrix from the local ellipsoidal coordinate system to the WGS84 space rectangular coordinate system. X_{GPS} is the coordinate vector of the GNSS antenna phase center in the space rectangular coordinate system.

Through this coordinate system transformation, the observed values of airborne LiDAR, including distance, scanning angle, sensor position, and attitude, can be transformed into 3D coordinates of laser points in WGS84.

3.1.3.2 IMU Boresight Angle Error Calibration

The airborne LiDAR system is composed of several components (GNSS, INS, laser rangefinder and scanning mirror, etc.). The point cloud coordinates calculated by Equation (3.1) do not fully eliminate the systematic errors. In order to improve the accuracy of airborne laser point cloud data, calibration must be carried out before flight operations. Among the systematic errors that affect the geometric positioning accuracy of airborne LiDAR, the IMU boresight angle error is the largest source of systematic error.

The error of the IMU boresight angle is shown in Figure 3.6, which is caused by the fact that the coordinate axis of the IMU reference coordinate system and the coordinate axis of the laser scanner coordinate system are not parallel. The IMU boresight angles are in roll, pitch, and heading directions of the coordinate axis. Their impact on ground laser point coordinates depends on the flight altitude and scanning angle. The IMU boresight angle error correction is designed to correct the laser point cloud data with geometric deviation to the correct position through concurrent and coplanar constraints. At present, IMU boresight angle error calibration mainly includes manual solutions and automatic solutions.

At the initial stage of commercial LiDAR equipment, the manual calculation method is usually used. For example, different routes (parallel routes and opposite routes)

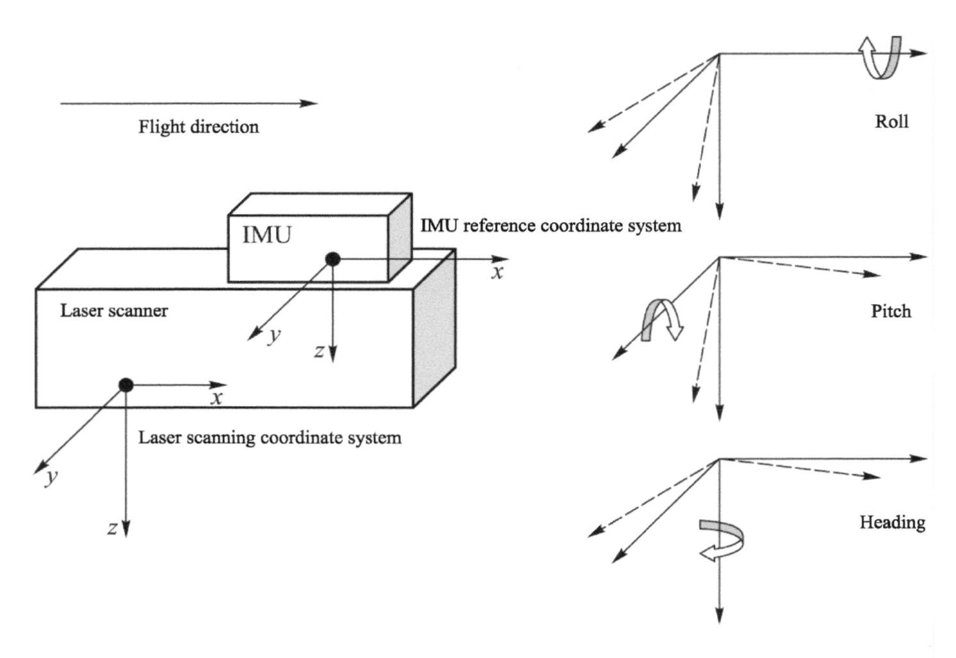

FIGURE 3.6 Schematic diagram of IMU boresight angle errors.

are selected to gradually separate the IMU boresight angle errors in roll, pitch, and heading directions through characteristic ground objects (spires, artificial platforms, straight roads, etc.). The deviation values are calculated according to the empirical formula through multiple iterations. Automatic calibration of IMU boresight angle errors takes the laser point cloud coordinate calculation equation as the mathematical model, takes the IMU boresight angle errors as the unknown parameters, solves the system parameters through adjustment, and eliminates the position error of the overlapping area. It can be understood as the registration process. That is, the point cloud with errors is registered to the reference position or the real position. Therefore, the key and difficult point of IMU boresight angle error calibration of the airborne LiDAR system is to establish appropriate "connection" conditions in adjacent strip point clouds.

Airborne LiDAR point cloud data rarely have real homonymous points. Hence, it is difficult to establish a "point-to-point" relationship in practice. Some scholars have carried out research on the "surface-to-surface" connection relationship, that is, the method of finding the homonymous plane. This method has strong anti-noise interference ability, and the results are the most reliable. The premise is that the point clouds of the same plane ground objects obtained from different routes should meet the coplanar condition. The airborne LiDAR coordinate solution equation is substituted into the plane equation, and the conditional error equation is established to solve the IMU boresight angle errors. Li et al. (2016a) carried out algorithm research on automatic extraction of connection planes and automatic matching of homonymous planes based on airborne LiDAR data and implemented the IMU boresight

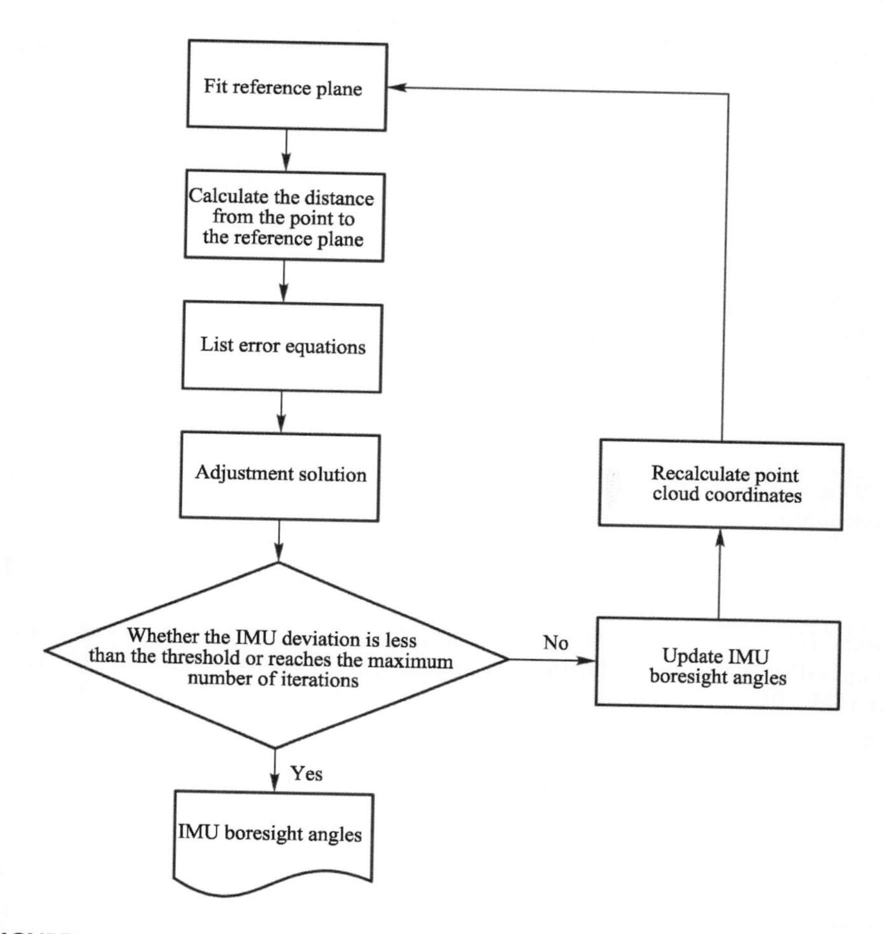

FIGURE 3.7 Calibration process of IMU boresight angle errors based on coplanar constraint.

angle error correction based on plane constraints. The process is shown in Figure 3.7. According to the airborne LiDAR point cloud coordinate solution equation, this method transformed the point cloud coordinates with error offsets to the correct position by the least square method through constraint conditions and calculated the IMU boresight angle error parameters. Assuming the IMU boresight angles in roll, pitch, and heading directions are α, β, and γ respectively, the errors in corresponding directions are $\Delta\alpha$, $\Delta\beta$, and $\Delta\gamma$ respectively, the rotation matrix $\Delta \boldsymbol{R}_M$ can be directly expressed as:

$$\Delta \boldsymbol{R}_M = \begin{bmatrix} 1 & -\Delta\gamma & \Delta\beta \\ \Delta\gamma & 1 & -\Delta\alpha \\ -\Delta\beta & \Delta\alpha & 1 \end{bmatrix} \tag{3.2}$$

Considering the boresight angle errors matrix, the airborne LiDAR laser footprint coordinates are changed from Equation (3.1) to Equation (3.3):

$$\boldsymbol{X} = \boldsymbol{R}_W \boldsymbol{R}_G \boldsymbol{R}_N \left[\Delta \boldsymbol{R}_M \boldsymbol{R}_M \boldsymbol{R}_L \begin{bmatrix} 0 \\ 0 \\ \rho \end{bmatrix} + \boldsymbol{P} \right] + \boldsymbol{X}_{GPS} \tag{3.3}$$

The distance (d) from the laser footprint to the homonymous plane can be expressed by the boresight angle parameters as:

$$d = d_0 + \frac{\partial d}{\partial \alpha}\Delta\alpha + \frac{\partial d}{\partial \beta}\Delta\beta + \frac{\partial d}{\partial \gamma}\Delta\gamma \tag{3.4}$$

Due to the influence of boresight angle errors, the residual distance between the laser footprint and the plane is $\boldsymbol{V}$, so the error equation is:

$$\boldsymbol{V} = \boldsymbol{B}\boldsymbol{X} + \boldsymbol{L} \tag{3.5}$$

where $\boldsymbol{B}$ is the unknown coefficients matrix, $\boldsymbol{L}$ is the distance matrix from the laser point to the plane, and $\boldsymbol{X}$ is the IMU boresight angle errors to be solved $[\Delta\alpha, \Delta\beta, \Delta\gamma]^{\mathrm{T}}$. When $\boldsymbol{V}^{\mathrm{T}}\boldsymbol{P}\boldsymbol{V}$ is the smallest, the IMU boresight angle errors can be obtained according to Equation (3.6):

$$\boldsymbol{X} = -(\boldsymbol{B}^{\mathrm{T}}\boldsymbol{P}\boldsymbol{B})^{-1}\boldsymbol{B}^{\mathrm{T}}\boldsymbol{P}\boldsymbol{L} \tag{3.6}$$

After the coordinates are solved according to Equation (3.1), the calibration field is calibrated with the boresight angle errors and the parameters are output, which can be substituted into Equation (3.3) to complete the high-precision coordinate solution of the whole survey area.

3.1.3.3 Strip Adjustment

Due to the limitation of altitude and scanning field angle, each flight strip can only cover a certain width of the ground. To complete a large range of operation area, multiple routes must be flown. These routes must maintain a certain degree of overlap (>20%). The error usually leads to systematic migration among the homonymous features of different flight strips of LiDAR data, which seriously affects the relative accuracy of point cloud data. The purpose of flight strip adjustment is to generate seamless products by eliminating or reducing the differences among overlapping areas of different flight strips, so as to provide quality assurance for the final geospatial products. Strip adjustment usually includes the following two aspects:

1. Automatic registration of point cloud data. The purpose of point cloud automatic registration in the strip adjustment is to determine the systematic offsets among strips. At present, the commonly used methods include the matching method based on regular grid, the least square matching method

based on triangulated irregular network (TIN), the least square 3D surface matching method, the iterative closest point (ICP) method, and their improved methods.

The matching method based on regular grid is the most widely used and the simplest surface matching technique. It resamples the two datasets into evenly spaced grids, so the vertical difference between them can be easily calculated. This expression is often referred to as 2.5-dimensional data; that is, any position (x, y) coordinate pair can only have one elevation value. Its advantage is that it can be directly processed by using the original standard image processing algorithms. However, the image matching scheme only provides a two-dimensional offset, while the LiDAR data processing needs to obtain the difference in the 3D direction. Therefore, we can consider using intensity image matching to directly obtain the 3D direction offsets of the corresponding range of data. In addition, the original point cloud grid would contain various errors, and the LiDAR point density itself is lower than the minimum spatial sampling distance required for complete surface sampling. Therefore, the homonymous feature points based on the regular grid need to be interpolated twice. The matching accuracy is easily affected by the errors caused by the second interpolation.

The least square matching method based on the TIN uses TIN as the data organization structure to construct irregular triangulation patches for the data of different overlapping areas of flight strips, so as to realize the matching. This data organization method can preserve the 3D information of the original point cloud at any location and provide interpolation of the surface defined by adjacent triangles at any location, so as to avoid the influence of secondary interpolation of point cloud data.

The principle of the ICP algorithm is to take the closest point on two free surfaces as the corresponding point, then establish the objective equation based on the principle of minimum sum of squares of distances between them, and iteratively solve the transformation parameters according to the principle of least square. The ICP algorithm is widely used in image matching and pattern recognition. It does not need to specially extract point features from point cloud data. Meanwhile, it can be combined for data with different features, such as line features, surface features, and corner features. However, it has many limitations, such as a large amount of computation, requiring initial values, and easy local convergence. Therefore, many scholars have improved and optimized this algorithm.

The least square 3D surface matching method takes the 3D feature surface as the splicing unit and achieves the complete splicing of the 3D surface according to the pregiven 3D surface template. It is the derivation of the least square matching in the 3D space, and is also a global registration method. The transformation parameters need to be compared with reasonable initial values.

During the adjustment of flight strips, areas with a uniform surface, such as roads and building surfaces, shall be selected. The vegetation area shall be avoided, because the vegetation area may contain multiple echo signals, which may affect the matching and adjustment results.

2. Selection and solution of the model for the strip adjustment. The strip adjustment model can be divided into the data-driven model and sensor calibration model. For the applications with low precision requirements, a simple data-driven model can meet the requirements. In contrast, the sensor calibration model has higher requirements for the original data. For the strip-shaped long measurement area, the sensor calibration model must be used because the overlapping area of the strip-shaped area is limited. Only a good calibration procedure can provide better overall data accuracy.

   The data-driven model is designed to establish the corresponding mathematical model according to the plane coordinates and elevation deviation of the same object in the adjacent flight strips, connect the overlapping areas by using the matching principle, and use the model to solve the parameters to correct the laser point coordinates. For example, seven-parameter conversion model is a typical data-driven model. It includes three spatial translation parameters, three spatial rotation parameters, and one scale factor. The spatial translation parameter $[\ \Delta x_{i,i+1}\ \ \Delta y_{i,i+1}\ \ \Delta z_{i,i+1}\ ]^{\mathrm{T}}$ represents the translation relative to the x, y, z. The spatial rotation parameter $[\ \alpha\ \ \beta\ \ \gamma\ ]^{\mathrm{T}}$ is the rotation relative to the axis. The scale factor m represents the relative scaling. Thus, the seven-parameter conversion model is constructed as follows:

$$\begin{bmatrix} x_{H_i} \\ y_{H_i} \\ z_{H_i} \end{bmatrix} = m \times \boldsymbol{R}_z \times \boldsymbol{R}_y \times \boldsymbol{R}_x \times \begin{bmatrix} x_{H_{i+1}} \\ y_{H_{i+1}} \\ z_{H_{i+1}} \end{bmatrix} + \begin{bmatrix} \Delta x_{i,i+1} \\ \Delta y_{i,i+1} \\ \Delta z_{i,i+1} \end{bmatrix} \tag{3.7}$$

   of which

$$\boldsymbol{R}_x = \begin{bmatrix} 1 & 0 & 0 \\ 0 & \cos\alpha & -\sin\alpha \\ 0 & \sin\alpha & \cos\alpha \end{bmatrix}, \boldsymbol{R}_y = \begin{bmatrix} \cos\beta & 0 & \sin\beta \\ 0 & 1 & 0 \\ -\sin\beta & 0 & \cos\beta \end{bmatrix}, \boldsymbol{R}_z = \begin{bmatrix} \cos\gamma & -\sin\gamma & 0 \\ \sin\gamma & \cos\gamma & 0 \\ 0 & 0 & 1 \end{bmatrix},$$

   where $[\ x_{H_i}\ \ y_{H_i}\ \ z_{H_i}\ ]^{\mathrm{T}}$ is the point coordinate of the reference flight strip, and $[\ x_{H_{i+1}}\ \ y_{H_{i+1}}\ \ z_{H_{i+1}}\ ]^{\mathrm{T}}$ is the point coordinate of the adjacent strip. This method is only applicable to the case where the systematic offsets among multiple flight strips are relatively consistent.

   The sensor calibration model calibrates the sensor through calibration parameters to achieve the goal of minimizing the systematic offsets among the flight strips. The model is established based on the airborne LiDAR equation. It considers the geometric positioning process of airborne LiDAR. The theory is rigorous. However, the established error model has the problem of strong correlation among parameters. Therefore, in practical applications, in order to ensure the accuracy and reliability of parameter solutions, the error equation model is often simplified, resulting in unknown residual errors after adjustment. In addition, due to the confidentiality of LiDAR hardware system, it usually only provides users with 3D coordinates rather than original observation values (such as distance and angle), which also brings difficulties to the application of the sensor calibration model.

The data-driven model is simple and feasible without original observation values, but it is not rigorous in theory. The similarity of the two models is the criteria for parameter adjustment.

The last step of the strip adjustment is to apply the error correction values determined in the previous step to LiDAR point cloud data. For data-driven models, 3D similarity transformation or a simpler method is usually used to directly apply the correction values to the original LiDAR point cloud data. For the sensor calibration model, LiDAR point cloud data should be completely reconstructed based on the sensor calibration model.

3.1.3.4 Radiometric Correction

The intensity information recorded by airborne LiDAR is not only related to the surface reflectance but also the transmission distance, atmospheric environment, equipment parameters, etc. The process of eliminating these effects is called relative radiation correction. The relative-corrected intensity is closely related to the parameters selected in the correction process. The calculation of the radiation parameters (reflectance, radar cross-sectional area, backscatter coefficient, etc.) of features by reference to the intensity and radiation parameters of the calibrated feature is called absolute radiation correction. The following section discusses the three factors considered in the relative intensity correction, i.e., observation geometry, atmospheric environment and equipment parameters, and the calculation of radiation parameters in the absolute radiation correction.

1. Observation geometry

   The physical quantities associated with the observation geometry include distance and incident angle, as shown in Figure 3.8. For a spot falling within the same target and assuming that the target conforms to Lambertian reflection, the LiDAR equation is expressed as Equation (3.8):

$$P_r = \frac{P_t D_r^2 \rho}{4R^2} \cos\theta \eta_{sys} \eta_{atm} \tag{3.8}$$

   where P_t is the laser power emitted by the laser, D_r is the antenna aperture diameter of the laser receiver, ρ is the reflectance of the scatterer, R is the distance from the laser sensor to the target ground feature, θ is the incident angle, η_{sys} is the influence coefficient of the sensor on signal power, and η_{atm} is the atmospheric influence coefficient. R is the distance between the laser scanner and the target, and θ is the scanning angle, which is equal to the incident angle for flat ground. When the ground has a certain slope, the incident angle is the angle between the laser and the normal of ground surface.

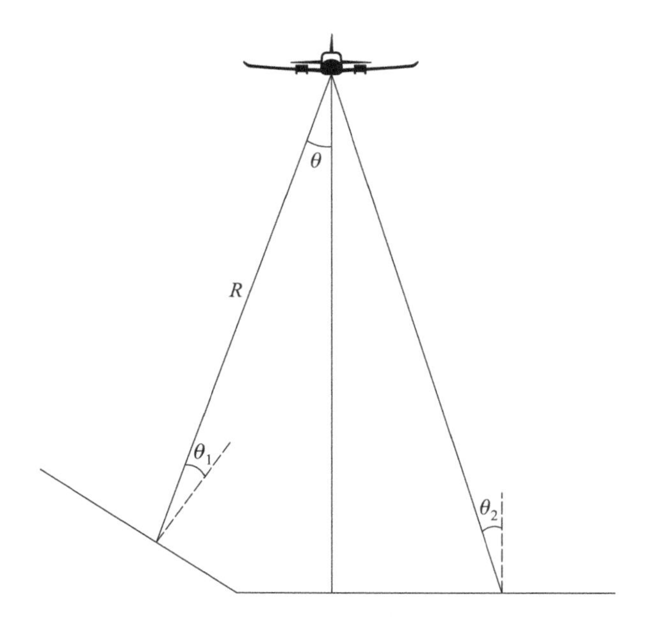

FIGURE 3.8 Schematic diagram of airborne LiDAR observation geometry.

It is observed that the intensity is inversely proportional to the square of the distance and proportional to the cosine of the incident angle. The distance is related to the altitude of the flight, the position of the target in the scanned airstrip, and the terrain relief. Generally, the average relative heading height is the reference distance and the intensity of all targets is normalized based on the intensity value at the reference distance. When the terrain is flat, the angle of incidence is equal to the scanning angle; when the terrain is undulating, the angle between the normal vector and the incident laser beam needs to be calculated. Considering the effects of the observation geometry, the correction formula is expressed as Equation (3.9):

$$I_c = I_{raw} \frac{R_i^2}{R_{ref}^2} \frac{1}{\cos\theta_i} \tag{3.9}$$

where I_{raw} and I_c are the intensity values before and after correction, R_i is the observation distance, R_{ref} is the reference distance, and θ_i is the incident angle.

2. Atmospheric environment and equipment parameters

Due to atmospheric scattering and absorption, laser energy is attenuated during transmission, with different attenuation coefficients for different meteorological conditions, sight distances, and wavelengths. Parameters of airborne LiDAR equipment that affect laser intensity include laser emission power, laser receiver aperture size, laser divergence angle, etc. The laser receiver aperture size and laser divergence angle are constants for the same airborne LiDAR sensor. Only when calibrating the intensity data for different sensors do differences in these parameters need to be considered. In most cases, the laser emission power and the signal power impact factor are mainly considered for the data acquired by the same device.

The set laser emission power is usually the same in the same survey area, whereas it might vary considerably from survey area to survey area, especially at different flight altitudes. Hence, during preflight mission planning, the corresponding flight software of the device shall calculate the laser pulse frequency and pulse power according to the terrain, point density, ground reflectivity, and other information input by the user.

A special example is that some airborne LiDAR systems have differences in the amplitude and width of the emitted pulses, i.e., the power of the emitted pulses varies from moment to moment. For this type of equipment, the amplitude and width of each pulse can be obtained by waveform decomposition. We assume that the amplitude of the transmit pulse only affects the amplitude of the return pulse and the pulse width of the transmit pulse only affects the pulse width of the return pulse. By taking a reference amplitude and pulse width, the intensity and pulse width can be corrected (Qin et al., 2011).

3. Absolute radiometric correction

After the relative radiometric correction to eliminate the effects of observation geometry, environment, and equipment parameters is carried out, the intensity is proportional to the reflectance. This intensity is related to factors such as the choice of reference distance and laser receiver aperture size, but does not correspond to what is customary for expressing the radiometric properties of a feature target. The radiation characteristics of a feature are usually expressed in terms of reflectance, backscatter coefficient, etc. A feature can be selected, and its reflectance measured by the spectrometer is used as a reference to solve for the reflectance of other targets, i.e., absolute radiation correction. There are two important assumptions in the process of absolute radiometric correction. One is that the feature target follows the Lambertian reflection, and the other is that the feature target is larger than the spot area, i.e., that there is only the same feature within a spot.

$$\frac{\rho}{\rho_{ref}} = \frac{I_c}{I_{ref}} \tag{3.10}$$

$$I_c = I_{raw} \frac{R_i^2}{R_{ref}^2} \frac{1}{\cos\theta} \frac{1}{\eta_{atm}} \frac{P_{ref}}{P_t} \tag{3.11}$$

$$\rho = \frac{\rho_{ref}}{I_{ref}} I_{raw} \frac{R_i^2}{R_{ref}^2} \frac{1}{\cos\theta} \frac{1}{\eta_{atm}} \frac{P_{ref}}{P_t} \tag{3.12}$$

where ρ and ρ_{ref} are reflectance of target and reference object respectively; I_{cor} and I_{ref} are the corrected intensity of the target and the reference object; P_t is the emitted pulse power; P_{ref} is the received pulse power of the reference object; and η_{atm} is the atmospheric transmission efficiency.

3.1.3.5 Quality Control

Quality control mainly includes point cloud density check, airstrip edge check, and point cloud elevation accuracy check.

1. Point cloud density check. Point cloud density is used to evaluate the quality of the overall point cloud data. Two indicators, average density and variance, are usually used. In general, a high-density point cloud gives more surface features and details.
2. Airstrip edge check. For overlapping areas of adjacent routes, the overlap of the adjacent strip point cloud is checked by pulling the profile. If the deviation of the overlapping point cloud is large, the system error parameter check and strip adjustment are required to be redone. For adjacent survey areas, a joint edge check of the overlapping areas is required.
3. Point cloud elevation accuracy check. Data located in a flat area are usually used to evaluate the accuracy of the point cloud elevation. Specifically, the elevations of a certain number of distinctive checkpoints are measured by GNSS and compared with the obtained point cloud elevations to calculate the root mean square error (RMSE) in elevation measurement.

3.2 TERRESTRIAL LiDAR DATA ACQUISITION

Terrestrial LiDAR includes vehicle mobile LiDAR, backpack LiDAR, and ground-based LiDAR. This section mainly introduces the data acquisition process of the ground-based LiDAR, which is generally divided into three stages: plan preparation stage, scan implementation stage, and data collection stage. Figure 3.9 shows its overall workflow.

3.2.1 Plan Preparation Stage

According to some instructional documents, such as technical specifications for terrestrial three-dimensional laser scanning (CH/Z 3017—2015), and existing related studies (Liu et al., 2010), the work in the plan preparation stage can be divided into defining task requirements, scan route and plot planning, data acquisition method selection, instrument and software configurations, operator allocation, safeguard measures formulation, and pre-scanning inspection.

3.2.1.1 Define Task Requirements

After receiving a scanning task, it is important to understand the background, content, objective, work region, workload, and deadline of the task, which are the main bases for formulating the scanning plan.

3.2.1.2 Scan Route and Plot Planning

Scan route and plot planning are the most important steps in the whole plan preparation stage. It is necessary to select plots and plan a reasonable route on the premise of considering representativeness and operability according to the extent and physical geography (topography, climate, etc.) of the survey area, the shape and spatial

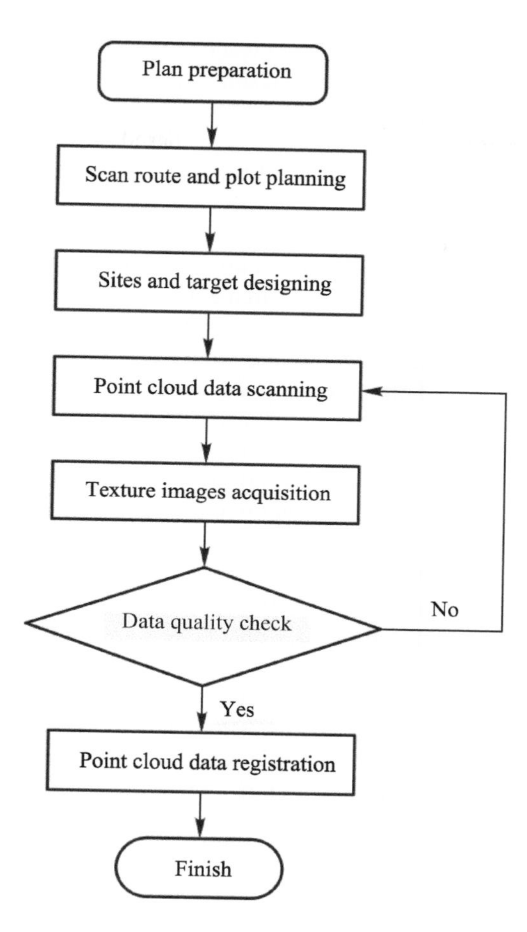

FIGURE 3.9 The workflow of terrestrial LiDAR data acquisition.

distribution of the scanned objects, and the measurement accuracy requirements. This is done in order to improve the efficiency of field data collection, avoid blind measurement, and reduce waste of manpower and material resources.

In order to ensure the rationality of the scan route and plot planning, it is necessary to survey the site and formulate a practical work scheme when the working range and distance allow. If the terrain of the survey area is complex or the distance is too far such that it is difficult to conduct a field survey, the preliminary design of the work scheme can be carried out on the drawings according to the existing data of the survey area and then flexibly adjusted in combination with the scanning object and the surrounding environment during field work (Xie et al., 2014).

3.2.1.3 Data Acquisition Method Selection

The data acquisition method directly affects the point cloud registration method and the field operation in the scanning implementation stage. Therefore, it is necessary to select the appropriate data acquisition method according to the plot location, the structural characteristics of the scanning object, the surrounding environment, and the specific application of the scanning results, so as to minimize the cumulative errors from multi-station data registration. There are three main data acquisition methods for terrestrial laser scanners: the method based on "stationary point + rear-view point", the target-based method, and the method based on point cloud automatic registration (Ou, 2014).

1. Data acquisition method based on "stationary point + rear-view point." This method is like the traditional measurement, which requires the scanner to be placed on the known control points, then the operations such as station setting, orientation, and scanning are completed in sequence. For a start, the control network shall be set up in advance, and the coordinates of each control point can be measured with instruments such as total station and GNSS. After the previous control survey, the coordinates of each station and the rear-view point (target point) are unified under the same spatial reference system in theory. The later point cloud registration process is just to complete the point cloud registration of different stations, without coordinate conversion. The stitching accuracy is relatively high. The method does not require overlapping areas between adjacent stations, as is usual for a belt survey or complex projects with a large survey area.
2. Target-based data acquisition method. The station and target should be placed in a position with good visibility and a wide field of vision within the survey area. The point cloud data registration of each station is achieved through the common target. In general, at least three non-collinear targets are required between adjacent stations. The method is simple, fast, and accurate, as is usual for scanning small independent objects.
3. Data acquisition method based on point cloud automatic registration. The method is like the target-based data acquisition method, but the difference is that this method does not require the assistance of the target and only needs to ensure at least 30% overlapping area between adjacent stations. Therefore, it is necessary to manually select the correspondence points with obvious characteristics in the overlapping area. The scanning results are highly dependent on the experience of the operator. The method is lower than the previous two methods in terms of data registration accuracy. It is suitable for projects with low accuracy requirements and obvious characteristics of the survey area.

3.2.1.4 Instrument and Software Configurations

Besides determining the laser scanner, GNSS, target, digital camera, portable computer, storage medium, and data processing software that meet the scanning needs, protective equipment such as a sunshade shall be configured. In special operation environments, the selected instruments and equipment shall meet the safety requirements.

1. Selection of terrestrial 3D laser scanners. There are many types of terrestrial 3D laser scanners; the primary manufacturers are Leica (Switzerland), Riegl (Austria), Trimble (United States), Faro (United States), Optech (Canada),

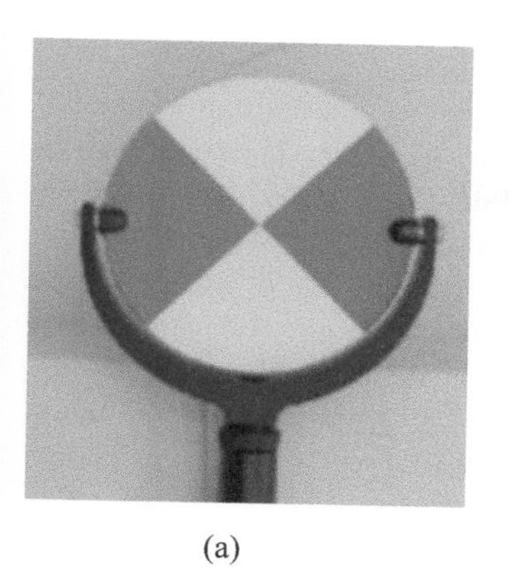
(a)

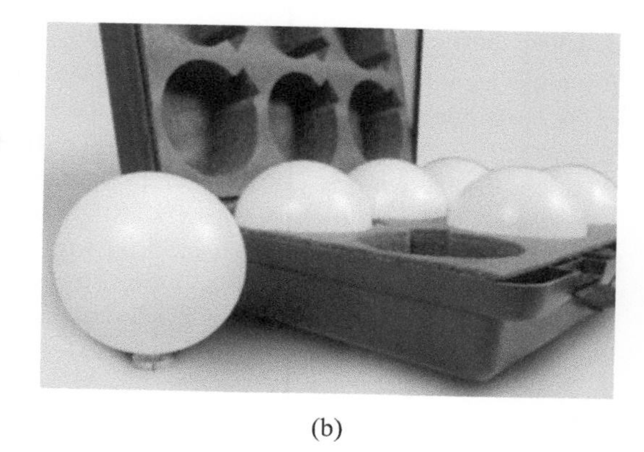
(b)

FIGURE 3.10 Well-known targets. (a) Plane target; (b) spherical target.

TOPCON (Japan), I-Site (Australia), Z+F (Germany), Hi-TARGET (China), Surestar (China), STONEX (China), etc. There are differences in laser wavelength, laser type, range, scanning speed, ranging accuracy, field of view, etc., among products of various manufacturers. For some equipment parameters, please refer to Tables 1.4 and 1.5 in Section 1.2 of this book. When selecting an instrument, the technical requirements of the task, the environment of the survey area, and other factors shall be considered first, and then the main technical parameters of the instrument shall be considered. Generally, one terrestrial 3D laser scanner can meet the operation requirements. However, in the case of large task volume, short construction period, or special requirements of the scanned object, multiple instruments are required to work together, and various brands and types of instruments may even be used (Xie & Gu, 2016).

2. Target selection. A target is an important symbol of point cloud registration in the process of data collation, which is directly related to the quality of point cloud registration. Well-known targets include the spherical target and plane target (Figure 3.10). The spherical target can obtain the coordinate of a spherical center from any direction and can help with the scanning inside and outside the building and at the corner. It is mainly used for the registration of multi-view point cloud data. The plane target is similar to the spherical target and has high registration accuracy. It can be used in conjunction with the total station. It is usually used for registration and coordinate transformation of strip-shaped and area-shaped objects.

3.2.1.5 Operator Allocation

The number of operators shall be allocated according to the operation content, which is generally four to five persons. Operators shall be trained to be familiar with all technical processes. Only after passing the training can they participate in the operation.

3.2.1.6 Safeguard Measures Formulation

It mainly includes safeguarding and schedules. The operating temperature requests of the instrument should be considered in the scanning operation. If the instrument is exposed to strong sunlight for a long time, the instrument should be shaded. In addition, the laser might cause a certain degree of damage to the human eye, so it is necessary to avoid looking directly at the laser transmitter. Sufficient operation space and stand area shall be ensured for high-altitude operation. The stability of the platform should be checked to ensure the safety of instrument and personnel. Finally, a reasonable schedule of the indoor data processing and field work should be formulated according to the task period.

3.2.1.7 Pre-scanning Inspection

First, check whether the parts and accessories of the scanning instrument are complete and matched and whether the parts of the instrument are connected tightly and stably. The instrument with laser alignment and dual-axes compensation shall also be functionally inspected. Second, check whether the scanner can be used normally after being powered on and whether the power capacity and memory capacity meet the operation time requests to avoid the battery being run down during scanning. Third, the built-in coaxial camera is checked for matching the image and point cloud to eliminate the error between them. The external camera should be calibrated for parameters such as camera host, principal point, distortion, and installation orientation. Finally, prepare vehicles, power supply equipment, digital camera, and other auxiliary facilities. Check whether the personnel are in place.

3.2.2 Scan Implementation Stage

During the scan implementation stage, the scanning personnel shall be required to standardize the operation as much as possible and select good weather for operation to obtain a high-quality point cloud. This section describes the target-based data acquisition method in detail, including the steps of station layout, target layout, scanner placement, scan implementation, and texture image acquisition.

3.2.2.1 Station Layout

The establishment of the station needs to comprehensively consider the integrity, accuracy, overlap, redundancy, and scanner security of the scanned data. The specific requests are as follows:

1. Select the appropriate station. In order to ensure the integrity of the scanned data, the site selection should follow the principle that all the measured objects can be scanned. In addition, the occlusion of the measured objects by branches and leaves should be considered to ensure that the complete 3D point cloud can be obtained.
2. Select an appropriate scanning distance. The accuracy of the data obtained by the three-dimensional laser scanner is inversely proportional to the scanning distance. Hence, the distance between the scanner and the measured object should not be too long. Try to ensure that the measured object is

within the 45° incident angle of the scanner. For example, if the station is placed directly in front of the dense area of the measured object, the data with high precision can be obtained with a small incident angle.

3. Set a reasonable overlap. To ensure the accuracy of subsequent data registration, the overlap degree between two adjacent stations shall not be less than 30%.
4. Reduce data redundancy. On the premise of ensuring data integrity, data acquisition shall be completed with the fewest number of stations and the appropriate scanning mode. The scanning of irrelevant objects shall be avoided as much as possible during the operation.
5. Make sure the scanner is safe. The scanner shall be placed in an area with a wide view and flat terrain.

3.2.2.2 Target Layout

The target is easily identified and measured in laser point cloud data, which can be used for point cloud data quality inspection and point cloud registration. Therefore, target layout is one of the keys in 3D terrestrial laser scanning operation. Four principles shall be followed for target layout. First, ensure that the target can be completely scanned by scanners at least two stations. Second, when the target is used as the same point in multi-station data registration, it should be ensured that at least three of the same points exist in two adjacent stations. Third, establish the targets uniformly and avoid establishing them in a narrow shape. For example, three targets can be arranged in the survey area with an approximate regular triangle shape. Fourth, the distance from the scanner must be suitable. Too short will lead to a large coordinate transformation error, and too long will decrease the recognition accuracy of the center of the target. Figure 3.11 shows an example of the distribution of stations and targets.

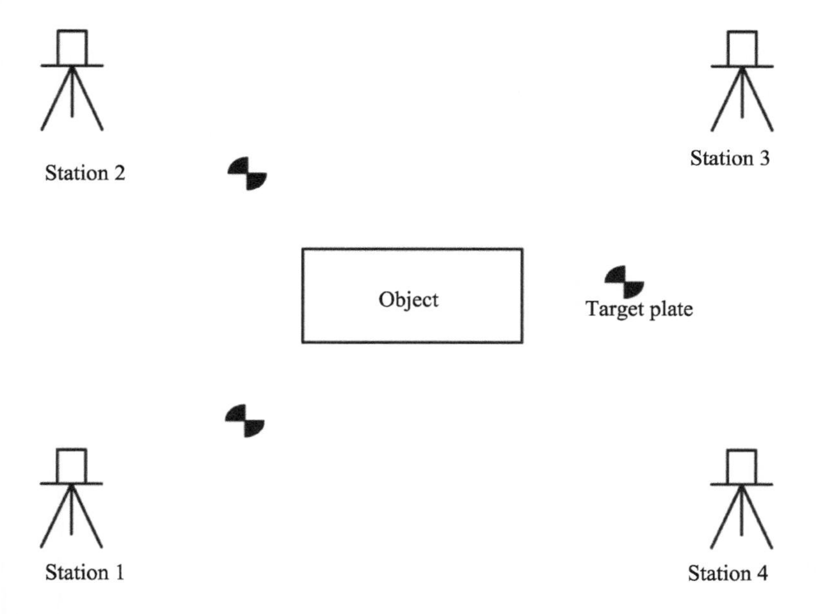

FIGURE 3.11 The distribution of stations and targets.

3.2.2.3 Terrestrial Laser Scanner Placement

Use a tripod to hold the terrestrial laser scanner at the selected station. When installing the instrument, keep the scanner horizontal and at a moderate height and then align, level, and turn on the instrument.

Since the suitable operating temperature of the laser scanner is 0°C–40°C and high-definition digital cameras and other sensitization devices are installed inside (or outside), direct sunlight should be avoided when installing the instrument. If scanning is conducted in a high-temperature environment, it is necessary to place the scanner in a cool place or use a parasol to shield it to avoid observation errors caused by the high temperature of the instrument.

3.2.2.4 Scan Implementation

After setting up the scanner, turn on the power and set the scanning parameters, sampling resolution, and other parameters. In addition, avoid glare from the object during scanning. The implementation consists of the following steps:

1. Coarse scan. Coarsely scan the object in low-resolution mode to obtain its general range and orientation. In this way, it can help to obtain accurate point cloud data of the object during the fine scan stage and reduce data redundancy.
2. Fine scan. Scan the object with a high-resolution mode according to the coarse scan result.
3. Target scan. Scan the target finely. Set a unique mark for the targets and determine their center. Use the total station or real-time kinematic (RTK) to obtain the geodetic coordinates of the targets and stations to prepare for subsequent multi-station data registration and coordinate conversion.
4. After the current station is scanned, move to the next station, and repeat steps 1–3 until all the objects are scanned.

There are a few precautions during scanning. First, avoid shaking the tripod to ensure the precise measuring. Second, avoid personnel or suspended solids within the scanning range to reduce noise points. Third, if the scanner suffers from a system halt, power off, position change, or other emergencies, the scanner shall be checked and then rescan after confirming that it is in good condition. Fourth, during the scanning, try to keep the measured object in a static state to avoid delamination and offset of the measured object.

3.2.2.5 Texture Image Acquisition

The texture image of a measured object acquired by the built-in (or external) camera of the 3D laser scanner can be used as a texture map or provide a reference for multi-station registration. When acquiring a texture image, ensure that the overlap of two adjacent images is not less than 30%. Additionally, the backlight and low light shall be avoided to cause image quality loss. After the operation, turn off the laser scanner and camera, and import the data into the computer in a timely manner. Check whether the point cloud or image conforms to the requests and whether the object is complete and available. If there are missing or abnormal data, supplementary measurement shall be conducted as soon as possible.

3.2.3 Data Collection Stage

After completing the data acquisition of the entire survey area, the collected data can be preprocessed and organized. It mainly includes point cloud registration and quality inspection. For multi-station scanning, the data of each station is under an independent local coordinate system. It is necessary to splice the data of multiple stations in the same survey area after the acquisition, which is called point cloud registration. In addition, due to the influence of the instrument itself or the surrounding environment, it is necessary to check the quality of the point cloud at the first time, such as point density, integrity, and overlap, to determine whether the object needs to be supplemented or remeasured. Finally, organize and store the dataset.

3.2.3.1 Point Cloud Registration

For terrestrial LiDAR, point cloud registration is done to unify the data with separate coordinate systems into the same relative coordinate system or real geographic coordinate system to obtain the complete point cloud data of the object (Dong et al., 2020; Li et al., 2016b). It is a 3D rigid transformation without deformation, which usually includes two steps: coarse registration and fine registration. The coarse registration is generally achieved by manually selecting feature points, of which registration accuracy reaches up to the millimeter level. The fine registration criterion automatically selects feature elements for matching, and its accuracy reaches the sub-millimeter level.

Coarse registration generally selects several similar points in the overlap area of adjacent stations and calculates the transformation matrix parameters (Zhang & Zhang, 1997). There are seven transformation matrix parameters, including three rotation angle parameters (ω, φ, κ), three translation parameters (ΔX, ΔY, ΔZ), and a scale coefficient (λ). The laser scanning data are measured at the original scale, and the scale coefficient is fixed at 1. Hence, there are actually only six unknown parameters. The calculation equation of the transformation matrix parameters is as in Equation (3.13).

$$\begin{pmatrix} X \\ Y \\ Z \end{pmatrix} = \boldsymbol{R}_{\varphi}\boldsymbol{R}_{\omega}\boldsymbol{R}_{\kappa}\begin{pmatrix} X_m \\ Y_m \\ Z_m \end{pmatrix} + \begin{pmatrix} \Delta X \\ \Delta Y \\ \Delta Z \end{pmatrix} = \boldsymbol{R}\begin{pmatrix} X_m \\ Y_m \\ Z_m \end{pmatrix} + \begin{pmatrix} \Delta X \\ \Delta Y \\ \Delta Z \end{pmatrix} \tag{3.13}$$

where $(X\ Y\ Z)^{\mathrm{T}}$ is the coordinate of the center of the base station, $(X_m\ Y_m\ Z_m)^{\mathrm{T}}$ is the coordinate of the corresponding point in the registration station, $(\Delta X\ \Delta Y\ \Delta Z)^{\mathrm{T}}$ is the translation, and R is the rotation angle matrix:

$$\boldsymbol{R} = \boldsymbol{R}_{\varphi}\boldsymbol{R}_{\omega}\boldsymbol{R}_{\kappa} = \begin{pmatrix} \cos\varphi & 0 & -\sin\varphi \\ 0 & 1 & 0 \\ \sin\varphi & 0 & \cos\varphi \end{pmatrix} \cdot \begin{pmatrix} 1 & 0 & 0 \\ 0 & \cos\omega & -\sin\omega \\ 0 & \sin\omega & \cos\omega \end{pmatrix} \cdot \begin{pmatrix} \cos\kappa & -\sin\kappa & 0 \\ \sin\kappa & \cos\kappa & 0 \\ 0 & 0 & 1 \end{pmatrix} = \begin{pmatrix} a_1 & a_2 & a_3 \\ b_1 & b_2 & b_3 \\ c_1 & c_2 & c_3 \end{pmatrix} \tag{3.14}$$

where $\boldsymbol{R}_\varphi$, $\boldsymbol{R}_\omega$, $\boldsymbol{R}_\kappa$ are the rotation matrices under the three rotation angle parameters ω, φ, κ respectively, and the calculation method is:

$$\begin{cases} a_1 = \cos\varphi\cos\kappa - \sin\varphi\sin\omega\sin\kappa \\ a_2 = -\cos\varphi\sin\kappa - \sin\varphi\sin\omega\cos\kappa \\ a_3 = -\sin\varphi\cos\omega \\ b_1 = \cos\omega\sin\kappa \\ b_2 = \cos\omega\cos\kappa \\ b_3 = -\sin\omega \\ c_1 = \sin\varphi\cos\kappa + \cos\varphi\sin\omega\sin\kappa \\ c_2 = -\sin\varphi\sin\kappa + \cos\varphi\sin\omega\cos\kappa \\ c_3 = \cos\varphi\cos\omega \end{cases} \tag{3.15}$$

The three rotation angle parameters ω, φ, and κ refer to the angles that the original coordinate system rotates counterclockwise around: the X-axis, Y-axis, and Z-axis, respectively.

$$\begin{cases} \varphi = -\arctan\left(\dfrac{a_3}{c_3}\right) \\ \omega = -\arcsin(b_3) \\ \kappa = -\arctan\left(\dfrac{b_1}{b_2}\right) \end{cases} \tag{3.16}$$

Equation (3.13) contains six unknown parameters. Theoretically, at least three sets of corresponding points are required to determine the unique values of the six parameters. Usually, four or more groups of corresponding points are selected. Then the high-precision transformation matrix is solved with the least square method.

The coarse registration methods include the control-point-based registration, target-based registration, and same-point-based registration in the overlap area, among which the target-based registration has the highest accuracy (Wan, 2015). During the scanning process, the center of the target is selected as the reference point to calculate the parameters of the rotation matrix. The accuracy of this method usually reaches ~3 mm. The same-point-based registration selects the feature points in the overlap area of the adjacent station as the reference points to calculate the parameters of the rotation matrix (Ge et al., 2010). The control-point-based registration method is designed to measure the latitude, longitude, and north direction of the center point of each station synchronously during the scanning process and then to register the point cloud data directly to the absolute coordinate system. Its registration accuracy is related to the positioning and orientation accuracy.

There are many algorithms for point cloud fine registration, such as the ICP algorithm, normal distribution transform (NDT) algorithm, and random sampling consensus (RANSAC) algorithm. The ICP and a variety of its improved algorithms are the most popular (Besl & McKay, 1992). The ICP algorithm is used to establish a mapping between the closest point pairs in the reference and target point clustering and calculate the optimal coordinate transformation function with the minimum sum of squared distance differences of all pairs as the constraint. The process of the traditional ICP algorithm is as follows:

Suppose that $P = \{p_i \mid p_i \in R^3, i = 1, 2, \cdots, N_p\}$ and $X = \{x_i \mid x_i \in R^3, i = 1, 2, \cdots, N_x\}$ is the set of the reference point cloud and to-be-registered point clouds in 3D space R^3, respectively.

① Input the reference point cloud P (with N_p point) and the to-be-registered point cloud X (with N_x point), $N_p \le N_x$.
② Initialize the same point set and the parameters of the transformation matrix: $P_0 = P$, $\overrightarrow{q_0} = [1\ 0\ 0\ 0\ 0\ 0\ 0]^{\mathrm{T}}$, $k = 0$.
③ Calculate the corresponding nearest neighbor point set C in the point set X according to the points in the point set P.
④ Solve for the registration parameter vector $\overrightarrow{q_{k+1}}$: $\overrightarrow{q_{k+1}} = (P_0, C)$. Once the parameter vector $\overrightarrow{q_{k+1}}$ is obtained, the sum of the square distance is calculated as f_{k+1}: $f(q) = \sum_{i=1}^{N_p} \|d_i\|^2 = \sum_{i=1}^{N_p} \left(\overrightarrow{d_i}^{\mathrm{T}} \overrightarrow{d_i}\right)$.
⑤ Transform P_0 based on the registration-based parameters: $P_{k+1} = \overrightarrow{q_{k+1}}(P_0)$.
⑥ Determine whether the operation converges. When the change of the sum of squared distances $(f_{k+1} - f_k)$ is less than the preset threshold, it is judged as converged, and the accurate registration is completed. Otherwise, return to step ③ to continue the iterative operation.
⑦ The coordinate of the source point cloud is transformed: $P' = \overrightarrow{q_k}(P)$; and the ICP fine registration is completed.

It can be seen that the iterative computation limits the registration efficiency of large area and high-density point clouds, and the quality of the initial value selection determines whether the iteration converges correctly. The improved ICP algorithm expands the selected object of the feature elements in comparison with the traditional ICP algorithm, such as the normal vector of surface feature points or face-to-face distance. Each convex or concave point of the surface is a feature point. The offset of all feature points to the normal vector is used as the control factor of the least squares for iterative computation to derive the optimal conversion matrix. Compared with the traditional ICP algorithm, the improved algorithms automatically select the distinctive features to participate in the calculation, which greatly improves the calculation efficiency without affecting the registration accuracy.

For the coarse and fine registrations of multi-station point cloud data, most LiDAR data processing software, such as CloudCompare, Riscan Pro, Meshlab, PCM, and LiDAR360, have built-in point cloud registration functions. Taking the software PCM v2.0 as an example, it generates the feature plane from the data of each station

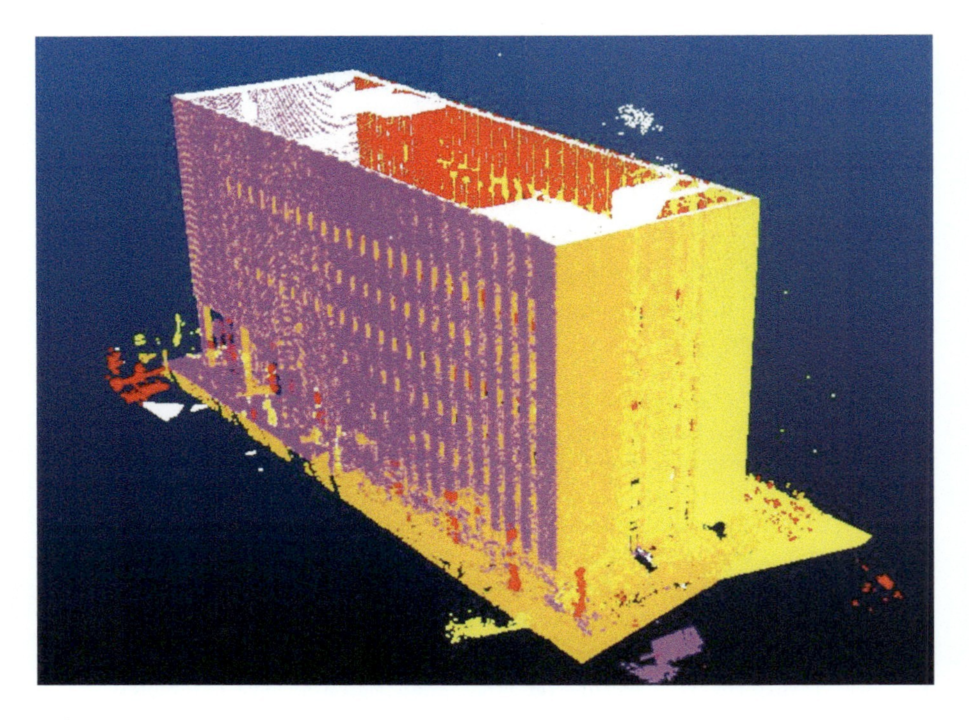

FIGURE 3.12 Point cloud data of building after registration.

before fine registration and then adopts an improved ICP algorithm based on the feature planes. The algorithm takes the minimum sum of squared face-to-face distance differences as a constraint and sets parameters such as search radius for iteration, to reduce the error rate of computation and registration. For example, taking one building as the experimental scenario, the Riegl VZ-1000 scanner was used to acquire the scanning data of multiple stations, and the PCM v2.0 was used to successively perform pairwise registration of point cloud data from adjacent stations. However, this method has transfer errors. Specifically, excessive cumulative errors would prevent the iterations from converging correctly or not, resulting in large registration errors or registration failures. In order to avoid this situation, the method of joint registration in blocks is adopted. That is, all station data are divided into several blocks, and the joint registration of multi-station data within each block is performed first, and then the blocks are merged under the condition that the registration error within the block is small, until the overall joint registration of all sites is completed. Figure 3.12 shows the point cloud data of the building after multi-station registration. Different colors represent the scanning data of different stations.

3.2.3.2 Point Cloud Quality Inspection

Terrestrial laser scanning is susceptible to occlusion, resulting in data voids, incompleteness, etc. Therefore, the secondary scanning plan needs to be developed based on data inspection results. The quality inspection should be carried out in time to confirm the scope that needs to be supplemented or remeasured. The key points for quality inspection are as follows:

1. Integrity and overlap of point cloud. In the process of terrestrial laser scanning, the laser beam cannot reach all parts of the object through one scan due to the mutual occlusion between the objects in complex scenes. Hence, multi-station scanning is required. However, the registration accuracy of multi-station data is closely related to the overlap of point clouds. The overlap degree of point cloud data between stations should be at least 20%–30% to meet the requirements of registration. Some complex buildings have more occlusions, and more station scanning is needed to ensure the integrity of the point cloud data.
2. Point cloud density. It is determined according to the accuracy requirements of the measurement project results. Specifically, the point cloud data are imported in the software according to the project design requirements. The coverage area and point number of the point cloud are derived to calculate the density of points within the measurement extent. The areas that do not meet the requirements are rescanned.
3. Noise points. During the scanning process, the points that are obviously not in the survey area are noise points. These are caused by clouds, instruments, terrain, or other factors. The noise points will interfere with the subsequent processing, analysis, and application of the point cloud. Therefore, it is necessary to check and analyze the distribution and number of noise points as well as the causes of their formation. For the noise points that obviously affect the data quality, it is necessary to delete and supplement the data. For the noisy points that are obviously far from the actual object point cloud, it can be removed by the point cloud denoising algorithms (introduced in Section 4.1).

3.3 SPACEBORNE LiDAR DATA ACQUISITION

The LiDAR sensors onboard the satellite platforms and the International Space Station (ISS) are collectively referred to as the spaceborne LiDAR. This section focuses on several major spaceborne LiDAR systems, including the platform, system parameters, data products, and download methods.

3.3.1 ICESat/GLAS Mission

The Ice, Cloud and Land Elevation Satellite (ICESat) is the world's first laser altimetry satellite for Earth observation. It was launched on January 13, 2003, at Vandenberg Air Force Base on the central coast of California. Its main payload is the Geoscience Laser Altimeter System (GLAS). The scientific goals of ICESat/GLAS are to measure changes in polar ice sheet elevation and sea ice, ice sheet mass balance, cloud and aerosol heights, and topographic and vegetation characteristic parameters (Schutz et al., 2005; Wang et al., 2011, 2015).

ICESat flew along a near-circular and near-polar orbit with an altitude of about 600 km and an orbital inclination of 94°. It collected data between 86°N and 86°S with a repetition period of approximately 183 days. The GLAS system was equipped with a Nd:YAG laser. GLAS generated footprints with a diameter of 70 m, spaced at 170-m intervals in the along-track direction. The distance between adjacent orbits varies with latitude: 15 km near the equator and 2.5 km at 80° latitude. The GLAS system emits infrared (1064 nm) and green (532 nm) laser pulses at a frequency of 40 Hz. The laser in the 1064-nm band is mainly used to measure ground and sea-level elevations, and the laser in the 532-nm band is used for atmospheric backscattering measurements. The spatial resolution of clouds and aerosol height distribution along the orbit reach 75–200 m. The horizontal resolution of the thick cloud layer measurement is 150 m. The key parameters of the GLAS are shown in Table 3.5.

GLAS data products are divided into three levels (level 0, 1, and 2), including 15 standard data products and auxiliary data (Table 3.6). Level 0 provides the raw telemetry data. Level 1A records the instrument parameters. Level 1B is the primary products. Level 2 is the application products for ice, ocean, geophysics, and atmosphere.

TABLE 3.5
Key Parameters of ICESat/GLAS Configuration

Parameters	Specifications
Launch time	January 2003
Platform	Satellite
Platform height	600 km
Wavelength	1064 nm/532 nm
Receiving caliber	0.709 m^2
Receiving field of view	450 μrad/150 μrad
Laser divergence angle	110 μrad
Pulse repetition frequency	40 Hz
Pulse width	5 ns
Pulse shape	Gaussian
Signal sampling frequency	1 GHz
Vertical resolution	15 cm
Footprint diameter	60–70 m
Pulse energy	72 mJ/36 mJ
Footprint spacing	
Along track	170 m
Across track	15 km (maximum); 2.5 km (minimum)
Repetition period	183 days
Design life	3 years (actually 6 years in orbit, expired in November 2009)
1064-nm ranging accuracy	13.8 cm (ice, land)
1064-nm horizontal accuracy	4.5 m (ice, land)
532-nm vertical resolution	75–200 m (cloud)
532-nm horizontal resolution	150 m (cloud)

TABLE 3.6
Overview of GLAS Data Products

Levels	Products	Descriptions
L1A	GLA01	Global altimetry data
	GLA02	Global atmosphere data
	GLA03	Global engineering data
	GLA04	Global laser pointing data
L1B	GLA05	Global waveform-based range corrections data
	GLA06	Global elevation data
	GLA07	Global backscatter data
L2	GLA08	Global planetary boundary layer and elevated aerosol layer height
	GLA09	Global cloud heights for multi-layer clouds
	GLA10	Global aerosol vertical structure data
	GLA11	Global thin-cloud/aerosol optical depths data
	GLA12	Antarctica and Greenland ice-sheet altimetry data
	GLA13	Sea-ice altimetry data
	GLA14	Global land-surface altimeter data
	GLA15	Ocean altimetry data

The ICESat/GLAS data are publicly released at the National Snow and Ice Data Center (NSIDC) and available for free download. There are three steps to download the GLAS data. The first step is to register an account, the second is to enter the time and geographic coordinate range of data acquisition, and the last step is to download related GLAS data products. The NSIDC official website provides three data download methods, namely Python Script, Order Files, and Large/Custom Order.

1. The Python Script method requires the Python 2 or Python 3 programming language to be installed in advance. The GLAS data are automatically downloaded to the specified directory through the downloaded Python source code.
2. Order Files provides a ZIP archive of GLAS data, as well as a list of individual file download links.
3. Large/Custom Order is used when the number of downloaded files exceeds 2000. It can obtain the download link for a single file, or it can be downloaded directly through the command line.

3.3.2 ICESat-2/ATLAS Mission

ICESat ceased operation in 2009. The National Aeronautics and Space Administration (NASA) launched the ICESat-2 satellite on September 15, 2018, from Vandenberg Air Force Base in California. The photon-counting laser altimeter — Advanced Topographic Laser Altimeter System (ATLAS) carried by ICESat-2 uses micro-pulse, multi-beam, photon-counting LiDAR technology, which is the first time the technology has been applied to the spaceborne platform (Markus et al., 2017; Neumann et al., 2019). It uses a more sensitive single-photon detector with a higher pulse repetition frequency, which acquires photon-counting data with a smaller footprint and a higher density, thereby achieving precise 3D surface measurements. The main scientific goals are (1) to quantitatively assess the contribution of polar ice sheets to current and near-term sea-level change; (2) to quantify the regional characteristics of ice sheet changes to assess their change-driving mechanisms and improve ice sheet prediction models; (3) to estimate the sea ice thickness and study the energy, material, and water exchange between sea ice/ocean/atmosphere; and (4) to measure the height of vegetation to reveal the status and variation of vegetation biomass over large areas (Markus et al., 2017).

The orbital height, orbital inclination, and observation coverage of ICESat-2 are 500 km, 92°, and 88°S–88°N, respectively. The repetition period is 91 days, and each period has 1387 orbits. There are two lasers in the ATLAS system. Generally, only one is in the working state. It emits pulses in the green band of 532 nm at a repetition rate of 10 kHz and pulse width of 1.5 ns. It acquires overlapping footprints spaced about 0.7 m apart along the track with a diameter of about 17 m. ATLAS emits a total of six laser beams, which are arranged in parallel in three groups along the track (Figure 3.13). Each group contains a strong beam and a weak beam, and the energy ratio of the two is about 4:1. The cross-track distance between groups is about 3.3 km, while the cross-track distance within the group is about 90 m (Neuenschwander & Pitts, 2019). Table 3.7 lists the key parameters of the ICESat-2/ATLAS platform and sensor (Markus et al., 2017).

ICESat-2/ATLAS provides 21 standard data products, ATL00–ATL21, which are divided into level 0, level 1, level 2, and level 3 (Markus et al., 2017) (Table 3.8). ATL00 is a level 0 product, providing raw telemetry data. ATL01 and ATL02 are level 1 products, providing telemetry data after format conversion and instrument error correction. ATL03 and ATL04 are level 2 products, of which ATL03 derives the geodetic position (i.e., latitude, longitude, and altitude) of the photons received by ATLAS based on photon round-trip time, laser position, and attitude angle. ATL06–ATL21 are level 3 products, providing the relevant information about glacier heights, ice cap heights, sea ice heights, vegetation heights, and inland water body elevations.

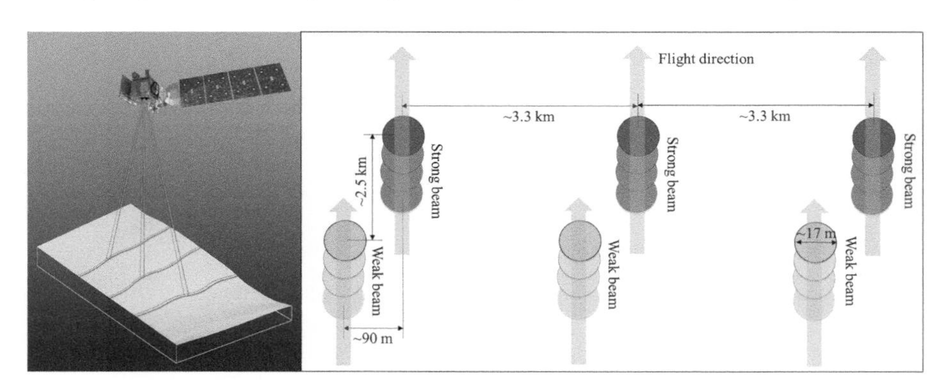

FIGURE 3.13 ICESat-2/ATLAS beam distribution.

TABLE 3.7
Key Parameters of ICESat-2/ATLAS Platform and Sensor

Parameters	Specifications	Parameters	Specifications	Parameters	Specifications
Platform height	500 km	Footprint diameter	~17 m	Spacing within beam group	~90 m
Orbital inclination	92°	Footprint spacing along the track	~0.7 m	Spacing between beam groups	~3.3 km
Coverage	88°S–88°N	Pulse width	~1.5 ns	Beam energy (strong)	175 ± 17 μJ
Repetition period	91 days	Number of beams	6 beams, arranged in 3 groups	Beam energy (weak)	45 ± 5 μJ
Transmit frequency	10 kHz	Beam energy ratio (strong: weak)	~4:1		
Laser wavelength	532 nm	Receiver aperture	0.8 m		

TABLE 3.8
Overview of ICESat-2/ATLAS Data Products

Levels	Products	Product Names	Descriptions
Level 0	ATL00	Telemetry Data	Raw ATLAS telemetry in packet format
Level 1	ATL01	Reformatted Telemetry	Parsed, partially reformatted into HDF5, generated daily, segmented into several-minute granules.
	ATL02	Science Unit Converted Telemetry	Photon time of flight, corrected for instrument effects. Includes all photons, pointing data, spacecraft position, housekeeping data, engineering data, and raw atmospheric profiles, segmented into several-minute granules.
Level 2	ATL03A	Atmospheric Delay Corrections	Correcting the laser footprint altitude as a result of atmospheric path delay, refraction, and other atmospheric properties.
	ATL03G	Received Photon Geolocation	General overview of steps/procedures required to provide geolocated heights for each ATLAS photon event, as well as a detailed geolocation algorithm implementation and processing flow specifically designed for the ICESat-2 mission.
	ATL03	Global Geolocated Photon Data	Precise latitude, longitude, and elevation for every received photon, arranged by beam in the along-track direction. Photons classified by signal vs. background, as well as by surface type (land ice, sea ice, land, ocean), including all geophysical corrections (e.g. Earth tides, atmospheric delay). Segmented into several-minute granules.
	ATL04	Uncalibrated Backscatter Profiles	Along-track atmospheric backscatter data, 25 times per second. Includes calibration coefficients for polar regions. Segmented into several-minute granules.

(Continued)

TABLE 3.8 *(Continued)*
Overview of ICESat-2/ATLAS Data Products

Levels	Products	Product Names	Descriptions
Level 3	ATL06	Land Ice Elevation	Surface height for each beam with along- and across-track slopes calculated for each beam pair. Posted at 40 m along-track; segmented into several-minute granules.
	ATL07	Arctic/Antarctic Sea Ice Elevation	Height of sea ice and open water leads at varying length scale based on returned photon rate for each beam presented along-track.
	ATL08	Land Water Vegetation Elevation	Height of ground including canopy surface posted at variable-length scales relative to signal level, for each beam presented along-track. Where data permit, include canopy height, canopy cover percentage, surface slope and roughness, and apparent reflectance.
	ATL09	Calibrated Backscatter and Cloud Characteristics	Along-track cloud and other significant atmosphere layer heights, blowing snow, integrated backscatter, and optical depth.
	ATL09, Pt. II – DDA	Density-Dimension Algorithm for Atmospheric Layers and Surface	Detection of clouds and other atmospheric layers, such as blowing snow and aerosols, and their boundaries in ATLAS data.
	ATL10	Arctic/Antarctic Sea Ice Freeboard	Estimate of sea ice freeboard over specific spatial scales using all available sea surface height measurements. Contains statistics of sea surface and sea ice heights.
	ATL11	Antarctica/Greenland Ice Sheet H(t) Series	Time series of height at points on the ice sheet, calculated based on repeat tracks and/or cross-overs.
	ATL12	Ocean Elevation	Surface height at specific length scale. Where data permit, include estimates of height distribution, roughness, surface slope, and apparent reflectance.
	ATL13	Inland Water Height	Along-track inland and near-shore water surface height distribution within water mask. Where data permit, include roughness, slope, and aspect.
	ATL14	Antarctica/Greenland Ice Sheet H(t) Gridded	Height maps of each ice sheet for each year based on all available elevation data.
	ATL15	Antarctica/Greenland Ice Sheet dh/dt Gridded	Height change maps for each ice sheet, for each mission year, and for the whole mission.
	ATL16	ATLAS Atmosphere Weekly	Polar cloud fraction, blowing snow frequency, and ground detection frequency.
	ATL17	ATLAS Atmosphere Monthly	Polar cloud fraction, blowing snow frequency, and ground detection frequency.
	ATL18	Land/Canopy Gridded	Gridded ground surface height, canopy height, and canopy cover estimates.
	ATL19	Mean Sea Surface (MSS)	Gridded ocean height product.
	ATL20	Arctic/Antarctic Gridded Sea Ice Freeboard	Gridded sea ice freeboard.
	ATL21	Arctic/Antarctic Gridded Sea Surface Height w/in Sea Ice	Gridded monthly sea surface height inside the sea ice cover.
	ATL22	Mean Inland Surface Water	Mean inland surface water products for each beam transect.

TABLE 3.9
ICESat-2/ATLAS Standard File Naming

Keywords	Implication
xx	Product number (02-21)
HH	Hemisphere ID, Northern Hemisphere = 01, Southern Hemisphere = 02
yyyymmdd	Data acquisition time: year, month, day
hhmmss	Data acquisition time: hours, minutes, seconds (UTC)
tttt	Referring to orbital numbers, the ICESat-2 mission has 1387 orbitals, numbered from 0001 to 1387
cc	Number of repeating orbital periods
ss	Track segment number, ATL02/ATL03/ATL06/ATL08 track number range is 01–14, ATL04/ATL07/ATL09/ATL10/ATL12/ATL13/ATL16/ATL17 is 01
vvv_rr	Version and revision number

The ICESat-2/ATLAS data have been publicly released at NSIDC website since May 2019. Similar to GLAS, the NSIDC website provides three identical methods for downloading the ATLAS data.

The HDF5 file naming of ICESat-2 data products follows a unified specification. Almost all the products are named ATLxx_yyyymmddhhmmss_ttttccss_vvv_rr.h5 except for ATL07 and ATL10. In contrast, ATL07 and ATL10 add a parameter named ATLxx-HH_yyyymmddhhmmss_ttttccss_vvv_rr.h5. The naming rules are detailed described in Table 3.9.

3.3.3 GEDI Mission

In November 2018, the United States carried the full-waveform LiDAR sensor Global Ecosystem Dynamics Investigation (GEDI) on the ISS with a design life of 2 years. The system has a total of three lasers that simultaneously acquire eight beams of full waveform data and will generate about 10 billion cloud-free ground observations during orbit (Dubayah et al., 2020; Duncanson et al., 2020). The main scientific goals include (1) high-resolution laser ranging observations of the 3D structure of the Earth; (2) accurate measurement of forest canopy height, canopy vertical structure, and surface elevation; and (3) research on carbon/water cycle processes, biological diversity, and habitat characteristics.

The wavelength of the three GEDI lasers is 1064 nm. One of the lasers is divided into two beams with weaker energy, so the three lasers generate a total of four beams and eight ground tracks through optical dithering. The distance between adjacent tracks is about 600 m. The scanning width is 4.2 km. The footprint diameter is 25 m, and the distance between the footprints along the track is approximately 60 m. GEDI records full-waveform LiDAR data with a vertical accuracy of 2–3 cm. The GEDI observations cover a range of 51.6°S– 51.6°N, including almost all tropical rain forests and temperate forests. Table 3.10 lists the key parameters of the GEDI system.

GEDI data products are divided into four levels: L1, L2, L3, and L4. The L1 product is the geolocated waveform data. L2 products mainly provide canopy height and profile metrics such as terrain elevations, relative height (RH) metrics, and leaf

TABLE 3.10
Key Parameters of GEDI System

Parameters	Specifications	Parameters	Specifications
Platform height	~400 km	Distance between tracks	~600 m
Coverage	51.6°S–51.6°N	Scan breadth	~4.2 km
Transmit frequency	242 Hz	Pulse width	~14 ns
Laser wavelength	1064 nm	Emitted pulse power	~10 mJ
Footprint diameter	~25 m	Number of ground orbits	8
Footprint space along track	~60 m	Operation hours	2 years

area index (LAI) for each footprint. L3 products provide gridded canopy height, coverage, and LAI data, which are obtained by spatial interpolation of L2 footprint-scale parameters. L4 products are footprint scale and gridded aboveground biomass (AGB) estimation, which is the highest level of GEDI products. It estimates the AGB at the footprint scale based on the canopy parameters of the L2 product and derives the average biomass of the 1-km grid using statistical theory. Table 3.11 provides the overview of GEDI products.

The products GEDI01_B, GEDI02_A, and GEDI02_B were released publicly in January 2020 and can be downloaded for free from its website. There are three data download methods, namely Data Pool, Earthdata Search, and GEDI Finder.

1. Data Pool directly provides a list of the acquisition time directory of the entire GEDI data, so that users can download GEDI data of a specific time period. This method cannot filter the data of a specific area.
2. Earthdata Search requires a registered account, but currently it is not possible to search and query by time and geographic coordinate.
3. GEDI Finder is used to filter the GEDI data in a specified area.

TABLE 3.11
Overview of GEDI Data Products

Levels	Product Names	Description	Resolution
L1	GEDI01_A-RX	Raw waveforms	25 m
	GEDI01_B	Geolocated waveforms	25 m
L2	GEDI02_A	Footprint-level ground elevation, canopy top height, relative height (RH) metrics	25 m
	GEDI02_B	Footprint-level canopy cover fraction (CCF), CCF profile, leaf area index (LAI), LAI profile	25 km
L3	GEDI03	Gridded CCF, CCF profile, LAI, LAI profile	1 km
L4	GEDI04_A	Footprint-level aboveground biomass	25 m
	GEDI04_B	Gridded aboveground biomass	1 km

TABLE 3.12
Key Parameters of China's Spaceborne LiDAR Sensors

Satellites	Launch Time	Detection Type	Data Recording	Beam Number	Transmit Pulse Width	Footprint Diameter	Track Spacing	Coverage Area
ZY3-02	2016	Linear	–	1	7 ns	~75 m	3.5 km	83° N/S
GF-7	2019	Linear	Full-waveform	2	7 ns	~17 m	2.5 km	83° N/S
TECIS-1	2022	Linear	Full-waveform	5	7 ns	~30 m	200 m	–

3.3.4 China's Laser Altimetry Satellite Program

Since the 2010s, China's LiDAR satellites have developed rapidly. A number of satellites equipped with LiDAR systems have been launched one after another. The ZY3-02 satellite launched by China in 2016 was equipped with an experimental laser altimetry, which laid the foundation for the development, operation, and application of the subsequent spaceborne laser altimetry (Huang et al., 2017; Liu et al., 2019). In November 2019, the GaoFen-7 (GF-7) satellite was successfully launched. It is equipped with a laser altimeter and a dual-line-array stereo camera, which is used for high-resolution stereo mapping, urban and rural construction survey, and evaluation (Xie et al., 2020). The GF-7 satellite has successfully undergone on-orbit geometry calibration. The on-orbit test shows that the GF-7 laser altimetry operates stably, with good quality and high precision, and it can provide high-precision elevation control points. In addition, China launched the Terrestrial Ecosystem Carbon Inventory Satellite (TECIS-1) in 2022, which is equipped with a full-waveform LiDAR for estimation and monitoring of carbon storage in global forest ecosystems (Du et al., 2020).

Table 3.12 lists the key system parameters of the LiDAR carried by China's three satellites. Among them, the designs of GF-7 and TECIS-1 are very similar to ICESat/GLAS. Both adopt linear detection systems and are equipped with full-waveform laser altimeters. Some indicators are also close to ICESat/GLAS.

3.4 SUMMARY

This chapter introduces the data acquisition of airborne, terrestrial, and spaceborne LiDAR in detail. For airborne and terrestrial LiDAR, the process mainly includes preparations before data acquisition, laser scanning tasks, and data preprocessing. For spaceborne LiDAR, this chapter introduces the system parameters, products, and data download methods of different LiDAR sensors.

EXERCISES

1. Please briefly describe the workflow of airborne LiDAR data acquisition.
2. Please briefly describe the data acquisition methods of terrestrial LiDAR systems and their advantages and disadvantages.

3. What are the specific contents in the quality inspection of airborne and terrestrial LiDAR point clouds?
4. What are the typical spaceborne LiDAR systems? Please briefly describe their similarities and differences in terms of platforms, systems, and data.
5. Please briefly describe the main scientific missions of ICESat/GLAS, ICESat-2/ATLAS, GEDI, and GF-7.

REFERENCES

Besl, P. J., & McKay, N. D. (1992). A method for registration of 3D shapes. *IEEE Transactions on Pattern Analysis and Machine Intelligence*, *14*(2), 239–256.

Dong, Z., Liang, F., Yang, B., Xu, Y., Zang, Y., Li, J., Wang, Y., Dai, W., Fan, H., Hyyppäb, J., & Stilla, U. (2020). Registration of large-scale terrestrial laser scanner point clouds: A review and benchmark. *ISPRS Journal of Photogrammetry and Remote Sensing*, *163*, 327–342.

Du, S., Liu, L., Liu, X., Zhang, X., Gao, X., & Wang, W. (2020). The solar-induced chlorophyll fluorescence imaging spectrometer (SIFIS) onboard the first Terrestrial Ecosystem Carbon Inventory Satellite (TECIS-1): Specifications and prospects. *Sensors*, *20*(3), 815.

Dubayah, R., Blair, J. B., Goetz, S., Fatoyinbo, L., Hansen, M., Healey, S., Hofton, M., Hurtt, G., Kellner, J., Luthcke, S., Armston, J., Tang, H., Duncanson, L., Hancock, S., Jantz, P., Marselis, S., Patterson, P. L., Qi, W. L., & Silva, C. (2020). The global ecosystem dynamics investigation: High-resolution laser ranging of the earth's forests and topography. *Science of Remote Sensing*, *1*, 100002.

Duncanson, L., Neuenschwander, A., Hancock, S., Thomas, N., Fatoyinbo, T., Simard, M., Silva, C. A., Armston, J., Luthcke, S. B., Hofton, M., Kellner, J. R., & Dubayah, R. (2020). Biomass estimation from simulated GEDI, ICESat-2 and NISAR across environmental gradients in Sonoma County, California. *Remote Sensing of Environment*, *242*, 111779.

Ge, X., Lu, X., Wang, Y., Lu, Y., & Li, T. (2010). Registration method of laser point cloud data in multi-stations (in Chinese). *Bulletin of Surveying and Mapping*, (11), 15–17.

Huang, X., Wen, D., Li, J., & Qin, R. (2017). Multi-level monitoring of subtle urban changes for the megacities of China using high-resolution multi-view satellite imagery. *Remote Sensing of Environment*, *196*, 56–75.

Lai, X. (2010). *Basic principle and application of airborne LiDAR* (in Chinese). Publishing House of Electronics Industry.

Li, D., Guo, H., Wang, C., Dong, P., & Zuo, Z. (2016a). Improved bore-sight calibration for airborne light detection and ranging using planar patches. *Journal of Applied Remote Sensing*, *10*, 024001.

Li, S., Wang, J., Liang, Z., & Su, L. (2016b). Tree point clouds registration using an improved ICP algorithm based on KD-tree. 2016 IEEE International Geoscience and Remote Sensing Symposium (IGARSS), IEEE, 4545–4548.

Liu, C., Chen, H., & Wu, H. (2010). *Data processing and feature extraction of laser three-dimensional remote sensing* (in Chinese). Science Press.

Liu, C., Huang, X., Zhu, Z., Chen, H., Tang, X., & Gong, J. (2019). Automatic extraction of built-up area from ZY3 multi-view satellite imagery: Analysis of 45 global cities. *Remote Sensing of Environment*, *226*, 51–73.

Markus, T., Neumann, T., Martino, A., Abdalati, W., Brunt, K., Csatho, B., Farrell, S., Fricker, H., Gardner, A., Harding, D., Jasinski, M., Kwok, R., Magruder, L., Lubin, D., Luthcke, S., Morison, J., Nelson, R., Neuenschwander, A., Palm, S., … Zwally, J. (2017). The ice, cloud, and land elevation satellite-2 (ICESat-2): Science requirements, concept, and implementation. *Remote Sensing of Environment*, *190*, 260–273.

Neuenschwander, A., & Pitts, K. (2019). The ATL08 land and vegetation product for the ICESat-2 Mission. *Remote Sensing of Environment, 221*, 247–259.

Neumann, T. A., Martino, A. J., Markus, T., Bae, S., Bock, M. R., Brenner, A. C., Brunt, K. M., Cavanaugh, J., Fernandes, S. T., Hancock, D. W., Harbeck, K., Lee, J., Kurtz, N. T., Luers, P. J., Luthcke, S. B., Magruder, L., Pennington, T. A., Ramos-Izquierdo, L., Rebold, T., … Thomas, T. C. (2019). The ice, cloud, and land elevation satellite-2 mission: A global geolocated photon product derived from the advanced topographic laser altimeter system. *Remote Sensing of Environment, 233*, 111325.

Ou, B. (2014). The study on method of terrestrial three dimensional laser scanning technology for outdoor data collection (in Chinese). *Geomatics and Spatial Information Technology, 37*(1), 106–108, 112.

Qin, Y., Li, B., Niu, Z., Huang, W., & Wang, C. (2011). Stepwise decomposition and relative radiometric normalization for small footprint LiDAR waveform (in Chinese). *Science China Earth Science, 41*(1), 103–109. https://doi.org/10.1007/s11430-010-4120-y

Schutz, B. E., Zwally, H. J., Shuman, C. A., Hancock, D., & DiMarzio, J. P. (2005). Overview of the ICESat mission. *Geophysical Research Letters, 32*(21), 97–116.

Wan, Y. (2015). Research on building reconstruction based on TLS technology (in Chinese). Institute of Remote Sensing and Digital Earth, Chinese Academy of Sciences.

Wang, C., Xi, X., Luo, S., & Li, G. (2015). *Data processing and application for space-borne LiDAR*. Science Press.

Wang, M., Sui, L., & Li, H. (2010). On strip mosaic for airborne LiDAR point cloud data (in Chinese). *Bulletin of Surveying and Mapping*, (7), 5–8.

Wang, X., Cheng, X., Gong, P., Huang, H., Li, Z., & Li, X. (2011). Earth science applications of ICESat/GLAS: A review. *International Journal of Remote Sensing, 32*(23), 8837–8864.

Xie, H., & Gu, F. (2016). *Technology and application of terrestrial three-dimensional laser scanning* (in Chinese). Wuhan University Press.

Xie, H., Gu, F., Li, Y., & Sun, M. (2014). *Application practice of three-dimensional modeling based on laser point cloud data* (in Chinese). Wuhan University Press.

Xie, J., Huang, G., Liu, R., Zhao, C., Dai, J., Jin, T., Mo, F., Zhen, Y., Xi, S., Tang, H., Dou, X., & Yang, C. (2020). Design and data processing of China's first spaceborne laser altimeter system for earth observation: GaoFen-7. *IEEE Journal of Selected Topics in Applied Earth Observations and Remote Sensing, 13*, 1034–1044.

Zhang, X. (2007). *Theory and method of airborne LiDAR measurement technology* (in Chinese). Wuhan University Press.

Zhang, Z., & Zhang, J. (1997). *Digital photogrammetry* (in Chinese). Wuhan University Press.

4 LiDAR Data Processing

4.1 LiDAR POINT CLOUD PROCESSING

The purpose of point cloud processing is to extract key features from the massive and disordered three-dimensional (3D) point clouds by specific algorithms and provide high-precision data and accurate information for subsequent digital surface model generation, 3D modeling of features, and other related applications (Yang et al., 2017). Point cloud denoising, point cloud filtering, and point cloud classification are the key steps in the point cloud processing, which are described in this section.

4.1.1 Point Cloud Denoising

In the process of point cloud acquisition, equipment errors, personnel operations, object reflections, and environmental interference usually produce a small amount of noise point clouds. According to their spatial distribution, noise points are simply divided into two categories: typical noise points and atypical noise points. Typical noise points refer to anomalous points or point clusters in the local range away from the scan target, such as birds and clouds. The atypical noise points refer to inconspicuous noise points mixed with the scan target, such as multi-path effects and internal factors of the system. The purpose of denoising is to remove the noise points and maximize the retention of local features of the scanned object. The commonly used denoising algorithms are mainly divided into three categories, i.e., statistical-based algorithm, frequency-domain-based algorithm, and surface-based algorithm.

The statistical-based algorithm considers the significant differences between the elevation and density of noise points and non-noise points and distinguishes noise and non-noise points by setting an adaptive threshold. For example, Han and Zuo (2012) used the smoothness constraint of the triangular mesh constructed by the point cloud as the noise differentiation criterion. The denoising algorithm based on frequency domain transforms the point cloud into the frequency domain and uses the frequency difference to remove the noise. For example, Somekawa et al. (2013) used the wavelet transform to convert the point cloud into the frequency domain and removed noise by filtering the abrupt signals. Its difficulty lies in the selection of the filter and transformation function. The surface-based point cloud denoising algorithm removes discontinuous noise points in 3D space by constructing local surfaces of non-noise points, such as morphological operations. The selection of window size and threshold is critical (Mongus & Zalik, 2012). The denoising algorithm based on multiscale curvature estimation maintains feature information by estimating the Gaussian curvature when computing point cloud features (Dutta et al., 2014) and then uses a moving least squares algorithm to construct local surfaces to remove noise. Its disadvantage is high time complexity and difficulty in applying it to large point cloud data volumes (Wu & Huang, 2017).

DOI: 10.1201/9781032671512-4

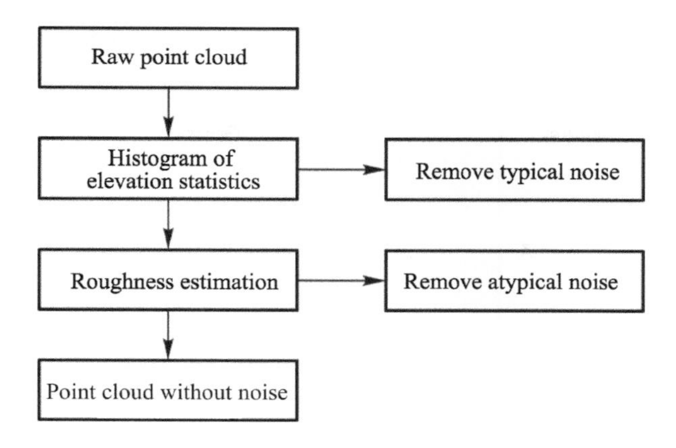

FIGURE 4.1 The procedure of point cloud denoising.

This section introduces two denoising algorithms for typical and atypical noise points. The technical flow is shown in Figure 4.1.

1. For typical noises, a denoising algorithm based on an elevation statistical histogram is used. First, set the sampling interval to draw a histogram of the elevation distribution. Then the elevation and frequency thresholds are set to remove abnormal high and low points.
2. For atypical noises, a denoising algorithm based on roughness estimation is used. The local Delaunay triangular network of the point cloud vertices is constructed to determine the neighborhood points of each vertex. The roughness r_v of the vertex P is estimated according to Equation (4.1).

$$r_v(P) = \left| 2\pi - \sum_{i=1}^{n} \theta_i \right| \tag{4.1}$$

   where n is the number of points in the neighborhood where the vertex P is located, and θ_i is the angle between the vertex P and each neighborhood point. When the vertex P is higher or lower in the triangle network, the value of roughness r_v is larger, which means the greater deviation of this vertex in the neighborhood and the higher possibility of P being a noise point.

The criterion for determining if a vertex P is marked as a noise point is based on the roughness $r_v(Q_i)$ and the elevation value $h_v(Q_i)$ for each vertex Q_i in the neighborhood of P. Specifically, the expectation $E_{r_v(P)}$ and the standard deviation $\sigma_{r_v(P)}$ of the roughness and the expectation $E_{h_v(P)}$ and the standard deviation $\sigma_{h_v(P)}$ of the elevation value are calculated. In the probability density function, 99.7% of the normally distributed data lie within 3 times the standard deviation. Hence, if both the roughness and elevation values of the vertex P exceed 3 times the standard deviation, they are marked as noise (Equations (4.2) and (4.3)). The pseudo-code of the point cloud denoising algorithm is shown in Table 4.1.

$$\left|r_v(P) - E_{r_v(P)}\right| > 3\sigma_{r_v(P)} \tag{4.2}$$

$$\left|h_v(P) - E_{h_v(P)}\right| > 3\sigma_{h_v(P)} \tag{4.3}$$

TABLE 4.1
Pseudo-Code of Point Cloud Denoising Algorithm

Input: ***P***, point cloud; ***d***, histogram sampling interval; ***f***, histogram frequency threshold.
Output: ***P_Result***, point cloud without noise
Algorithm:

```
#import point cloud
for p_i in P
        save p_i.z to array E;
endfor;
# Remove typical noise
create Histogram with interval d using array E;
for bin in Histogram
        if bin.points_number < f
                for p in bin
                        p.is_noise = TRUE;
                endfor;
        else
                save p in bin into P_New;
        endif;
endfor;
# Remove atypical noise
construct Delaunay triangulation using P_new;
for p in P_New
        p_neighbor = FindNeighborInTriangulation(p);
        compute roughness p.r for p using p_neighbor;
        save p.r to array R;
        save p.z to array Z;
endfor;
compute mathematical expectations E_r and standard deviation S_r for array R;
compute mathematical expectations E_z and standard deviation S_z for array Z;
for p in P_New
        if p.z - E_z > S_r * 3 AND p.r - E_r > S_r * 3
                p_i.is_noise = TRUE;
        else
                save p to P_Result;
        endif;
endfor;
return P_Result
```

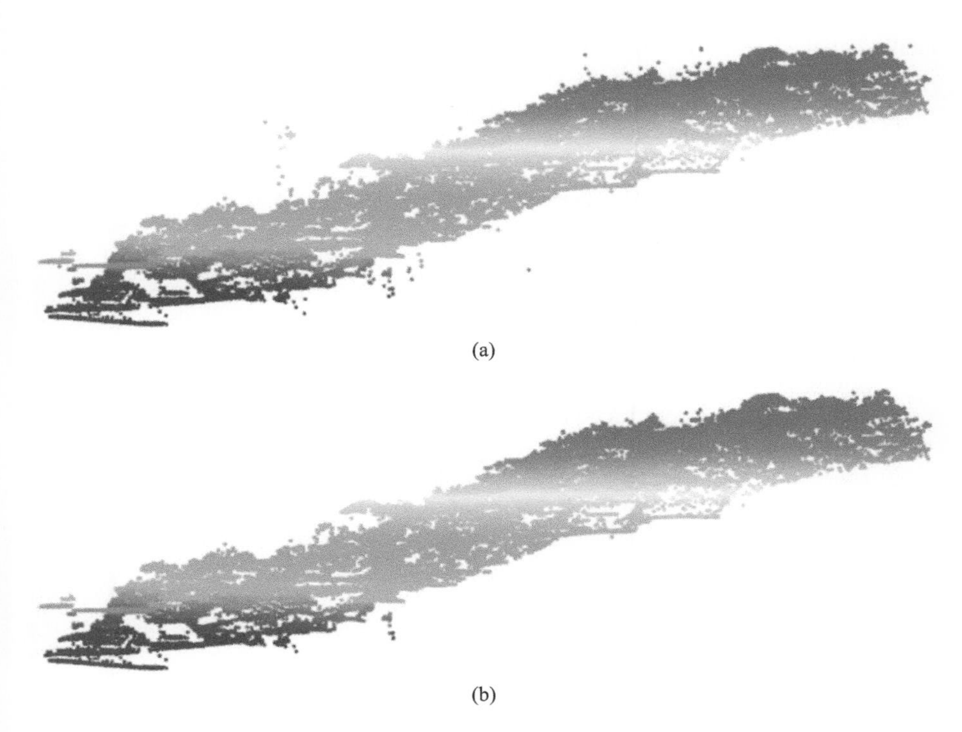

FIGURE 4.2 Example of airborne LiDAR point cloud denoising: (a) raw data; (b) data after denoising.

In this section, the point cloud denoising algorithm is applied to a public LiDAR dataset released by the International Society for Photogrammetry and Remote Sensing (ISPRS) in 2003. Figure 4.2 shows the raw data with noise points and the data after denoising. It can be seen that the noise points are accurately removed.

4.1.2 Point Cloud Filtering

Point cloud filtering is the process of separating ground points and non-ground points from the original point cloud, usually based on certain conditional rules or prior knowledge. For example, in a certain area, non-ground points are always higher than ground points, terrain slope changes are always within a certain range, and ground objects have specific geometric structures, such as buildings and vegetation. Three classic point cloud filtering algorithms are introduced next, i.e., mathematical morphology point cloud filtering, multi-level moving surface point cloud filtering, and progressive densification triangulation filter.

4.1.2.1 Mathematical Morphology Point Cloud Filtering

Mathematical morphology is an image analysis method based on grid theory and topology. Erosion and dilation are two basic operators in mathematical morphology

(Kilian et al., 1996). The point cloud data are usually processed based on the opening operation in morphology (erosion followed by dilation). The main steps include rasterization, morphological filtering, and elevation threshold determination.

1. Rasterization. Gridding the messy point cloud data mainly includes three steps. First, set the grid size and divide the point cloud data into a grid. The size of the grid should be determined according to the density of the point cloud, generally set to 1–2 m. Second, traverse all grids. For each grid, if it contains multiple points, the point with the lowest elevation is selected as the attribute value of the grid; if only one point is included, the point is directly used as the grid attribute value; if the grid does not contain any points, mark it as "no data." Third, traverse all grids marked as "no data" and derive the Z coordinate of these grids using the nearest neighbor interpolation method.
2. Morphological filtering of raster data. The morphological opening operation (erosion followed by dilation) is used to process the rasterized data to derive the raster ground points. The erosion operation and dilation operation are shown in Equations (4.4) and (4.5), respectively.

$$E\big(Z(i,j)\big) = \min\big(Z(i',j')\big) \tag{4.4}$$

$$D\big(Z(i,j)\big) = \max\big(Z(i',j')\big) \tag{4.5}$$

where i' and j' are the row and column numbers of the adjacent grids of the grid whose row and column numbers are (i, j), respectively. For a grid with row and column number (i, j), its Z coordinate becomes the smallest Z value in all grids in its neighborhood after the erosion operation and becomes the maximum Z value in all grids in its neighborhood after the dilation operation.

The specific steps of raster data filtering include:

i. Set an initial neighborhood search window and perform the erosion operation on all grids.
ii. Perform the dilation operation for the eroded raster data with the same window size.
iii. Traverse all grids, calculate the difference between the Z values before and after the dilation of each grid, and determine the ground point by setting the adaptive threshold dh_k [Equation (4.6)]. If the difference is less than the threshold, the point is determined to be a ground point; otherwise, it is determined to be a non-ground point.

$$dh_k = \begin{cases} dh_0, & \textit{if } W_k \le 3 \\ \min\left[dh_{\max},\ S\cdot\left(W_k - W_{k-1}\right) + dh_0\right], & \textit{if } W_k > 3 \end{cases} \tag{4.6}$$

where W_k is the window size of the k-th filtering; dh_0 and $dh_{\max}$ are the minimum and maximum height difference thresholds, respectively;

and S is the terrain slope parameter. It can be seen that the height difference threshold increases with the filter window size, and the increase is determined by the terrain slope S.

iv. Expand the neighborhood search window step by step [see Equation (4.7)]. Repeat steps (ii) and (iii) until the maximum window is reached.

$$W_k = 2 \times b^k + 1 \tag{4.7}$$

where k is the number of iterations, k = 0, 1, 2, ..., M. b is the initial window size, which is generally set to 2. The maximum window size is equal to $2 \times b^M + 1$, which is generally set as the largest building size in the area.

3. Elevation threshold determination. The ground points processed by the previous steps have been rasterized, rather than the original ground point cloud data obtained by scanning. All ground points in the original data are further filtered by determining the elevation threshold. The specific steps are as follows. Traverse the original point cloud. Then use the inverse distance weighting (IDW) method to interpolate the real ground elevation value directly below the point. That is, search for the ground grid points within a certain neighborhood of the point and perform the IDW interpolation. Finally, the height difference threshold determination is performed between the elevation value of the point and the elevation value obtained by interpolation. The threshold value is generally set to 0.5 m. If the difference is less than the threshold value, it is determined as a ground point.

Figure 4.3 shows the morphological filtering effect of unmanned aerial vehicle (UAV) laser point cloud data in a high-voltage transmission line corridor.

4.1.2.2 Multi-Level Moving Surface Fitting Point Cloud Filtering

The filtering algorithms based on the surface constraint usually fit the surface with a local point cloud. The traditional filtering algorithm based on moving surface fitting uses a 2 × 3 moving window to obtain a rough fitting surface by finding the lowest point of each window and then determines the height difference of the points in the six moving windows and the fitted surface; finally, all points that have a height difference lower than the given threshold are marked as ground points (Zhang & Liu, 2004). However, fitting the surface with only six points may produce inaccurate fitted surfaces because the influence of the surrounding terrain is not taken into account.

This section introduces a point cloud filtering algorithm based on multi-level moving surface fitting of continuous terrain surfaces (Zhu et al., 2018b). By increasing the window neighborhood and reducing the grid size by multiples, different window neighborhoods and height difference thresholds are automatically set when changing the grid size. In this way, the ground point is better filtered. The algorithm

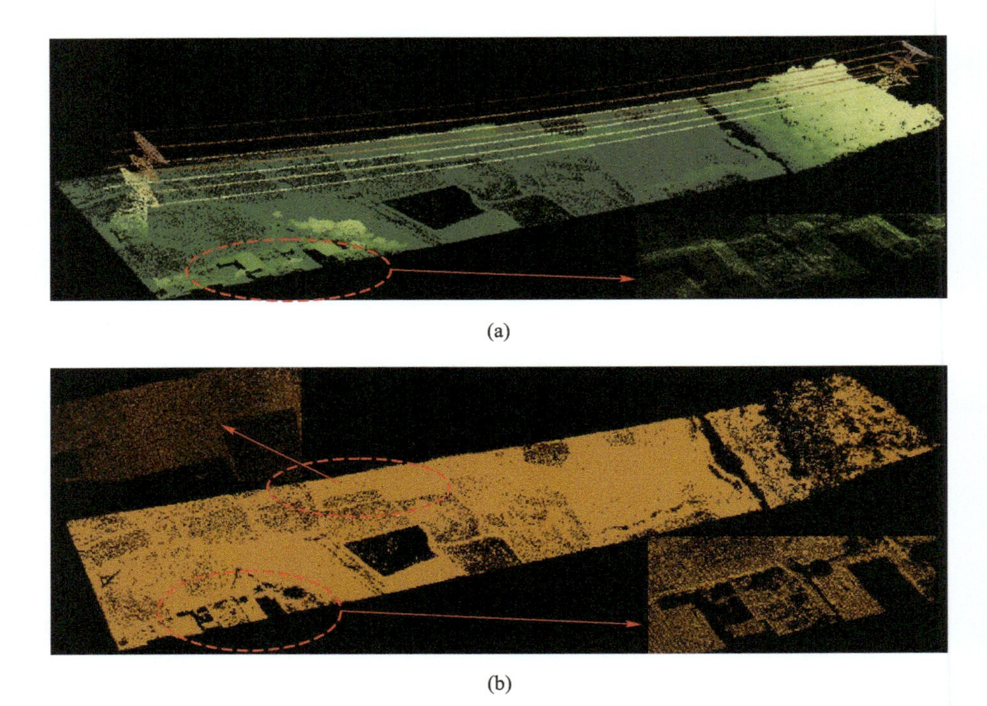

(a)

(b)

FIGURE 4.3 Example of morphological point cloud filtering. (a) Original point cloud. (b) Ground points after filtering.

flow is shown in Figure 4.4 and the pseudo-code is shown in Table 4.2. The specific steps are as follows:

1. Grid segmentation and indexing

   This step converts the point cloud to the corresponding grid according to the plane coordinates. To ensure that the lowest point in each grid is the real ground point the initial setting of the grid size should be larger than the size of the largest building, and then gradually decrease by multiples until most ground objects are eliminated and the best-fitting terrain surface is derived. Each grid records four properties (Figure 4.5). (i) Point set attribute: record the point number and calculate the grid corresponding to each point. (ii) Lowest point ID (ID_of_low): record the ID of the lowest point of each grid. (iii) Boundary coordinates: the relative coordinates (point cloud coordinates minus the boundary coordinates of the lower left corner of the grid) of the point cloud in the grid is used when fitting the surface, so the boundary coordinates (X_{min}, Y_{min}) of each grid need to be recorded. (iv) Labeling attribute: label each grid with the row and column numbers (i, j) of the two-dimensional (2D) matrix, which is conducive to the rapid storage of point clouds and improves the efficiency of point cloud filtering and interpolation.

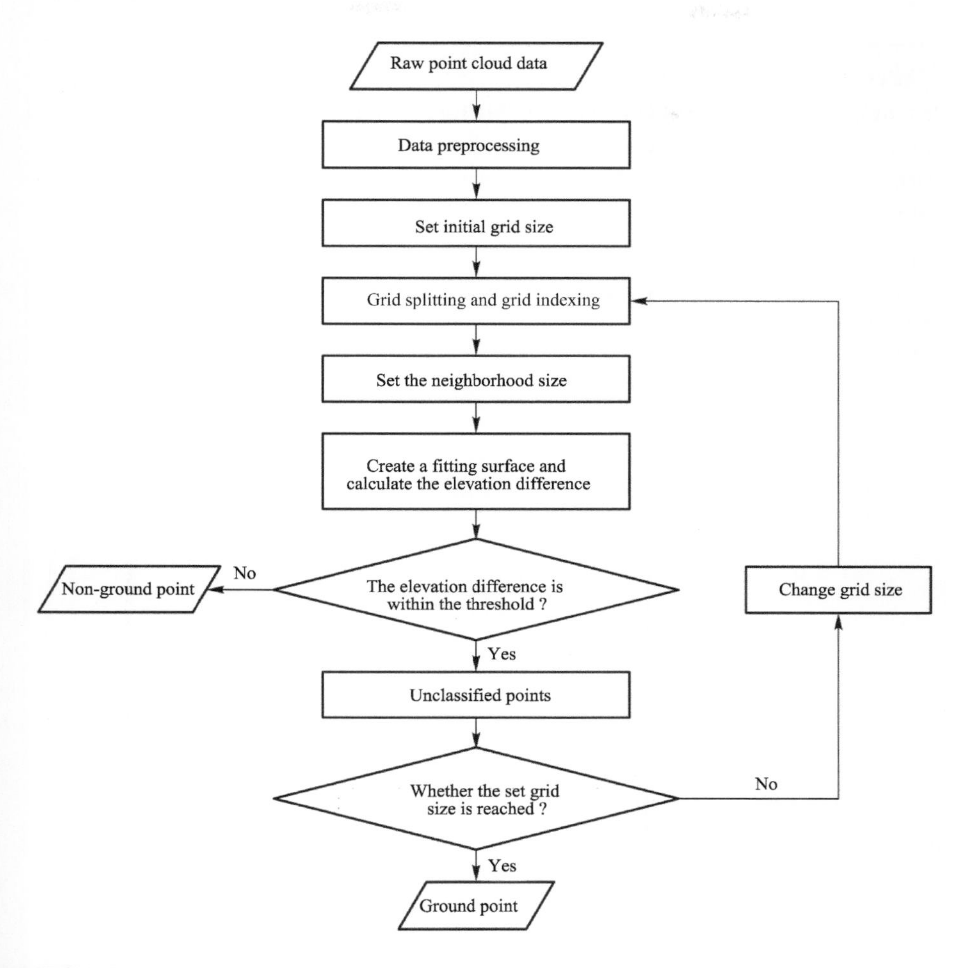

FIGURE 4.4 Flowchart of moving surface filtering algorithm.

2. Create a fitting surface

   Assuming that the terrain surface is a complex space surface, the local surface elements can be approximated by quadric surfaces, as in Equation (4.8).

$$f(X_i, Y_i) = a_0 + a_1 X_i + a_2 Y_i + a_3 X_i^2 + a_4 Y_i^2 + a_5 X_i Y_i \tag{4.8}$$

The fitted surface of the real ground points reflects the degree of ground fluctuation. Take the lowest point of the neighborhood window of each grid as the real ground point of the grid. Use the lowest point in all grids to fit the surface. Then calculate the difference between the fitted elevation and the real elevation of the point in the grid. Distinguish whether the point is a ground point by the height difference threshold. The key to fitting the

TABLE 4.2
Pseudo-Code of Multi-Level Moving Surface Filtering Algorithm

Input: ***P***, point cloud; ***s_max***, max grid size; ***s_min***, min grid size.

Output: ***P_Ground***, ground point; ***P_Object***, non-ground points.

Algorithm:

```
import point cloud P;
# Initialization
s = s_max: initialize grid size as s_max;
# Iterative calculation
while s > s_min
        construct grid G of point set P using grid size s;
        for g_j in G
                find the lowest point p_i in the grid g_j;
                g_j.h = p_i.z;
        endfor;
        for g_j in G
                slope = CompueSlopeWithNeighbor3x3(g_j. h);
                if slope < 0.176
                        n_s = 3;
                else if slope < 0.087
                        n_s = 5;
                else
                        n_s = 7;
                endif;
        endfor;
        fit surface with n_s × n_s neighbor of g_j;
        h_fit = ComputeFitHeightOnSurface(g_j, surface);
        # detect non-ground points, h_t is height threshold calculated by 3-σ law
        for p in g_j
                if p.z - h_fit > ht
                        p.is_ground = FALSE;
                endif;
        endfor;
        s = s/2;
endwhile;
for p_i in P
        if p_i.is_ground
                save p_i to P_Ground;
        else
                save p_i to P_Object;
        endif;
endfor;
return P_Ground, P_Object;
```

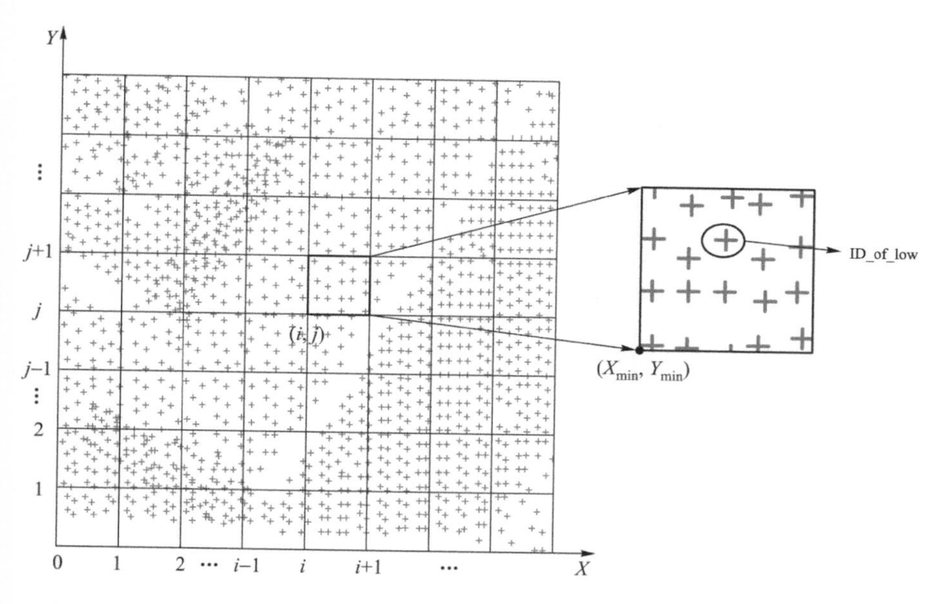

FIGURE 4.5 Grid division schematic diagram.

surface is the automatic setting of the grid neighborhood size and the solution of the parameters $(a_0, a_1, a_2, a_3, a_4, a_5)$ of the fitting surface, as follows:

i. Setting of grid neighborhood size

The quadratic surface used to fit the terrain surface has a total of six parameters. To ensure that the number of the lowest points is greater than 6 and the grid is in the middle position, the grid neighborhood size is generally set to 3×3, 5×5, etc. The 3×3 neighborhood of the gird (i, j) is the lowest point in the nine grids to fit a surface. The fitted surface is used to determine the ground points in the grid (i, j).

The selection of neighborhood size can be set according to the terrain slope. In this section, based on a large number of experiments, two slope thresholds, 5° and 10°, are selected. First calculate the average slope tangent of each grid of the 3×3 neighborhood grid, as in Equation (4.9). If there is a grid whose neighborhood grid with an average slope tangent value exceeding 0.176 [tan(10°)], the neighborhood size is set to 3×3; if the average tangent of the slope of the neighborhood grid exceeds 0.087 [tan(5°)], set the neighborhood grid size to 5×5; otherwise, set it to 7×7.

$$Slope_{ave} = \frac{1}{8}\sum_{k=-1}^{1}\sum_{t=-1}^{1}\frac{Z_{\min}(i,j) - Z_{\min}(i+k, j+t)}{\sqrt{(X_{\min}(i,j) - X_{\min}(i+k, j+t))^2 + (Y_{\min}(i,j) - Y_{\min}(i+k, j+t))^2}} \tag{4.9}$$

where $Z_{\min}(i, j)$ is the elevation value of the lowest point of the grid at row i and column j.

ii. Solution of surface parameters

The method increases the number of solving parameters; that is, the number of equations is greater than the number of parameters to be solved. Hence, the six parameters of the surface ($a_0, a_1, a_2, a_3, a_4, a_5$ in Equation 4.8) is solved by using the generalized inverse matrix and least squares method.

3. Determination of height difference threshold

This step sets different height difference thresholds when changing grid size and setting different window neighborhoods. Bartels and Wei (2010) proposed a point cloud distribution assumption: a large number of discrete ground point clouds are normally distributed in the natural state, while non-ground points affect the normal distribution. Compared with the ground points, the nonground points are regarded as error points whose standard deviation is greater than 3σ, thereby determining the height difference threshold. However, each grid may contain different features of the ground objects, and the same threshold cannot be directly selected for the entire point dataset. Considering the layering phenomenon of ground points and non-ground points, the idea of pedigree clustering is used to classify the fitted elevation difference. The height difference threshold is determined according to different classification numbers. However, the layering of ground points and non-ground points in some grids is not obvious. Hence, we combine the two methods. Specific steps are as follows:

i. Determination of overall height difference threshold. Assuming that there are N points, calculate the elevation difference between every point and its corresponding fitting surface equation, and arrange the N elevation difference data in ascending order, set $x_1 < x_2 < \ldots < x_N$. Calculate the height differences $l_{i(1 \le i \le N-1)}$ between adjacent data and their expected value μ and standard deviation σ. Set the overall height difference threshold as $\mu + 3\sigma$.

ii. Determination of the height difference threshold for each grid. Take one of the grids as an example, assuming that there are n points in the grid, calculate the elevation difference between two adjacent data points according to the same steps as the overall height difference threshold and arrange them in the order of $l_1 \ge l_2 \ge l_3 \ge \ldots \ge l_{n-1}$. Use l_1 to divide all data A in the grid into two categories, A_1 and A_2. The diameters of the two categories (the difference between the largest and smallest data in the same category) are Q_1 and Q_2, respectively. The center of the two categories (the arithmetic mean value of the largest and smallest data points in the same category) are O_1 and O_2, respectively. Once these parameters meet one of the following conditions, they are divided into two categories; otherwise, they are not classified.

a. $l_1 \ge \max\{Q_1, Q_2\}$. The distance between classes is relatively large. A_1 and A_2 are different classes.

b. $l_1 < \max\{Q_1, Q_2\}, l_1 > \min\{Q_1, Q_2\}, \text{and} (O_1 + O_2)/2 \in A_1$ or A_2. That is, the arithmetic mean value of the centers of the two categories falls into one of the two categories, indicating that this part is concentrated near the arithmetic mean of the two centers, while the other part is the opposite. The two parts of data have different characteristics.

The setting of the height difference threshold is related to the number of classifications and is specified as follows. First, let the first and the second largest distance between two adjacent data in A be l_1 and l_2, and $l_1 \geq l_2$:

a. When A is not classified, the stratification between the data is not obvious, and the height difference threshold is set as the overall height difference threshold.
b. A is divided into two parts, and the maximum diameter of the two categories is Q^*. When $Q^* \leq l_1 \leq 2Q^*$, the height difference threshold value is the fitting height difference value at l_2; in other cases, the height difference threshold value is the value with the fitted height difference at l_1.

This algorithm automatically sets the height difference threshold of each grid based on the actual situation of each grid itself in the case of multi-level filtering, which avoids as much as possible the excessive or incomplete filtering caused by manually setting the same height difference threshold.

4.1.2.3 Progressive TIN Densification Filtering

The idea of the progressive triangulated irregular network (TIN) densification (PTD) filtering algorithm is to subdivide the entire terrain surface into a connected triangular network and to remove non-ground points by searching for the points that conform to the ground characteristics to construct triangles, which is an iterative densification process (Zhang & Liu, 2004). It mainly includes three steps. First, construct a grid index for the point cloud data. The size of the grid generally depends on the size of the largest building in the data. Select the lowest point in each grid to construct an initial triangular network, that is, a rough ground model. Second, search for the points that fall on the triangular surface. When the height difference d between the point and the triangular surface and the terrain angles $\alpha_1, \alpha_2, \alpha_3$ (parameters are shown in Figure 4.6) satisfy the threshold conditions, the triangle is added to form

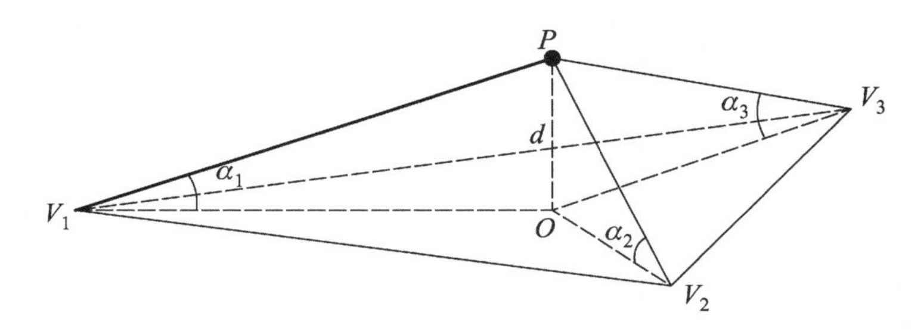

FIGURE 4.6 Parameters used in progressive TIN densification filtering algorithm. V_1, V_2, and V_3 are the three vertices of a triangular surface; P is one point in 3D space; O is the projection of point P onto the triangular plane.

a new triangular network. Third, perform iterative densification until all points are classified as ground points or non-ground points or reach preset constraints, such as the upper limit of iterations and the lower limit of the side length of triangular faces.

Although the PTD algorithm has good terrain universality, there are still some points that need to be optimized in complex areas with large terrain fluctuations and dense vegetation coverage (such as transmission corridors in mountainous areas). For example, the filtering effect depends on the selection of the initial ground point and the effect of densification order on filtering accuracy.

Here we introduce an improved progressive triangulation densification filtering algorithm (Liu et al., 2020). In the initial irregular triangulation construction stage, the constraints for the selection of ground seed points are added. The lowest point extracted by the extended local minimum method is set as the pending seed point. Extend the thin plate spline (TPS) interpolation method commonly used in 2D deformation detection to 3D, so as to achieve high-precision fitting interpolation of the 3D terrain. Determine the ground seed point by comparing the difference between the interpolation elevation and the real elevation. In the process of triangulation network densification, elevation sorting is used to optimize the point cloud densification process. The process of the improved PTD filtering algorithm is shown in Figure 4.7 and its pseudo-code is shown in Table 4.3. The main steps are as follows.

1. Extract pending seed points

   The discrete point cloud is evenly divided by a regular virtual grid. Two types of grids with and without a point cloud are obtained. For grids with a point cloud, the extended local minimum method is used to select seed points grid by grid, rather than directly taking the lowest point as the seed point, so as to avoid introducing too many noise points or non-ground points into the pending seed point set. The principle of the extended local minimum method is as follows. Select the first lowest and second lowest points in the grid. Determine whether the elevation difference between the two points is less than the set distance threshold. Once the difference is more than the threshold, consider the second lowest point as the lowest elevation point and compare first lowest and second lowest points again. Iterate the process until the elevation difference is less than the threshold. The lowest elevation point when iteration stops is used as the pending ground seed point. For grids that do not contain point clouds, the nearest neighbor method is used to find the elevation value of the nearest pending ground seed point as the elevation of the grid, so as to avoid the impact of data missing on the final seed point determined later.

2. Determine the final seed point

   The pending ground seed points selected in the previous step are close to the real ground points, but there are still a small number of atypical points (noise or ground objects) erroneously added to the point set. Hence, such points need to be removed from the pending seed point set to determine the final seed point (Figure 4.8).

   Usually, the TPS function, which performs well on local terrain simulation, is used to determine the final seed point (Chen et al., 2013).

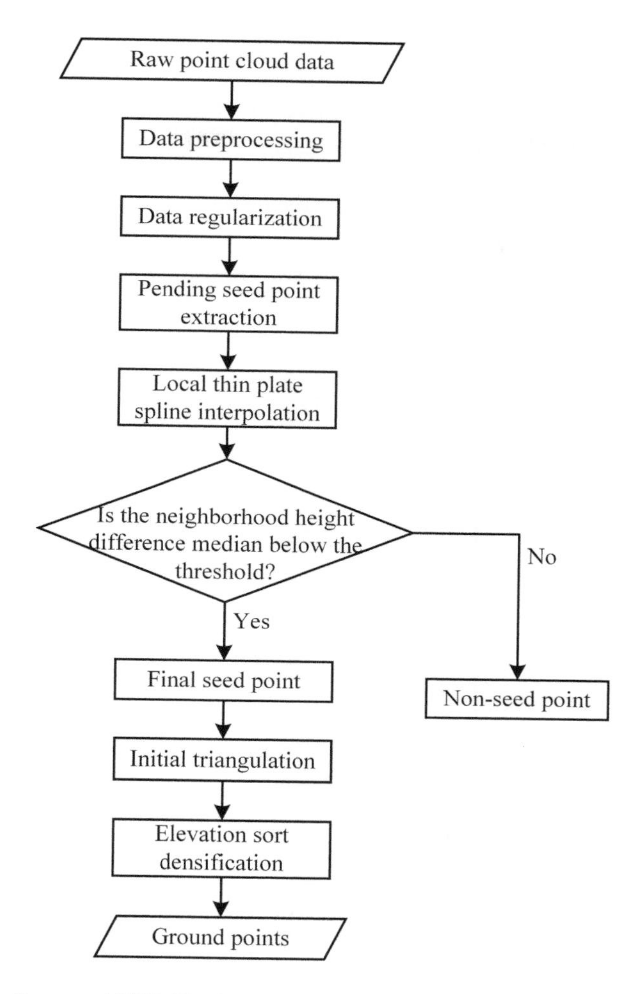

FIGURE 4.7 Improved PTD filtering algorithm flowchart.

Specifically, the minimum curvature surface is used to fit the control point and approximate the local surface where the control point is located, as follows.

Assuming that the local terrain surface is approximated by the function $z = f(x, y)$, the estimation of TPS is expressed by Equation (4.10).

$$f(x, y) = a_1 + a_2 x + a_3 y + \sum_{i=1}^{n} w_i u(r_i) \tag{4.10}$$

where a_1, a_2, and a_3 are the coefficients of the trend function $a_1 + a_2 x + a_3 y$, n is the number of control points for calculating the TPS function, w_i is the weight, r_i is the distance between the point to be interpolated and

TABLE 4.3
Pseudo-Code of Improved PTD Filtering Algorithm

Input: ***P***, point cloud; ***s***, grid size; ***ht1***, height difference threshold for seed points extraction; ***ht2***: height difference threshold for seed points optimization

Output: ***P_Ground***, ground point; ***P_Object***, non-ground points.

Algorithm:

```
import point cloud P;
# Extract seed points
construct grid G with P using grid size s;
for g_j in G
        sort all points in g_j in ascending order of z;
        for p in g_j
                p_n = GetNeighbor(p);
                if p.z - p_n.z < ht1
                        save p to P_seed;
                        break;
                endif;
        endfor;
endfor;

# Optimize seed points P_seed with TPS
for g_j in G
        if Exists p_seed in g_j
                h_fit = TPSInterplotation(g_j);
                Δh = p_seed.z - g_j;
        endif;
endfor;
for g_j in G
        if Exists p_seed in g_j
                Δh_m = GetNeighborMedian(g_j);
                if p_seed.z - Δh_m > ht2
                        remove p_seed from P_seed;
                endif;
        endif;
endfor;
construct triangulation network DT using P_seed;
sort all non-seed points P_others in ascending order of z;
for p in P_others
        insert p into DT follow the rules of classical PTD algorithm;
endfor;
for p_i in P
        if p_i inserted into DT
                save p_i to P_Ground;
        else
                save p_i to P_Object;
        endif;
endfor;
return P_Ground, P_Object;
```

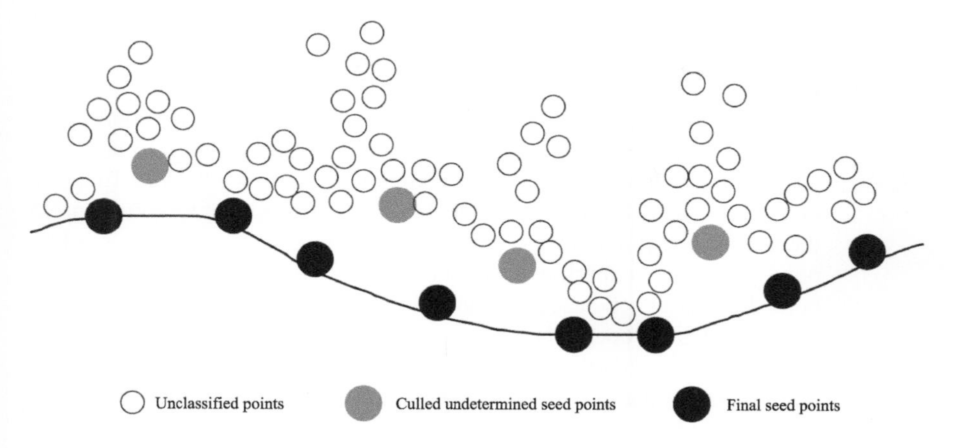

FIGURE 4.8 Schematic diagram of pending seed points.

the control point i, and $u(r)$ is the radial basis function, derived from Equation (4.11).

$$u(r) = r^2 \ln r^2 \tag{4.11}$$

There is a constraint relationship between the weight coefficient and the coordinates of the control point as in Equation (4.12):

$$\sum_{i=1}^{n} w_i = \sum_{i=1}^{n} w_i x_i = \sum_{i=1}^{n} w_i y_i = 0 \tag{4.12}$$

Based on the previous equations, a linear equation set is formed. Its matrix representation is shown in Equation (4.13).

$$\begin{pmatrix} \boldsymbol{K} & \boldsymbol{P} \\ \boldsymbol{P}^{\mathrm{T}} & \boldsymbol{O} \end{pmatrix} \begin{pmatrix} \boldsymbol{w} \\ \boldsymbol{a} \end{pmatrix} = \begin{pmatrix} \boldsymbol{z} \\ 0 \end{pmatrix} \tag{4.13}$$

where $\boldsymbol{K} = \begin{pmatrix} s_{11} & \cdots & s_{1n} \\ \vdots & \ddots & \vdots \\ s_{n1} & \cdots & s_{nn} \end{pmatrix}$, $\boldsymbol{P} = \begin{pmatrix} 1 & x_1 & y_1 \\ \vdots & \vdots & \vdots \\ 1 & x_n & y_n \end{pmatrix}$, $\boldsymbol{w} = \begin{pmatrix} w_1 \\ \vdots \\ w_n \end{pmatrix}$, $\boldsymbol{z} = \begin{pmatrix} z_1 \\ \vdots \\ z_n \end{pmatrix}$,

$\boldsymbol{a} = \begin{pmatrix} a_1 \\ a_2 \\ a_3 \end{pmatrix}$, s_{ij} is the distance between the control point i and the control point j, and z is the elevation value of the control point i. According to Equation (4.13), all the coefficients of the TPS function are obtained by using the LU decomposition algorithm.

The specific steps of determining the final seed point with TPS are as follows:

i. Local TPS interpolation

The TPS is used to calculate the fitted elevation of the pending ground seed points in the grid. The more control points used to fit the terrain surface, the higher the time complexity of solving the TPS function coefficients. Therefore, the local TPS method is used to search for the ten nearest points in the neighborhood range of the pending seed point (with many experiments, the number 10 is better, which considers both the search efficiency and interpolation accuracy) as the control point to solve the coefficient. The control point search method is as follows. First, by using the nearest neighbor search algorithm, a KD-tree of all points of the grid where the pending seed point is located is constructed and helps to quickly find the neighboring points of the pending seed point. Then, substitute the plane coordinates of the pending seed point into Equation (4.10) for which all coefficients have been solved and derive the fitting elevation h of the pending seed point.

ii. Extract the final point set

Select a 3×3 neighborhood window of the grid where the pending seed point is located, as shown in Figure 4.9. For grids without a point cloud, the grid elevation is used as the fitting elevation. For grids with a point cloud, the fitting elevation of pending seed points in the grid is calculated based on local TPS interpolation. The final seed points are further screened according to the following constraint criteria. The real elevation of the pending seed point in the central grid is known to be z_0. First calculate the difference between the real elevation and the fitted elevation h, obtain Δh_0, Δh_1,..., Δh_8, respectively. Then calculate the

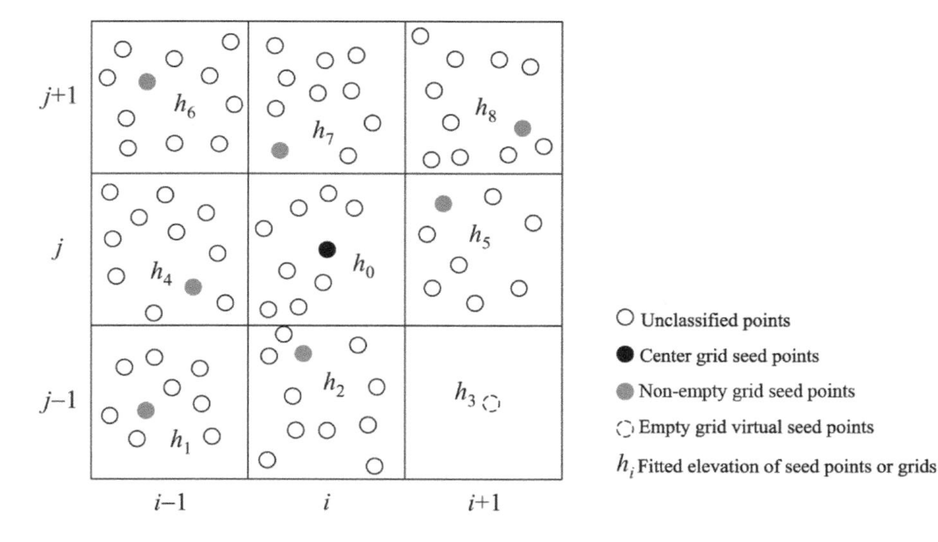

FIGURE 4.9 Ground seed point selection based on local TPS.

absolute value of Δh_i and get the median of $|\Delta h_i|_{mid}$. If the difference between $|\Delta h_i|_{mid}$ and z_0 is less than the set threshold, the pending seed point in the central grid will be selected as the final ground seed point; otherwise, it will be eliminated. Finally, the final ground point set is determined by traversing the pending seed point set grid by grid.

iii. Densification optimization based on elevation sorting

The densification sequence is a part that is often ignored in the PTD algorithm. If ignored, the real ground points might be misclassified or missed. Since the iterative densification process of the PTD algorithm has a high time complexity, here we adopt the densification optimization algorithm of sorting by elevation and preferentially densify the points near the identified ground seed points.

Take the airborne LiDAR point cloud of a power transmission corridor in a mountainous area of Guizhou as an example. The data cover vegetation, power towers, wires, buildings, and other objects. The length is about 300 m, and the number of points is 2,366,928, as shown in Figure 4.10(a). The parameters to be set in the experiment include grid division size, angle threshold, and distance threshold. The grid size is used to divide the data into regular grids. If the grid is too large, the terrain fitting accuracy will decrease; if too small, the calculation efficiency will decrease. The angle threshold and distance threshold are the parameters of the densified triangulation. With many experiments,

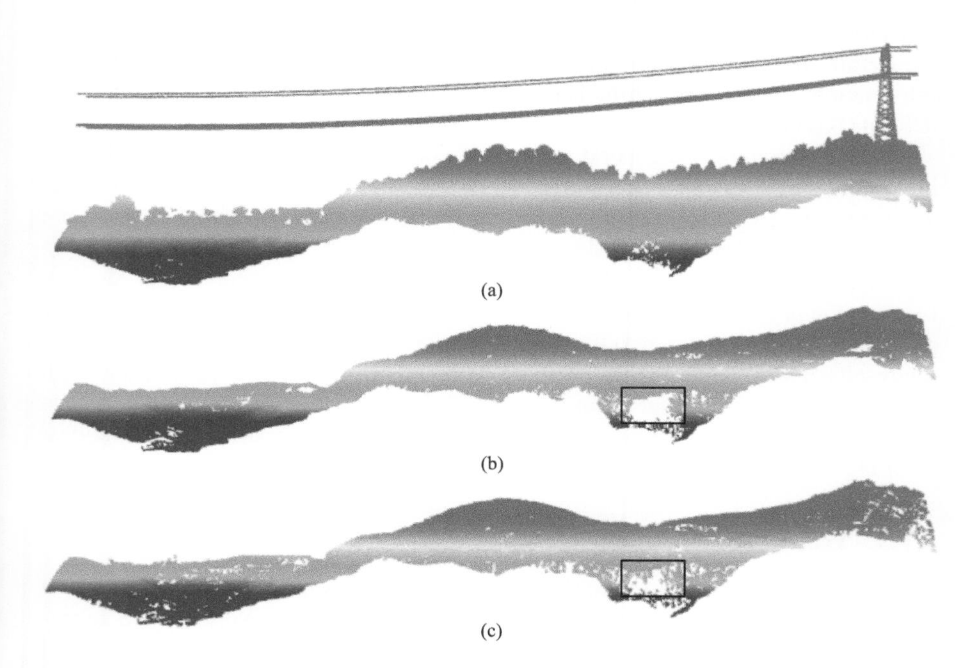

FIGURE 4.10 Comparison of filtering results. (a) Experimental data; (b) classified ground point of PTD algorithm; (c) classified ground point of improved PTD algorithm.

the grid size is set to 10 m, the angle threshold is 6°, and the distance threshold is 1 m. The filtering results of the PTD algorithm and the improved PTD algorithm are shown in Figure 4.10(b) and (c), respectively. It can be seen that both the PTD algorithm and the improved PTD algorithm can effectively separate ground points and non-ground points. Most vegetation, towers, power lines, and other obvious objects are filtered out. Compared with the PTD algorithm, the improved PTD algorithm retains more ground point clouds in areas with dense vegetation and large slopes, as shown in the box in Figure 4.10(c).

4.1.2.4 Other Point Cloud Filtering Algorithms

In addition to the previous three classical filtering algorithms, many scholars in the world have proposed some other filtering algorithms. For example, Zhang et al. (2016) proposed a cloth simulation filter (CSF). This algorithm constructs a "spring and mass point" model and iteratively simulates the up-and-down moving physical process of each terrain "mass point" based on the gravity of the mass point itself and the internal force between the mass points. The physical process reaches the final position, thereby determining the ground point set. The CSF algorithm has the advantages of few parameter settings, strong adaptability, and good filtering performance in urban areas. However, this algorithm has the possibility of misclassifying ground points into non-ground points in steep slope areas. Ding et al. (2019) proposed a point cloud filtering algorithm based on virtual grid and improved slope, which effectively avoids the loss of accuracy caused by point interpolation or smoothing. The shortcoming is that the slope threshold is difficult to determine according to the terrain. In addition, there are some filtering algorithms based on multiple echo information or various machine learning classifiers, e.g., conditional random fields (CRFs) (Bassier et al., 2017).

Scholars throughout the world are working on exploring better point cloud filtering algorithms. Generally speaking, the improvement of the point cloud filtering algorithm mainly focuses on three aspects. First, improve the automation degree of the algorithm. The existing algorithms cannot achieve high filtering accuracy in all complex scenes, and manual editing is still required in the follow-up. Second, improve the intelligence of the algorithm. Most filtering algorithms involve complex parameter settings, which require sufficient prior knowledge to achieve better filtering results. Third, improve the efficiency of the algorithm. High efficiency and high accuracy of the filtering algorithm often cannot be satisfied at the same time. In most cases, there are loop iterations and a large number of neighborhood calculations, resulting in high time complexity of the algorithm.

4.1.3 Point Cloud Classification

The point cloud classification in this section refers to assigning semantic labels to all points in the scene. Point cloud classification is the basic work for the subsequent applications. The existing point cloud classification algorithms are divided into two categories: one is based on semantic rules and the other is based on machine learning.

4.1.3.1 Point Cloud Classification Based on Semantic Rules

According to the spatial distribution characteristics of various ground objects, a series of semantic rules are set, such as elevation thresholds, linear constraints, and spatial position relationship constraints. Then, the scene objects are extracted one by one. The point cloud classification algorithms based on semantic rules mainly include the model-fitting-based algorithm and the clustering-based algorithm.

1. Point cloud classification algorithm based on model fitting

   The model-fitting algorithm can effectively segment point clouds that conforms to the model geometry, such as lines, planes, and cylinders. This method uses the mathematical model of the original geometric shape as prior knowledge to segment the point cloud and classifies the point cloud with the same mathematical expression into the same geometric region, such as power lines (Wang et al., 2017) and building roofs (Suveg & Vosselman, 2002). The method is based on mathematical principles, mainly including the Hough transform (Duda & Hart, 1972) and the random sampling consensus (RANSAC) (Fischler & Bolles, 1981) algorithm.

   The Hough transform utilizes the dual relationship between points and lines to express a straight line in a plane image as a point in the parameter space. By expanding the parameter space to 3D space, the planes can be detected from a 3D point cloud. Hough transform uses triples $\{\theta,\varphi,\rho\}$ to define a plane, where θ,φ, and ρ represent the three parameters of the plane normal vector passing through the origin of the coordinate axis. Calculate the position of the sample points in the 3D space and find the triplet containing the most points so as to determine the optimal plane parameters.

   The basic idea of the RANSAC algorithm is to randomly select a set of points from the point cloud to fit a predefined model (line, plane, sphere, etc.) and test other points with this model. If enough points fit the model, the model is considered more reasonable; otherwise, re-estimate the model and evaluate its accuracy. This process is iterated several times until the model with the highest accuracy is selected as the best model for the point set.

   This section presents the RANSAC linear fitting algorithm as an example. The linear mathematical model is:

$$y = ax + b \tag{4.14}$$

   The specific process is as follows:

   i. Project the point cloud into the 2D plane (*XOY/XOZ/YOZ* plane). Select two points randomly as inliers and use their projected planar coordinates (x_i, y_i) to establish a linear model $y = a_{\text{iteration}}x + b_{\text{iteration}}$ $(\text{iteration} = 0,1,\ldots k)$, with k the threshold of the number of iterations to calculate the optimal model.
   ii. Use the line model obtained in step (i) to test all other points. Calculate the distance from the point to the line model. If the distance satisfies the threshold requirement, the point is regarded as a point on the same line.

iii. If the number of points on the line exceeds the number threshold, then the line model is constructed.

iv. The line model is evaluated by the least squares fitting algorithm. The rationality of the model is evaluated by the number of points on the model and the accuracy.

v. Repeat steps (i)–(iv) several times. The model with the highest number of points on the line and satisfying the accuracy requirement is selected as the optimal model.

The algorithms based on model fitting have strong robustness. They can extract point clouds that conform to the specified geometry. However, such algorithms tend to extract point clouds that are mathematically correct but not real objects, and the algorithms are inefficient.

2. Point cloud classification algorithm based on clustering

The point cloud classification algorithm based on clustering is implemented by analyzing the local features of the point cloud and dividing the points with the same features into homogeneous regions, generally using an inter-class distance threshold or a predetermined number of classes as the iteration termination condition. The commonly used clustering algorithms are spectral clustering (Ng et al., 2002), *K*-means clustering (Hartigan & Wong, 1979), density-based spatial clustering of applications with noise (DBSCAN) clustering (Ester et al., 1996), and mean shift clustering (Comaniciu & Meer, 2002). The advantages and disadvantages of these algorithms are summarized in Table 4.4.

This section takes the DBSCAN algorithm as an example for a detailed introduction. The DBSCAN algorithm clusters samples of different

TABLE 4.4
Advantages and Disadvantages of Four Clustering Algorithms

Algorithm	Spectral Clustering	*K*-means	DBSCAN	Mean Shift
Measured value	Similarity	Euclidean distance	Density	Density/Euclidean distance
Number of clusters	Heuristic	Predefined	Automatic	Automatic
Advantages	Insensitive to noise; can handle high-dimensional data efficiently	Simple and fast; high scalability and efficiency for large datasets	Fast clustering and efficient handling of noise points; cluster shapes are unbiased	Automatically determine the number of clusters; insensitive to noise
Disadvantages	Sensitive to similarity matrices	Sensitive to noise and outliers; sensitive to initial values; the number of clusters must be given	Large memory consumption for high-dimensional data; sensitive to parameters	Large memory consumption for high-dimensional data

categories based on the density features of the point cloud. The centroids with neighborhood points greater than a threshold are used as cores and continuously expanded outward. The expansion process is mainly affected by two parameters: the neighborhood radius (eps) and the minimum number of points (min_pts) contained by core points. If a core point is in the neighborhood of another core point, then the two core points are called directly density-reachable. If two core points are connected by multiple "directly density-reachable" core points, they are called density-reachable. The core point and all points reachable by it (including core points and non-core points) form a cluster. As in Figure 4.11, the circle represents the neighborhood. The min_pts is set 4. Point *A* and other red points have more than four points in the neighborhood, so they are all core points. All core points in the figure are directly density reachable for each other, which form a cluster. Point *B* and point *C* are non-core points, but they can be reached by point *A* through other core points, so they belong to the same cluster as point *A*. Point *N* is an extra point; it is neither a core point nor reachable by other points.

The specific process of the DBSCAN algorithm is as follows. An unprocessed point P_i is randomly selected. If the point belongs to the core point, that is, the number of neighbor points is greater than min_pts, all points with the density reachable to the point P_i are found, and these points form a point cluster and are marked as the same class. If the point does not belong to the core point, the next core point is reselected, and the previous steps are repeated until all points are processed. The DBSCAN is very sensitive to the parameter settings. Different parameter settings would bring different classification results. The pseudo-code of the DBSCAN algorithm is shown in Table 4.5. Figure 4.12 shows the results of the DBSCAN algorithm for segmenting a building point cloud.

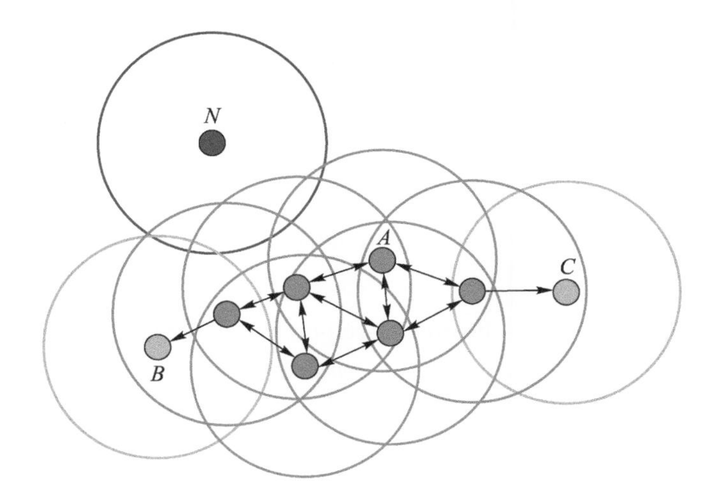

FIGURE 4.11 Schematic diagram of the DBSCAN algorithm.

TABLE 4.5
Pseudo-Code of DBSCAN Algorithm

Input: ***P***, point cloud data; **eps**, search radius; **min_pts**, minimum number of points in the neighborhood.
Output: $P_{DBSCAN\_Cluster}$, cluster points
Algorithm:
import point cloud ***P***;
for each point P_i in ***P***
    search all points within the circle with center of P_i and radius of **eps**;
    if the number of points within the radius of **eps** >**min_pts**
        create a new class $\mathbf{Class}_i$ and mark all points as $\mathbf{Class}_i$;
        for all points in the neighborhood of P_i
            if the number of points within the radius **eps** >**min_pts**
                classify the searched points into $\mathbf{Class}_i$;
            endif;
        endfor;
    else
        Mark the point as an outlier;
    endif;
endfor;

(a) (b)

FIGURE 4.12 Example of the DBSCAN algorithm. (a) Original point cloud (rendered by elevation); (b) clustering results.

4.1.3.2 Point Cloud Classification Based on Machine Learning

According to different feature extraction methods, point cloud classification algorithms based on machine learning are divided into classical machine learning based and deep learning based. The former extracts the custom features of the objects and then uses support vector machine (SVM) (Cortes & Vapnik, 1995), random forest (RF) (Breiman, 2001), Bayesian, and other classifiers to achieve classification. The technical process is shown in Figure 4.13. The point cloud classification algorithm

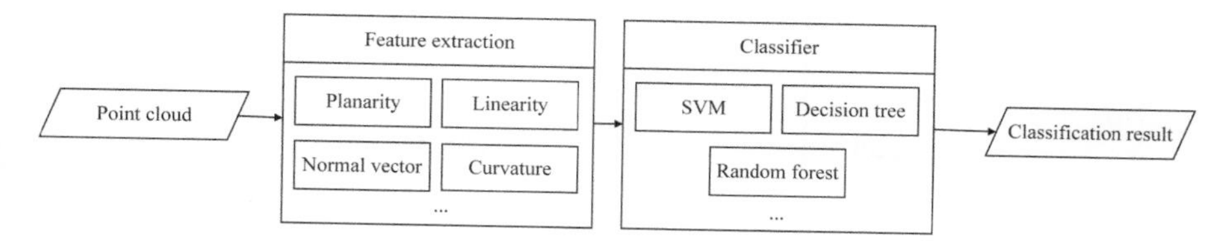

FIGURE 4.13 Point cloud classification process based on classical machine learning.

based on deep learning does not require manually defined features, but directly inputs the original point cloud into the deep neural network, automatically extracts features, and establishes a classification model to achieve point cloud classification.

1. Point cloud classification algorithms based on classical machine learning

   The algorithms design and extract features according to the characteristics of the objects and construct a multi-dimensional feature vector. The feature vector is input into the classifier for training to construct a point cloud classification model. The features of the test data are extracted and input into the point cloud classification model. Finally, the test data are classified. The specific steps are as follows.

   i. Feature extraction. Point cloud feature design and extraction is the basis of point cloud classification, which is straightforwardly related to the final classification accuracy. In view of the large variations in laser penetration rate, surface roughness, physical size, volume, and other surface characteristics of various types of objects, the point cloud features are designed according to the spatial distribution characteristics and the surface characteristic differences of the point cloud. According to the type of primitives used to calculate the features, there are two categories, i.e., point primitive-based feature and object primitive-based feature extraction methods. The former retains the original features of a single point to the maximum extent but requires point-by-point calculation, which is time-consuming. In addition, due to the complexity of spatial distribution and surface morphology of features, point primitive-based features are more sensitive to changes in the object shapes. The object primitive-based feature extraction method overcomes the sensitivity to the spatial distribution of point clouds, but it depends on the quality of object primitives and spends long time in the preprocessing stage.

      The concept of a neighborhood is usually involved in the point primitive-based feature extraction method. The single point neighborhood refers to the set of surrounding points with a point as the center point and searching its neighborhood. The choice of neighborhood size and type would affect the point cloud feature calculation and classification results. Neighborhood types include cylindrical neighborhood, spherical neighborhood, etc. (Figure 4.14). Constructing a cylinder based on a given point and a search radius r is called the cylindrical neighborhood

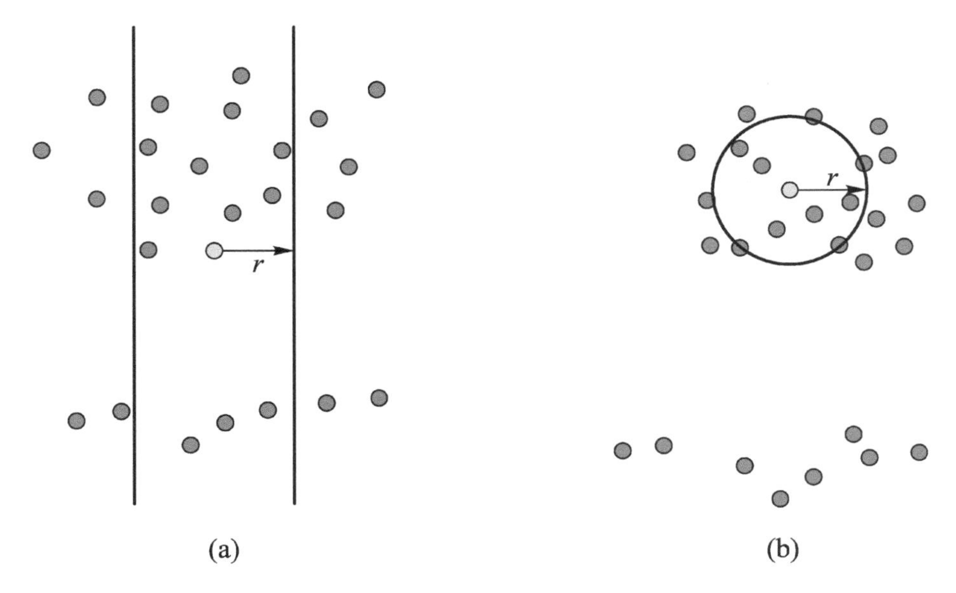

FIGURE 4.14 Neighborhood definition of point cloud features. (a) Cylindrical neighborhood; (b) spherical neighborhood.

of the current point. All points contained within the cylinder are called the cylindrical neighborhood points of the point. If constructing a sphere, the points contained within the sphere with radius r are called the spherical neighborhood points. The commonly used point cloud features extracted based on point primitives are mainly the features based on eigenvalue, density, and height (Table 4.6) (Peng et al., 2019).

The eigenvalues and eigenvectors of the point cloud provide important decision information for classification. The covariance matrix of the center point is calculated based on its neighborhood points to find the eigenvalues of the center point $\lambda_1, \lambda_2, \lambda_3$ ($\lambda_1 \geq \lambda_2 \geq \lambda_3$). The different relationships between the eigenvalues reflect the structure of the current point. If $\lambda_1 \approx \lambda_2 \approx \lambda_3$, the point is discrete. If $\lambda_1, \lambda_2 \gg \lambda_3$, the point has planar characteristics. If $\lambda_1 \gg \lambda_2, \lambda_3$, the point is in a linear structure. The point cloud of the building roof surface shows a planar distribution with a significant planarity feature, that is, $\lambda_1, \lambda_2 \gg \lambda_3$. The power lines and the edges of the building have an obvious linear structure. The linearity feature of these points is high, that is, $\lambda_1 \gg \lambda_2, \lambda_3$. The vegetation points have no tendency for the distribution of direction, that is, $\lambda_1 \approx \lambda_2 \approx \lambda_3$.

The point cloud density feature reflects the spatial distribution of the point cloud. The number of points per unit area is related to the surface characteristics irradiated by the laser beam. For example, the point cloud density of objects with multiple echoes (such as vegetation) is usually larger than that of impenetrable objects (such as buildings).

TABLE 4.6
Description of Point Cloud Features

Class	Full Name (Short Name)	Feature Calculation Equation	Description
Eigenvalues	Sum (SU)	$\lambda_1 + \lambda_2 + \lambda_3$	Sum of eigenvalues
	Omnivarance (OM)	$(\lambda_1 * \lambda_2 * \lambda_3)^{1/3}$	Full variance
	Eigenentropy (EI)	$-\sum_{i=1}^{3} \lambda_i * \ln(\lambda_i)$	Feature entropy
	Anisotropy (AN)	$(\lambda_1 - \lambda_3)/\lambda_1$	Anisotropy
	Planarity (PL)	$(\lambda_2 - \lambda_3)/\lambda_1$	Planarity
	Linearity (LI)	$(\lambda_1 - \lambda_2)/\lambda_1$	Linearity
	Surface variation (SUV)	$\lambda_3/(\lambda_1 + \lambda_2 + \lambda_3)$	Surface roughness
	Sphericity (SP)	λ_3/λ_1	Sphericity
Density	Point density (PD)	$\frac{3}{4} * \frac{N_{3D}}{\pi r^3}$	Point density in spherical neighborhoods
	Density ratio (DR)	$DR_{3D/2D} = \frac{N_{3D}}{N_{2D}} \cdot \frac{3}{4r}$	Ratio of the point density of the sphere to the point density in the circular area projected onto the horizontal plane. N_{3D} is the number of points in the sphere, and N_{2D} is the number of points in the circular area projected to the horizontal plane.
Height	Vertical range (VR)	$Z_{max} - Z_{min}$	Height difference between the highest point Z_{max} and the lowest point Z_{min} in the cylindrical neighborhood.
	Height above (HA)	$Z - Z_{min}$	Height difference between the current point Z and the lowest point Z_{min}.
	Height below (HB)	$Z_{max} - Z$	Height difference between the highest point Z_{max} and the current point Z.
	Sphere variance (SPV)	$\sqrt{\frac{\sum_{i=1}^{n}(z_i - z_{ave})^2}{n-1}}$	Standard deviation of the height difference in the spherical neighborhood; z_{ave} is the average height of the neighborhood.
	Cylinder variance (CV)	$\sqrt{\frac{\sum_{i=1}^{n}(z_i - z_{ave})^2}{n-1}}$	Standard deviation of height differences in the cylindrical neighborhood

The height-based features are used to distinguish objects with different elevations. Taking the power transmission corridors as an example, the vertical range of objects such as towers and trees are large; the standard deviation of elevation differences within the cylindrical neighborhood is low for the objects with little elevation variation, such as power lines with regular linear structures and buildings with planar structures.

ii. Classifier classification. The extracted feature vectors are input into the classifier to construct the classification model. The RF is one more

TABLE 4.7
Random Forest Algorithm Flow

Traverse ***J*** decision trees. For decision tree ***j***:

① Using the **bootstrap sample** method, randomly and with replacement, extract ***N*** samples from the training set ***D*** as the training set of the tree.

② Build decision trees using binary recursive classification.

a. Start training from the root node.
b. Repeat the following steps recursively for each unsplit node until the stopping condition is met:
   i. Randomly select ***m*** feature subsets from ***p*** features, where $m \ll p$;
   ii. Find the optimal split among all binary splits of ***m*** features, and record this feature as the feature with the best classification effect;
   iii. Using the features in (ii), split the node into two leaf nodes.

③ Input new sample data x into the trained RF model. Calculate the number of times that the predicted value $\hat{h}_j(x)$ of all decision trees is equal to class y. Determine the class $\hat{f}(x)$ with the most votes as the final attribution class [Equation (4.15)], where Equation (4.16) is the indicative function that the sample belongs to class y.

$$\hat{f}(x) = \arg\max_y \sum_{j=1}^{J} I\left(\hat{h}_j(x) = y\right) \tag{4.15}$$

$$I\left(\hat{h}_j(x) = y\right) = \begin{cases} 1, \hat{h}_j(x) = y \\ 0, \hat{h}_j(x) \neq y \end{cases} \tag{4.16}$$

commonly used RF classification algorithm. It is a kind of ensemble learning, also a combined classifier (Breiman, 2001). Its basic unit is a decision tree, and the decision trees are independent of each other. Each decision tree is trained based on a subset of training samples and a subset of randomly selected features. Multiple decision trees form a random forest. When new sample data are input, each tree makes a classification decision to produce its own classification result. The RF integrates these voting results. The class with the highest number of votes is output as the final class. The detailed RF algorithm is shown in Table 4.7.

Machine learning–based algorithms can automatically find appropriate thresholds according to the input features and then classify. They are more automated and less time-consuming. However, the algorithm accuracy is closely related to the effectiveness of the features. The commonly used features are designed manually and rely on a priori knowledge.

2. Point cloud classification algorithm based on deep learning

Deep learning algorithms automatically learn features from data instead of manual design and are widely used in image processing (LeCun et al., 2015). However, traditional deep learning models require the input data to have a regular structure (for example, an image is a regular matrix

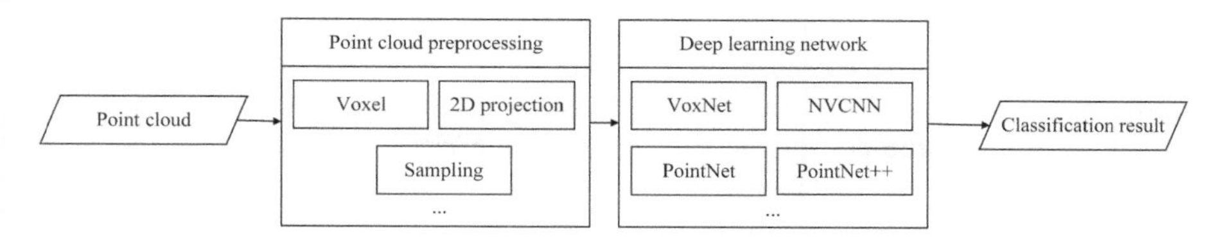

FIGURE 4.15 Point cloud classification process based on deep learning.

composed of pixels). In contrast, point clouds are unstructured, disordered, and discrete. The traditional deep learning models cannot be used for the point cloud classification directly. In the 2010s, many scholars attempted to improve deep learning techniques and apply them to point cloud classification (Figure 4.15). They proposed a variety of point cloud regularization methods, that is, converting discrete point clouds into regular input formats, such as meshes and 3D voxels. Voxel-based algorithms (such as VoxNet) are extremely complex and take a lot of time to train the model. The algorithms that project to 2D images, such as multi-view convolutional neural networks (MVCNN), reduce the network complexity but lose the point cloud 3D information. In addition, some algorithms directly input the original point cloud without regularization, such as the PointNet algorithm (Qi et al., 2017a). It solves the problem of point cloud disorder and retains high-precision 3D information, but only considers global features. The PointNet++, which is optimized on the basis of PointNet, takes into account the local features of adjacent points, and the classification result is better than some other algorithms (Qi et al., 2017b).

The point cloud classification model based on PointNet++ consists of two parts: hierarchical point set feature learning and point feature propagation. The main idea is to input the original point cloud, extract the features with different resolutions by hierarchical point set feature learning, then obtain the features of all original points by point feature propagation, and finally get the score of each point belonging to each class by classifier. The specific steps are as follows:

i. Hierarchical point set feature learning. Hierarchical point set feature learning consists of several "set abstraction" modules. It aims to divide the input point cloud into several local point clouds and extract the global features of each local point cloud. The combination of multiple set abstraction modules allows the features to be continuously abstracted to obtain higher dimensional features. The set abstraction module contains the following three basic units.
   a. Sampling layer. This layer uses the iterative farthest point sampling (FPS) algorithm (Kamousi et al., 2016) to complete the construction of the local area centroid set. First, a point is randomly selected as the initial point. Then the farthest point from the point is added to the centroid set and used as the starting point to continue iterating until the number of selected centroid points satisfies a given number N'.

b. Grouping layer. The point cloud is divided into multiple local point clouds based on the set of centroid points. The size of the input data is $N \times (d + C)$, where N is the number of input point clouds, d is the coordinate dimension of the point (usually 3), and C is other attribute dimensions such as intensity and echo. Using the centroid points in the sampling layer as the circle center and the predefined search radius r, all points in the spherical neighborhood are extracted as N' local point clouds, and each has a size of $k \times d$, where k is the number of point clouds in each local area.
c. PointNet layer. This layer learns the global features of each local area point cloud, that is, the holistic features of the abstract area point cloud. The input data are multiple sets of local area point clouds obtained by the grouping layer, and the size is $N' \times k \times d$. Each local area point cloud is learned by a PointNet model to derive a set of C'-dimensional feature vectors, so the final output data of this layer have a size of $N' \times (d + C')$.

ii. Point feature propagation. Hierarchical pointset feature learning downsamples the point cloud and retains the centroid points in each point cloud abstraction module. In contrast, the scene classification aims to assign corresponding class labels to the original points, so the points in set abstraction modules need to be restored to the original data. When the size of the stratified features is $N_l \times (d + C')$, the inverse distance weighting method is used to interpolate the layered features of N_{l-1} points before sampling that layer. In addition, in order to reduce the loss of details in the process of sampling and interpolation, the features obtained from the current interpolation are fused with the features of the point set abstraction layer to generate data of size $N_{l-1} \times (C_{l-1} + C')$. However, the feature dimensions of the data are too high. The correlation between some features would affect the training efficiency and model robustness. Hence, the data need to be processed by a unit PointNet to reduce the feature dimensions, reduce the computations, increase the nonlinearity of the network, and improve the generalization ability of the model. Interpolation and unit PointNet processing are done repeatedly until the number of output points is the number of original points.

Figure 4.16 shows the results of the PointNet++ algorithm and RF algorithm for the classification of the power transmission corridor point cloud, where the classification accuracy of the PointNet++ algorithm is 87.14% and the classification accuracy of the RF algorithm is 75.99%. The RF algorithm only performs better than the PointNet++ algorithm in vegetation point cloud extraction. In contrast, the classification of the building class produced by the RF algorithm is not satisfactory. The buildings are not retained integrally. Also, there are more missed and wrong points in the classification of towers. The RF algorithm cannot accurately classify the tower points, vegetation points, and power line points in comparison with the PointNet++.

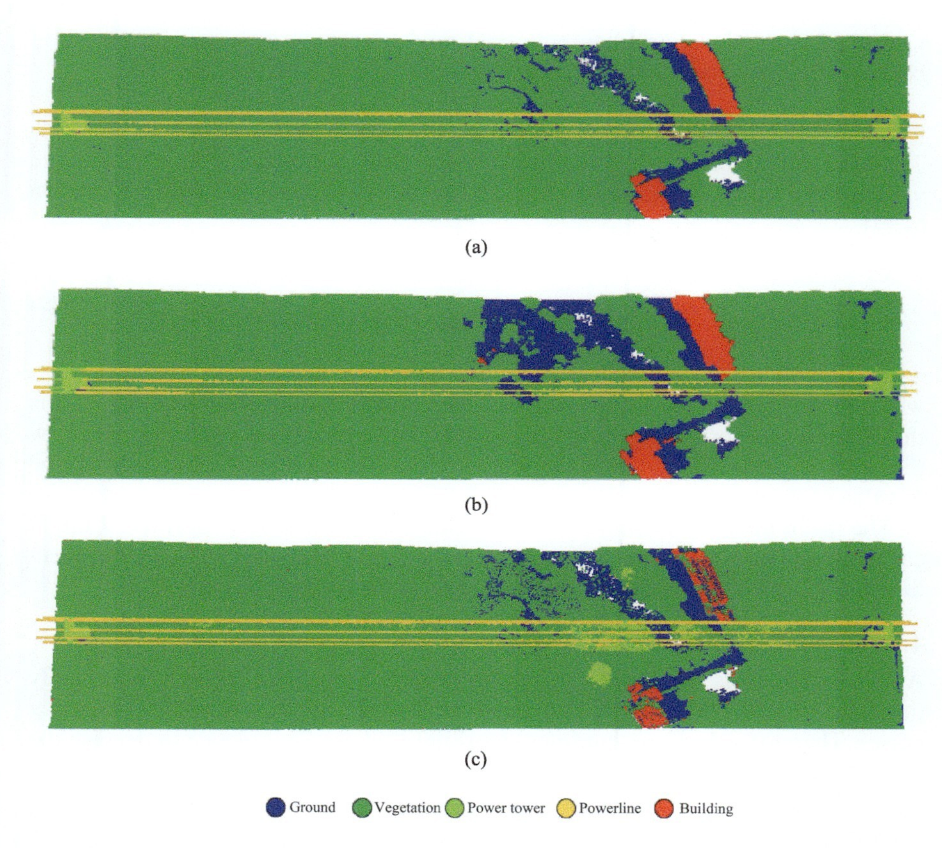

FIGURE 4.16 Classification results with different classification algorithms. (a) True values (manual classification); (b) PointNet++ classification; (c) random forest classification.

4.2 LiDAR WAVEFORM PROCESSING

The purpose of LiDAR waveform processing is to separate the returned signals of target objects from the original waveform and then extract the structure information of the objects (Mallet & Bretar, 2009). There are several key contexts for LiDAR waveform processing, including waveform denoising, waveform decomposition, waveform deconvolution, and waveform feature parameter extraction.

4.2.1 Waveform Denoising

The acquisition of a LiDAR waveform is usually affected by various factors such as external environment (sunlight, atmosphere, etc.) and internal systems, resulting in abundant noise in the acquired data. Waveform denoising is an important step in LiDAR waveform processing, aiming to provide a high-quality effective signal for subsequent waveform decomposition, waveform convolution, and feature parameter extraction. There are four kinds of commonly used waveform denoising algorithms, i.e., mean filter, Gaussian filter, empirical mode decomposition soft (EMD-soft)–based filter, and wavelet filter.

4.2.1.1 Mean Filter

The idea of a mean filter is to replace the original signal with the average value of the signal in the filtering window (i.e., neighborhood), as in Equation (4.17):

$$g(t) = \frac{1}{N} \sum_{i=t-N/2}^{t+N/2} f(i) \tag{4.17}$$

where N is the window size (usually an odd number), t is the sampling bin, $f(i)$ is the original waveform, and $g(t)$ is the filtered waveform. Figure 4.17(a) and (b) show the original waveform and the denoised waveform with a mean filter by setting the N value of 9.

The mean filter algorithm removes the Gaussian white noise. However, it might cause excessive smoothing of the waveform and loss of a valid signal since the mean filter assigns the same weight to each bin of the waveform in the window.

4.2.1.2 Gaussian Filter

All kinds of noises are generated during the data acquisition process, but mainly the Gaussian white noise with normal distribution. Most of them are high-frequency noises. Therefore, the Gaussian filter algorithm is chosen in practice to denoise the original waveform. The basic idea of a Gaussian filter is to denoise the original waveform by convolving the Gaussian kernel function with the original waveform. The one-dimensional Gaussian function is expressed by Equation (4.18):

$$g(t,\sigma) = \frac{1}{\sqrt{2\pi\sigma}} \exp\left(-\frac{t^2}{2\sigma^2}\right) \tag{4.18}$$

Its first derivative is a Gaussian filter, as shown in Equation (4.19):

$$g^{(1)}(t,\sigma) = \frac{-t}{\sqrt{2\pi\sigma^3}} \exp\left(-\frac{t^2}{2\sigma^2}\right) \tag{4.19}$$

The result of the convolution operation between the original waveform $f(t)$ and the Gaussian filter $g^{(1)}(t,\sigma)$ is the denoised waveform, as shown in Equation (4.20):

$$s(t,\sigma) = f(t) * g^{(1)}(t,\sigma) \tag{4.20}$$

where * is the convolution operator and σ is the standard deviation of the Gaussian filter. By changing the Gaussian standard deviation, we can adjust the smoothness of the filtered waveform. Figure 4.17(c) shows the denoised waveform with a Gaussian filter by setting the σ value to 3.

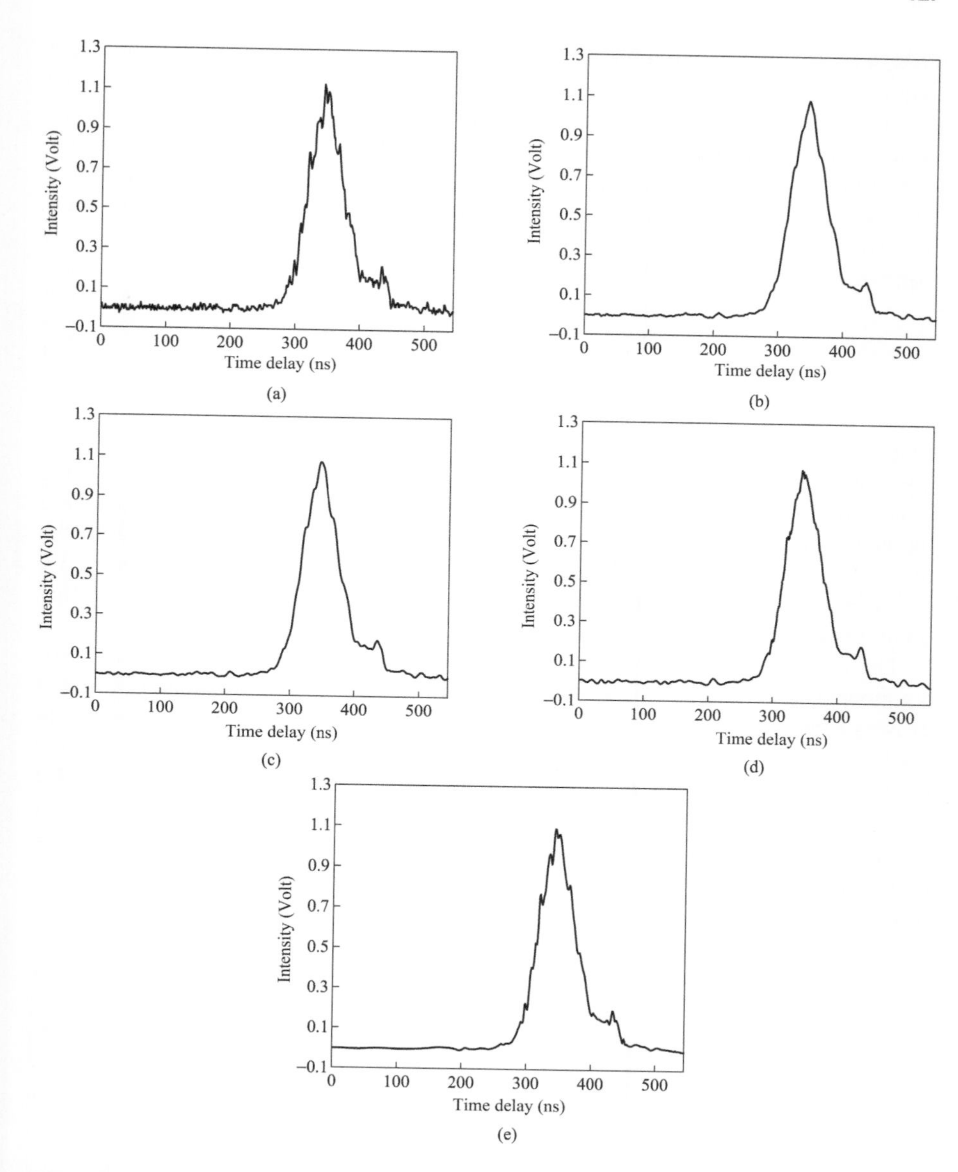

FIGURE 4.17 Waveform denoising. (a) Original waveform; (b) denoised waveform with mean filter; (c) denoised waveform with Gaussian filter; (d) denoised waveform with EMD-soft-based filter; (e) denoised waveform with wavelet soft-threshold filter.

4.2.1.3 EMD-Soft Filter

EMD is an adaptive signal decomposition method. Its essence is a process of decomposing the original signal from high frequency to low frequency in turn (Huang et al., 1998). The EMD algorithm assumes that the signal is composed of finite intrinsic mode functions (IMFs). There are two key steps to perform the EDM filter.

First, a series of IMF components with different frequencies (from high to low) are extracted from the original signal. Then each IMF component is further processed or filtered. Each IMF must satisfy two conditions: (1) the number of local extreme points and zero-crossing points must be equal or differ, at most, by one in the time range of the IMF and (2) at any bins in time, the average value of the local maximum envelope (upper envelope) and the local minimum envelope (lower envelope) must be zero. The relationship between the IMF components and the original waveform is expressed by Equation (4.21):

$$X(t) = \sum_{j=1}^{n} IMF_j(t) + r_n(t) \tag{4.21}$$

where $X(t)$ is the original waveform; $IMF_j(t)$ is the j-th IMF after EMD decomposition; and $r_n(t)$ is the remaining term after decomposition and also the monotonic trend term of the signal. The IMF component is derived by repeatedly subtracting the mean value of the upper and lower envelopes from the original signal until the residual meets the two conditions of the IMF component. Then, repeat the previous steps to derive more IMFs until the final monotonic trend item remains. At that point, the EMD process is completed.

The core idea of the EMD filter algorithm is to take the high-frequency IMF components derived by EMD as the background noises and remove them. Given that directly removing high-frequency IMFs might cause the loss of effective signals, the EMD-soft filter is usually chosen (Boudraa et al., 2013). First, extract useful information from the discarded p high-frequency IMFs by soft threshold after the EMD process is completed, as in Equation (4.25). The threshold is calculated based on Equations (4.22)–(4.24). Then, add the useful information, the remaining IMF components, and the monotonic trend term to obtain the final filtered signal, as shown in Equation (4.26).

$$MAD_j = \text{Median}\left| IMF_j(t) - \text{Median}\left[IMF_j(t) \right] \right| \tag{4.22}$$

$$\sigma_j = MAD_j / 0.6745 \tag{4.23}$$

$$\tau_j = \sigma_j \sqrt{2 \ln L} \tag{4.24}$$

$$s_j(t) = \begin{cases} IMF_j(t) - \tau_j, IMF_j(t) \geq \tau_j \\ 0, \left| IMF_j(t) \right| < \tau_j \\ IMF_j(t) + \tau_j, IMF_j(t) \leq -\tau_j \end{cases} \tag{4.25}$$

$$F(t) = \sum_{j=1}^{n} s_j(t) + \sum_{j=p+1}^{n} IMF_j(t) + r_n(t) \tag{4.26}$$

where $p \leq j$, σ_j is the noise estimate for $IMF_j(t)$, L is the array length of $IMF_j(t)$, and $F(t)$ is the waveform with the EMD-soft filter. Figure 4.17(d) shows the denoised waveform with EMD-soft filter working on the first two IMF components with abundant high-frequency noise.

4.2.1.4 Wavelet Filter

The wavelet filter first performs multi-scale wavelet transformation on the original signal with noise to obtain wavelet coefficients at each scale. Then extract the wavelet coefficients of the signal as much as possible and remove the wavelet coefficients belonging to the noise. Finally, the signal is reconstructed by inverse transformation based on the processed wavelet coefficients, so as to achieve the waveform denoising. This algorithm can not only retain the effective frequency components but also suppress high-frequency noise. Additionally, it overcomes the waveform overlapping phenomenon and maintains the effective information of the waveform. Here we introduce a waveform denoising algorithm based on the wavelet soft threshold. The key steps are as follows:

1. Wavelet decomposition. The LiDAR waveform is a one-dimensional signal of discrete sampling $f(t)$. The wavelet coefficients are obtained according to the Mallat fast algorithm, as in Equation (4.27).

$$\begin{cases} C_k^{j-1} = \sum_N a_{N-2k} C_N^j \\ d_k^{j-1} = \sum_N b_{N-2k} C_N^j \end{cases} \quad k = 1,2,\ldots,N \tag{4.27}$$

   where C_k^{j-1} are d_k^{j-1} are low-frequency and high-frequency wavelet coefficients at the scale of $j-1$, $\{a_{N-2k}\}$ and $\{b_{N-2k}\}$ are the decomposition sequences, N is the number of sampled bins, and j is the number of decomposition layers. Here, the sym6 wavelet, similar to the Gaussian function, is selected to perform wavelet transformation on the original signal, and the number of wavelet decomposition layers is set to 5 to calculate the wavelet coefficients $W_{j,k}$. The low-frequency wavelet coefficients indicate the degree of similarity between the wavelet and the original signal. The larger the coefficient, the greater the similarity between the wavelet and the original signal and vice versa.
2. Threshold processing of high-frequency coefficients. The low-frequency coefficients derived by the wavelet decomposition are reserved to ensure that the overall shape of the signal remains unchanged, and the high-frequency coefficients are quantized by threshold. The threshold λ is set as $\lambda = \sigma\sqrt{2\log n}$, where σ is the noise standard deviation and n is the signal length. When the wavelet coefficient $W_{j,k}$ is less than the threshold λ, it is considered to be caused by the noise, and it is set to 0; when the wavelet coefficient is greater than or equal to the threshold, it is considered to be caused by the signal. The wavelet coefficient caused by the signal is

processed using the soft threshold method. The soft threshold is determined based on Equation (4.28).

$$W_{j,k} = \begin{cases} \operatorname{sgn}(W_{j,k})(|W_{j,k}| - \lambda), & |W_{j,k}| \geq \lambda \\ 0, & |W_{j,k}| < \lambda \end{cases} \tag{4.28}$$

3. One-dimensional wavelet signal reconstruction. The reconstruction step is exactly the inverse process of the decomposition step. The wavelet reconstruction of the one-dimensional signal is conducted based on the low-frequency coefficients of the j-th layer and the modified high-frequency coefficients of the first layer to the j-th layer of the wavelet decomposition, as in Equation (4.29):

$$C_k^j = \sum_{k=1}^{N} p_{N-2k} C_k^{j-2} + q_{N-2k} C_k^{j-1} \tag{4.29}$$

where $\{p_{N-2k}\}$ and $\{q_{N-2k}\}$ are reconstruction sequences, N is the number of sampling bins, and j is the number of decomposition layers. In practical applications, it is necessary to select an appropriate wavelet filter threshold according to the specific situation. If the effective signal has a wide bandwidth and distributed from low frequency to high frequency, the effective signal might be suppressed when using the wavelet transform to suppress high-frequency noise. Figure 4.17(e) shows the denoised waveform with wavelet soft-threshold filter. It is seen that the wavelet filter preserves the waveform details well and reduces the loss of effective signals.

4.2.2 Waveform Decomposition

Waveform decomposition decomposes the original LiDAR waveform into multiple Gaussian sub-waveforms (Hofton et al., 2000; Wagner et al., 2006). The mathematical expression of the Gaussian function is as follows:

$$f_i(t) = A_i e^{-\frac{(t-\mu_i)^2}{2\sigma_i^2}} \tag{4.30}$$

where A_i, μ_i, and σ_i are the maximum amplitude, gravity center location, and standard deviation of the i-th Gaussian function $f_i(t)$, respectively.

The LiDAR waveform can be approximated by the mixed Gaussian waveform function. Each Gaussian sub-waveform is the result of the laser-target interaction. Specifically, the LiDAR waveform is expressed as in Equation (4.31).

$$P(t) = \varepsilon + \sum_{i=1}^{N} f_i(t) \tag{4.31}$$

where $P(t)$ is the amplitude of the LiDAR waveform at the time t, N is the number of Gaussian sub-waveforms, $f_i(t)$ is the i-th Gaussian sub-waveform function, and ε is the background noise mean.

The input of the waveform decomposition is the denoised waveform, and the output is the Gaussian function derived from the decomposition algorithm. It mainly goes through two steps: initial parameter estimates and waveform nonlinear fitting. The details are as follows.

4.2.2.1 Initial Parameter Estimates

We introduce two methods to estimate the initial values of parameters of the mixed Gaussian function: inflection point method and wavelet transform method.

1. Inflection point method

   The inflection point method is a classic algorithm of estimating the initial parameters for Gaussian decomposition. By solving the first-order and second-order derivatives of the waveform data and setting them as 0, the initial peak positions and inflection points of the waveform are determined. Take as an example Equation (4.32).

$$f_i(t) = \varepsilon + A_i \mathrm{e}^{\frac{-(t-\mu_i)^2}{2\sigma_i^2}} \tag{4.32}$$

   Its first-order and second-order partial derivatives are expressed as Equations (4.33) and (4.34), respectively.

$$\frac{\partial f_i(t)}{\partial t} = -A_i \frac{-(t-\mu_i)^2}{2\sigma_i^2} \mathrm{e}^{\frac{-(t-\mu_i)^2}{2\sigma_i^2}} = -f_i(t)\frac{(t-\mu_i)^2}{\sigma_i^2} \tag{4.33}$$

$$\frac{\partial^2 f_i(t)}{\partial t^2} = f_i(t)\left(\frac{(t-\mu_i)^2}{\sigma_i^4} - \frac{1}{\sigma_i^2}\right) \tag{4.34}$$

   Let $\frac{\partial^2 f_i(t)}{\partial t^2} = 0$, then $\frac{(t-\mu_i)^2}{\sigma_i^4} = \frac{1}{\sigma_i^2}$, so we derive:

$$\sigma_i = |t-\mu_i| \tag{4.35}$$

   One Gaussian function has two inflection points, i.e., $t = \mu_i + \sigma_i$ and $t = \mu_i - \sigma_i$. n Gaussian distributions would have $2n$ inflection points. Once the inflection points of the waveform curve are derived, the gravity center position, half-width, and maximum amplitude of the Gaussian function can be obtained by matching two parity-adjacent inflection points. After that, we further evaluate the extracted initial parameters of each Gaussian component. The steps are as follows:

   First, most of the inflection points extracted by using the aforementioned method are from random fluctuations of background noise rather than valid signals, so useless inflection points should be eliminated. We set the minimum Gaussian amplitude ($A_{\min}$) as:

$$A_{\min} = \varepsilon + k \cdot \sigma_{\text{noise}} \tag{4.36}$$

where ε is the mean of background noise of the LiDAR waveform, σ_{noise} is the standard deviation of background noise, and k is determined according to experience and is generally set as 3. The Gaussian peaks with amplitude less than $A_{\min}$ are removed. Only keep the Gaussian peaks larger than $A_{\min}$.

Second, we merge the Gaussian peaks whose adjacent peak interval is smaller than the transmitted pulse width. The specific steps are: (1) sort the Gaussian components by the area and remove the components whose area is less than or equal to 5% of the maximum component area and (2) merge the Gaussian component with the smallest area to its nearest component, until the number of Gaussian peaks is less than or equal to the set number of Gaussian components (usually set to 6). The commonly used method of merging Gaussian components is to merge based on the area weight, as in Equation (4.37).

$$\begin{cases} \text{Area}_{\text{new}} = \text{Area}_1 + \text{Area}_2 \\ A_{\text{new}} = \max(A_1, A_2) \\ \sigma_{\text{new}} = w_{t_1} * \sigma_1 + w_{t_2} * \sigma_2 \\ t_{\text{new}} = w_{t_1} * t_1 + w_{t_2} * t_2 \\ w_{t_1} = \dfrac{\text{Area}_1}{\text{Area}_2} \\ w_{t_2} = 1 - w_{t_1} \end{cases} \tag{4.37}$$

where w_{t_1} and w_{t_2} are the area weights of two components, and A_{new}, σ_{new}, and t_{new} are the amplitude, standard deviation, and gravity center position of the combined Gaussian peak, respectively.

The disadvantage of the inflection point method is poor performance in decomposing a superimposed waveform or weak waveform. Specifically, the superimposed waveform is composite signal from several objects with similar heights, which might be decomposed incorrectly into one object by using the inflection point method. The weak waveform signal is from the object with small cover or low reflectance within the footprint. With the inflection point method, it is difficult to distinguish weak signals from background noise, resulting in the lack of target components.

2. Wavelet transform method

Wavelet transform is a time-frequency, localized signal analysis method with a fixed window size and variable shape (Wang et al., 2013). It is regarded as the convolution process of the waveform and the wavelet function.

The continuous wavelet transform can be expressed as in Equations (4.38) and (4.39):

$$W_{P(a,b)} = \int_{-\infty}^{\infty} P(t)\psi_{a,b}(t)\mathrm{d}t \tag{4.38}$$

$$\psi_{a,b}(t) = \frac{1}{\sqrt{a}}\psi\left(\frac{t-b}{a}\right) \tag{4.39}$$

where $\psi(t)$ is the basic wavelet or mother wavelet, $\psi_{a,b}(t)$ is the wavelet function, and a and b are the scale factor and the position factor, respectively. $W_{P(a,b)}$ is the high-frequency approximation of the waveform $P(t)$ under the scale a, also called the wavelet coefficient. It indicates the similarity of the wavelet function at the position b under the scale a with the original waveform. The larger the wavelet coefficient, the higher the similarity.

Since the LiDAR waveform is a discrete signal, the discrete wavelet basis *sym6*, similar to the Gaussian shape, is used to perform wavelet transformation. Specifically, the original signal is decomposed into a low-frequency approximation signal and a high-frequency detail signal. The initial parameters (amplitude, standard deviation, gravity center location, etc.) of the decomposed Gaussian waveform can be derived from the reconstructed low-frequency approximation waveform through multiscale analysis. Figure 4.18 shows the wavelet transform results with different scales applied to the LiDAR waveform. It is seen that the small-scale wavelet transform derives two peaks and can correctly reconstruct the superimposed waveform, while the large-scale wavelet transform incorrectly decomposes the superimposed waveform and loses some detail information.

Compared with the inflection point method, multi-scale wavelet transform can effectively separate a superimposed waveform and identify a weak waveform. However, the wavelet transform is very sensitive to signal mutation, which might cause "misdetection" or "overdetection." In order to

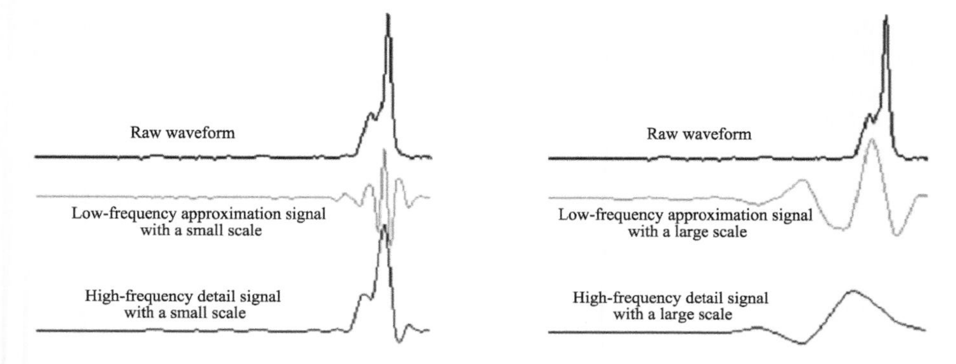

FIGURE 4.18 Wavelet transform results with different scales applied to the LiDAR waveform.

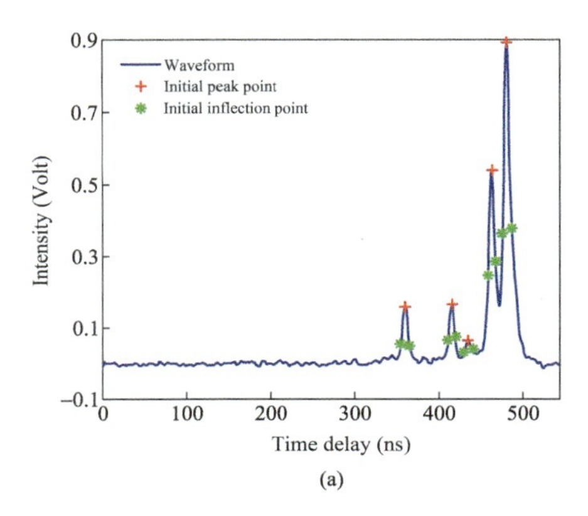

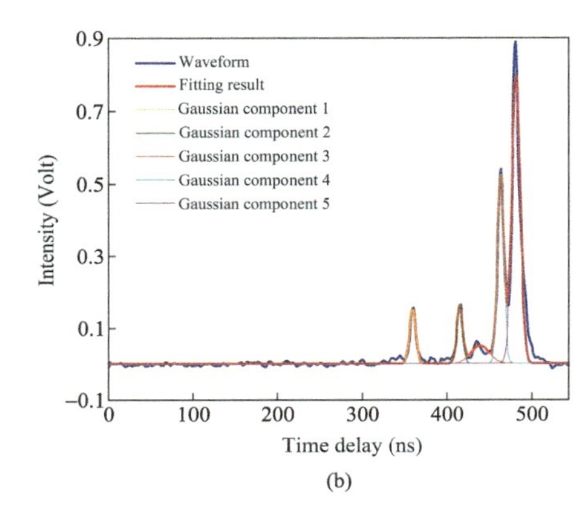

FIGURE 4.19 Waveform decomposition. (a) Initial parameter estimates; (b) LM nonlinear fitting.

avoid false detection of background noise as weak signals, it is necessary to denoise the waveform fully before the wavelet transform. Also, in the process of extracting object components, the detected components should be screened by setting a series of restrict conditions. For example, the components must be within the effective signal extent, and the component width must be greater than the emitted pulse width. For the detection of ground signal, in order to avoid overdetecting false components (targets that do not actually exist), constraints should also be put forward in the process of wavelet transform to filter out false components. Figure 4.19(a) displays the identified initial parameter of the LiDAR waveforms.

4.2.2.2 Nonlinear Fitting

To improve the fitting accuracy of the mixed Gaussian function, the estimated initial parameters need to be further optimized. The existing nonlinear fitting algorithms include Levenberg-Marquardt (LM), expectation maximization (EM), and least squares, among others. The LM method proposed by Levenberg is a non-linear fitting method between the inverse Hessian method and the gradient descent method (Marquardt, 1963), commonly used for waveform fitting. It has the advantages of both the gradient descent method and Gauss-Newton method and is similar to neural networks. The specific process is as follows.

Set the model to be fitted as:

$$y = y(x; a) \tag{4.40}$$

where $a = a_1, a_2, \ldots, a_m$, m is the number of parameters and y and x are variables with length of N. The best-fit parameters of the model can be derived by minimizing the value of χ^2:

$$\chi^2(a) = \sum_{i=1}^{N} \left[\frac{y_i - y(x_i; a)}{\sigma_i} \right] \tag{4.41}$$

The iteration rule is expressed as in Equation (4.42).

$$\boldsymbol{x}_{i+1} = \boldsymbol{x}_i - (\boldsymbol{H} + \lambda \boldsymbol{I})^{-1} \nabla \chi^2(a_i) \tag{4.42}$$

where $\boldsymbol{I}$ is the unity matrix, λ is the damping factor, $\boldsymbol{H} = \nabla^2 \chi^2(a_i)$ is the Hessian matrix, and $\nabla \chi^2(a_i)$ is the Jacobian matrix of partial differentiation of χ^2 with respect to vector a_i. In each iteration process, λ, affected by the estimated error χ^2, is continuously updated. In order to solve the classic problem of an "error valley," the LM algorithm replaces the unity matrix of the iteration rule [Equation (4.42)] with the diagonal matrix of the Hessian matrix $\boldsymbol{H}$, as in Equation (4.43), so that the estimated model value and gradient are used to calculate $\nabla^2 \chi^2(a_i)$ and $\boldsymbol{H}$.

$$\boldsymbol{x}_{i+1} = \boldsymbol{x}_i - (\boldsymbol{H} + \lambda \text{diag}[\boldsymbol{H}])^{-1} \nabla \chi^2(a_i) \tag{4.43}$$

In the iteration process, λ decreases with χ^2. As $\lambda \rightarrow 0$, Equation (4.43) is approximate to the Gauss-Newton equation of second-order local convergence. That is, as χ^2 is close to the minimum value, the LM method is converted to the Gauss-Newton method. As $\lambda \rightarrow \infty$, Equation (4.43) approximates the gradient descent equation of linear global convergence.

Assuming that the initial estimates of the fitting parameters a are known, the detailed process of the LM algorithm is as follows:

i. Calculate χ^2.
ii. Set the initial value for λ, such as =0.001.
iii. Solve the previous equations for parameter estimation and calculate χ^2 again.
iv. If $\chi^2(a_{i+1}) \geq \chi^2(a_i)$, increase λ with the appropriate factor multiples (such as 10), return to step (iii) to continue, or go to the next step.
v. If $\chi^2(a_{i+1}) < \chi^2(a_i)$, reduce λ with the appropriate factor multiples (such as 10) until $a_{i+1} \rightarrow a_i$; otherwise, return to step (iii).

For the LiDAR waveform decomposition, we substitute the estimated initial parameters into the LM algorithm. Then the mixed Gaussian function is fitted by minimizing the value of χ^2. Finally, the best-fitting parameters are obtained. To improve the fitting accuracy, we can set some constraints in the fitting process according to the actual situation. For example, the amplitude of the component must be greater than the minimum Gaussian amplitude; the fitted pulse width of the Gaussian component is not less than the emitted pulse width; and the minimum distance between adjacent peaks is 1.5 m. Additionally, to make the result meet the requirement of fitting accuracy, each Gaussian component can be added one by one according to the importance degree from high to low, until reaching the fitting accuracy. The pseudo-code of waveform decomposition algorithm based on inflection point method and LM nonlinear fitting is shown in Table 4.8. Figure 4.19(b) shows the LiDAR waveform decomposition result, in which the blue line is the original waveform and the red line is the fitted Gaussian waveform.

TABLE 4.8
Pseudo-Code of Waveform Decomposition Algorithm Based on Inflection Point Method and LM Nonlinear Fitting

Input:$w(t)$, LiDAR waveform; N_{max}, maximum number of Gaussian components,
Output: N, number of Gaussian components; amplitudes $\{A_i\}$, center locations $\{t_i\}$, and standard deviations $\{\sigma_i\}$ of component i; ε, average background noise,.
Algorithm:

```
ε_initial= NoiseMean(w(t));
std = NoiseStandardDeviation(w(t));
A_min =ε_initial + k × std;
# Recognize the turning points.
{turningPointLoc} = TurningPointLocation_Recognition(w(t));
# Generate the initial values of Gaussian parameters
n_initial = Count({turningPointLoc})/2;
for i = 1: n_initial
        t_i _initial = turningPointLoc_{2i-1}+ turningPointLoc_{2i};
        A_i _initial = w(t_i _initial);
        σ_i _initial = (turningPointLoc_{2i-1} - turningPointLoc_{2i})/2;
        if A_i _initial < A_min
                delete A_i _initial, t_i _initial, σ_i _initial;
        endif;
endfor;
# Merge Gaussian components
while Count(A_i _initial) > N_max
        MergeTwoComponents({A_i _initial},{t_i _initial},{σ_i _initial });
endwhile;
N = Count(A_i _initial)
# Nonlinear fitting
f(t) = sum(A_i×exp(-(t-t_i)×(t-t_i)/2/σ_i/σ_i))+ε;
{A_i, t_i, σ_i, ε} = LevenbergMarquardt(w(t), f(t), {A_i _initial}, {t_i _initial}, {σ_i _initial }, ε_initial);
```

4.2.3 Waveform Deconvolution

Waveform deconvolution is a waveform processing method that removes system effects to obtain the vertical distribution of the ground feature. In the time domain, the received waveform can be represented as a convolution of the transmit pulse and the vertical distribution of the ground feature (Wu et al., 2011). The mathematical expression is as in Equation (4.44):

$$w(t) = h(t) * w_x(t) \tag{4.44}$$

where $w_x(t)$ is the transmit pulse, $h(t)$denotes the surface vertical distribution, $w(t)$ is the received waveform, and * denotes the convolution operator.

Since the sensor itself records the received waveform and the transmit pulse, the surface vertical distribution can be obtained by waveform deconvolution.

Transformed to the frequency domain, Equation (4.44) is expressed as Equation (4.45).

$$W(f) = H(f)W_x(f) \tag{4.45}$$

where $W(f)$, $H(f)$, and $W_x(f)$ are the representations of the received waveform, the surface vertical distribution, and the emitted waveform in the frequency domain, respectively.

Without system noise, the surface vertical distribution can be derived according to the previous equation. However, the noise exists in both the transmitted and received waveforms. The noise would affect the estimation of the surface vertical distribution. To solve this problem, noise filtering must be performed first. Wiener filtering is one of the methods, as shown in Equation (4.46) (Jutzi & Stilla, 2006):

$$W_i(f) = \frac{|W(f)|^2}{|W(f)|^2 + |N(f)|^2} \tag{4.46}$$

The surface vertical distribution is derived by Equation (4.47).

$$H(f) = \frac{W(f)}{W_x(f)} \times W_i(f) \tag{4.47}$$

where $N(f)$ is the representation of the noise in the frequency domain and $H(f)$ is the representation of the surface vertical distribution in the frequency domain. The surface vertical distribution in the time domain is derived by inverse fast Fourier transform. Figure 4.20 shows an example of waveform deconvolution.

Compared to waveform decomposition, the waveform deconvolution does not require a priori knowledge and does not require an estimation of initial values. However, it is very sensitive to noise and requires high waveform quality.

4.2.4 Waveform Feature Parameter Extraction

The waveform feature parameters are metric parameters indicating the vertical distribution characteristics of the signal, which is calculated from the LiDAR waveform. The extraction process is that, based on the waveform decomposition result, the waveform components are determined to correspond to objects within the footprint, and then the waveform feature parameters are extracted to further estimate the Earth's surface features. The commonly used waveform feature parameters include waveform quantile height, waveform height indices, and waveform energy indices.

4.2.4.1 Waveform Quantile Height

To calculate the waveform quantile height requires one to first calculate the total effective signal energy between the beginning of the signal and the ground peak position. Then accumulate the signal energy from the ground surface position. Calculate the ratio of the accumulated energy at a certain position to the total effective signal energy. Determine the height of the position. The distance between the position and the ground surface is just the waveform quantile height. The position of the

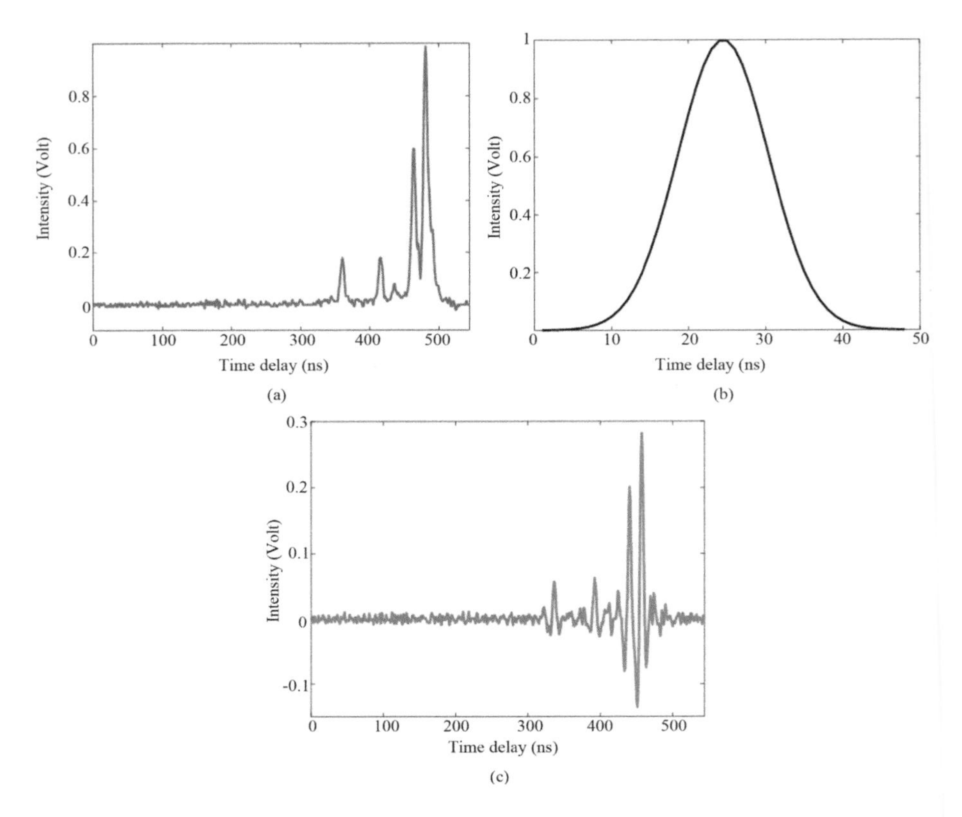

FIGURE 4.20 Waveform deconvolution. (a) Raw received waveform; (b) emitted waveform; (c) deconvoluted waveform.

ground surface is determined by the last waveform peak position derived from the waveform decomposition. Sun et al. (2008) recorded the height of the position from the ground surface when the accumulated energy reaches 25% of the total effective signal energy as H_{25}, which represents the quantile height of 25%. The quantile heights H_{25}, H_{50}, H_{75}, H_{100} are as shown in Figure 4.21. H_{100} is the height between the beginning of the signal and the last Gaussian peak, usually used to estimate the maximum tree height within the footprint. H_{50}, also named height of median energy (HOME), is sensitive to canopy vertical distribution and canopy closure. It has been used as an important parameter for forest biomass retrieval (Drake et al., 2002).

4.2.4.2 Waveform Height Indices

The waveform height indices include the waveform height, waveform distance, peak distance, leading edge extent, trailing edge extent, etc.

1. Waveform height

   All bins of the LiDAR waveform are not all valid signals. The initial part of the signal is from system noise, solar energy, clouds, and fog and does not

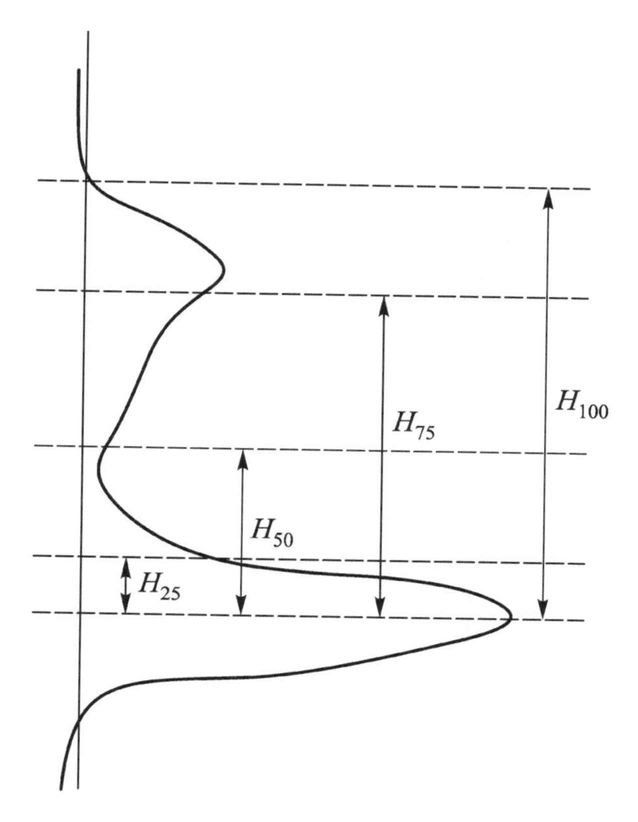

FIGURE 4.21 Waveform quantile height parameters.

contain the returns of surface objects. Before processing and analyzing the LiDAR waveform, it is necessary to determine the valid signal extent (Harding & Carabajal, 2005). That is the waveform height, w_{Echo} in Figure 4.22.

In practice, the beginning position of the valid signal (p_{Beg}) is considered the first bin whose voltage value exceeds a certain threshold, counting from the first bin of the LiDAR waveform in positive order. The threshold is usually set as the mean value of background noise plus N times the standard deviation of the noise. The end position of the valid signal (p_{End}) is the first bin whose voltage value exceeds the threshold, counting from the last bin of the LiDAR waveform in reverse order. The waveform height is calculated by the beginning and end positions of the valid signal, as in Equation (4.48):

$$w_{\text{Echo}} = p_{\text{End}} - p_{\text{Beg}} \tag{4.48}$$

2. Waveform distance

Waveform distance (d_{Echo}) is the distance from the beginning position of the valid signal to the last Gaussian peak (Sun et al., 2008) (Figure 4.22).

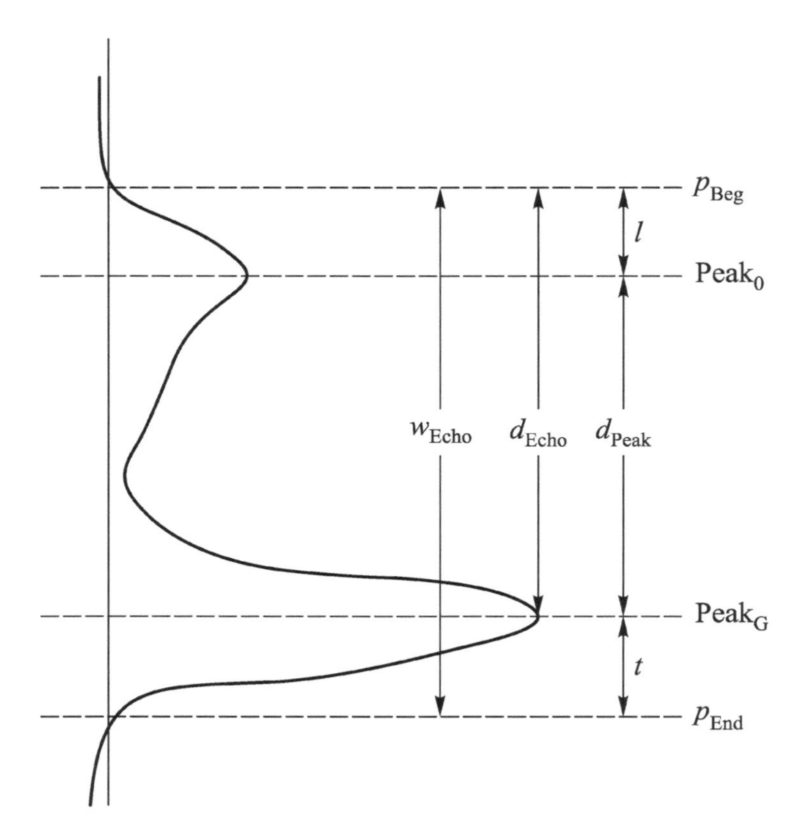

FIGURE 4.22 Waveform height indices.

The waveform distance can characterize the canopy height. It is calculated as Equation (4.49).

$$d_{\text{Echo}} = \text{Peak}_{\text{G}} - p_{\text{Beg}} \tag{4.49}$$

where Peak_{G} is the position of the last Gaussian peak and p_{Beg} is the beginning position of the valid signal.

3. Peak distance

   The peak distance (d_{Peak}) is the distance between the first and last Gaussian peaks after the LiDAR waveform is decomposed (Wang et al., 2013), as in Figure 4.22. The peak distance generally has a strong correlation with the vegetation height. The calculation equation is:

$$d_{\text{Peak}} = \text{Peak}_{\text{G}} - \text{Peak}_{0} \tag{4.50}$$

where Peak_{G} is the center position of the ground Gaussian peak (usually regarded as the last peak) and Peak_{0} is the center position of the first Gaussian peak.

4. Leading edge extent

   The leading edge extent (l) is the distance between the first Gaussian peak and the beginning position of the valid signal (Lefsky et al., 2007), as in "l" in Figure 4.22. Its calculation equation is:

$$l = \text{Peak}_0 - p_{\text{Beg}} \tag{4.51}$$

   where Peak_0 is the center position of first Gaussian peak after waveform decomposition and p_{Beg} is the beginning position of the valid signal. This parameter indicates the coupling influence of vegetation canopy and terrain fluctuations on the LiDAR waveform.

5. Trailing edge extent

   The trailing edge extent (t) is the distance between the ground Gaussian peak and the end position of the valid signal (Lefsky et al., 2007), as in "t" in Figure 4.22. Its calculation equation is:

$$t = p_{\text{End}} - \text{Peak}_{\text{G}} \tag{4.52}$$

   where Peak_{G} is the center position of the last Gaussian peak after waveform decomposition and p_{End} is the end position of the valid signal. This parameter indicates the influence of terrain slope and roughness on the LiDAR waveform.

4.2.4.3 Waveform Energy Indices

The total returned waveform energy (e_{Echo}) represents the waveform area from the beginning position of the valid signal to the end position (Figure 4.23). The relative waveform energy (r_{Echo}) is the ratio of the returned waveform energy to the emitted laser energy. It characterizes the surface reflectivity within the footprint. The ground echo energy (e_{Ground}) represents the accumulated value of the ground waveform energy. The canopy echo energy (e_{Canopy}) is the part of the echo energy minus the ground energy (Figure 4.23).

The surface echo energy ratio (r_{Ground}) is the ratio of the ground echo energy to the canopy echo energy (Drake et al., 2002), as in Equation (4.53). The canopy echo energy ratio (r_{Canopy}) is the proportion of the canopy echo energy to the total echo energy [Equation (4.54)]. These indicators represent the canopy closure within the footprint to a certain extent. They can be used to explore the quantitative relationships between LiDAR waveform parameters and forest canopy characteristics (Harding & Carabajal, 2005).

$$r_{\text{Ground}} = \frac{e_{\text{Ground}}}{e_{\text{Canopy}}} \tag{4.53}$$

$$r_{\text{Canopy}} = \frac{e_{\text{Canopy}}}{e_{\text{Echo}}} \tag{4.54}$$

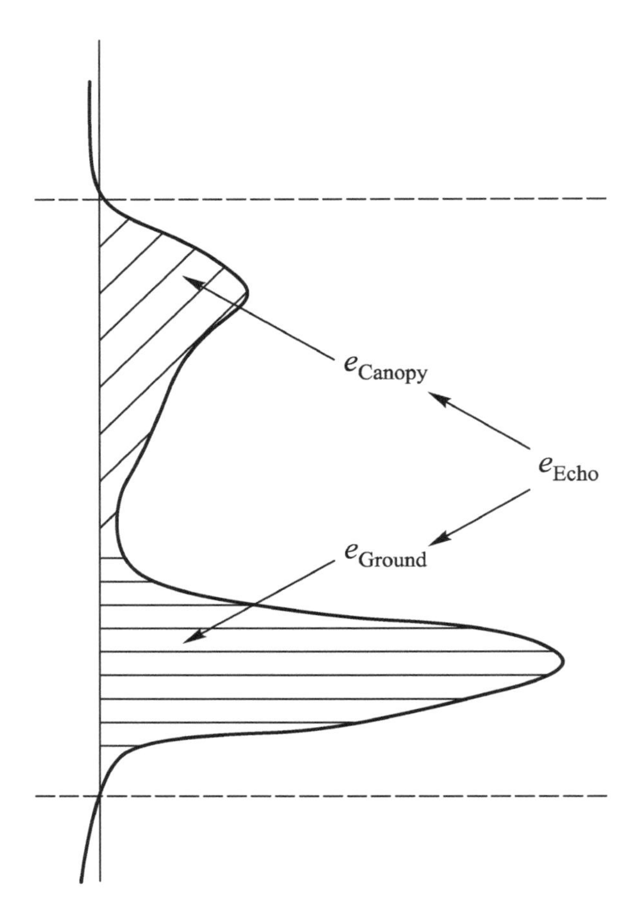

FIGURE 4.23 Waveform energy indices.

4.3 LiDAR PHOTON-COUNTING PROCESSING

The transmitted and received signals of the photon-counting LiDAR system are weak signals, which are greatly affected by noise (solar background noise, system noise, atmospheric scattering noise, etc.). Figure 4.24 shows the spatial distribution of ICESat-2/ATLAS photon-counting data for different land covers (forest, glaciers, lakes). It is seen that the spatial distribution of photon noise is random and wide, which is a huge challenge for photon-counting LiDAR data processing and applications. Photon denoising and photon classification are two key steps in photon counting LiDAR data processing. In areas such as glaciers, sea ice, and lakes, the signal points obtained by the photon denoising algorithm are ground signal photons. In contrast, in the vegetation-covered area, the denoised photon points need to be further classified to distinguish the ground and non-ground photon points.

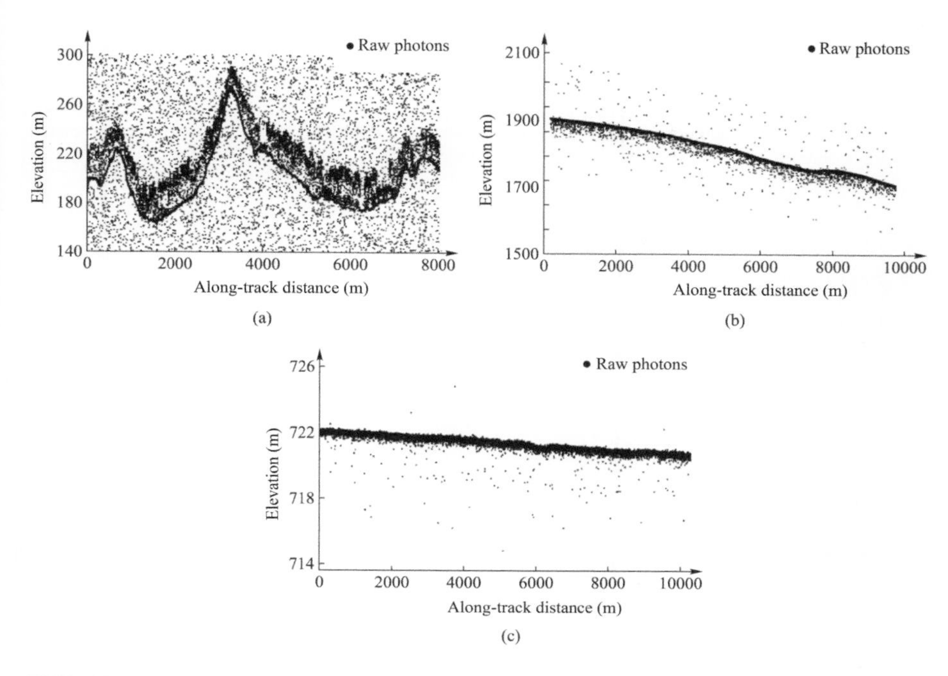

FIGURE 4.24 Examples of ICESat-2/ATLAS photon-counting distribution in (a) a forest area of California in the United States, (b) the Greenland ice sheet, and (c) Qinghai-Tibet Plateau lake.

4.3.1 Photon-Counting Denoising

This section introduces an effective noise removal algorithm based on localized statistical analysis (Zhu et al., 2018a). There are three key steps to implement this algorithm. The first step is to build an elevation frequency histogram for removing the apparent noise photons. The second step is to use the Ordering Points to Identify the Clustering Structure (OPTICS) algorithm to calculate the density of each photon and to achieve fine denoising based on the rule that the signal photon points are densely distributed and the noise photon points are sparsely distributed. The final step is to remove residual noise photons based on the photon elevation distribution features. An overview of the three steps is shown in Figure 4.25.

4.3.1.1 Coarse Denoising Based on Elevation Frequency Histogram

Coarse denoising roughly removes noise and determines the approximate elevation range of the photon signal. The photon noise is distributed in the entire vertical section [Figure 4.26(a)]. In order to initially determine the elevation range of the photon signal and reduce the data amount, a coarse denoising method based on the elevation frequency histogram is used. First, the original photon points are divided into grids, and an elevation frequency histogram is established based on a certain window size along the track direction. Then, the peak position of each histogram is obtained and used as the center height of the photon signal. Finally, a buffer centered

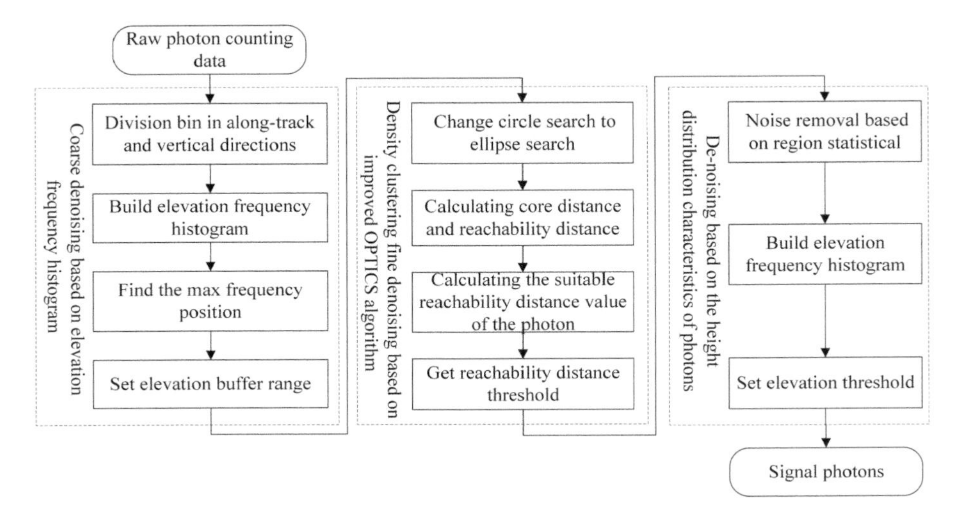

FIGURE 4.25 Photon-counting denoising flowchart.

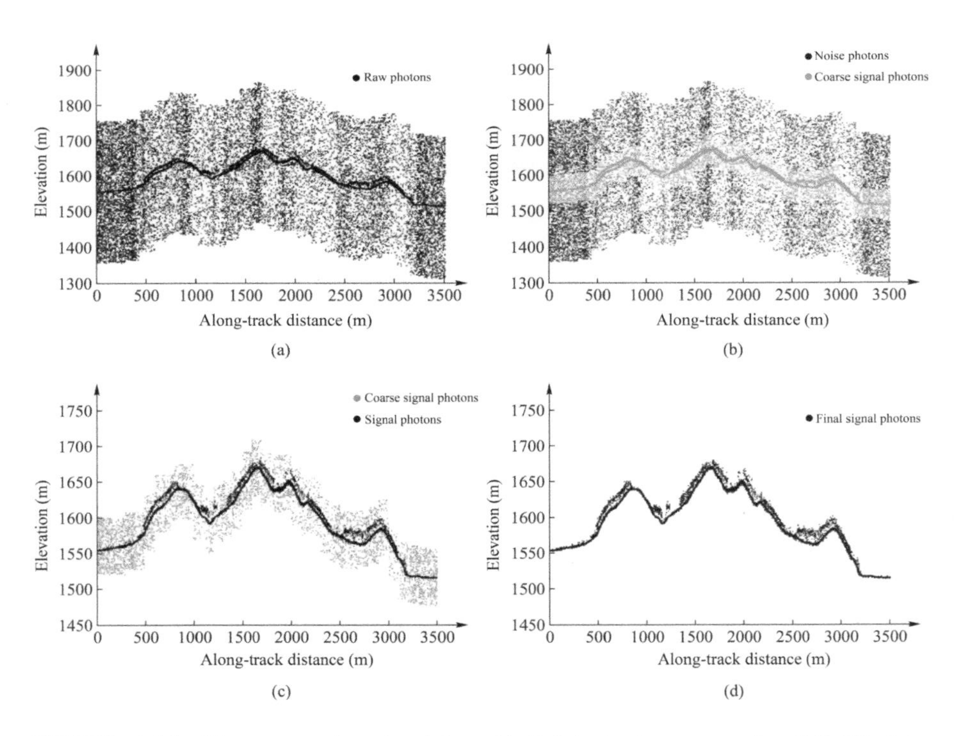

FIGURE 4.26 Photon-counting denoising. (a) Original photon-counting LiDAR data; (b) coarse denoising result; (c) fine denoising result; (d) final signal photons.

on the center height and whose length is the elevation threshold (depending on the maximum height of the objects in the study area) is established to remove most of the noise points. The coarse denoising result is shown in Figure 4.26(b).

4.3.1.2 Fine Denoising Based on Improved OPTICS Density Clustering

The coarse denoising method based on the elevation frequency histogram can remove most of the noise photons, but there are still many noise photons remaining in the photon-counting data. In order to further eliminate the noise photons, it needs to be finely denoised. Here we introduce a fine denoising method based on the improved OPTICS density clustering (Zhu et al., 2020).

The OPTICS density clustering algorithm (Ankerst et al., 1999) clusters the data in the space according to the density distribution. Its idea is very similar to the DBSCAN algorithm (Ester et al., 1996). The difference is that OPTICS does not explicitly generate data clustering, but sorts the objects in the dataset to obtain an ordered list of objects. Through this list, the clusters with an arbitrary density can be obtained. Therefore, the algorithm is not sensitive to the input parameter neighborhood radius (ε), as long as the value of the minimum number of points in the neighborhood (MinPts) is determined. The steps of fine denoising based on the improved OPTICS density clustering algorithm are as follows:

1. Given that there is an apparent difference in the distribution of photons in the horizontal and vertical directions, the signal photons and noise photons cannot be effectively distinguished by the circular search. Hence, the circular search in the OPTICS algorithm is improved to an elliptical search. The equation for calculating the distance between points $q(x_q, z_q)$ and $o(x_o, z_o)$ is:

$$\text{dist}(q,o) = \sqrt{\frac{(x_q - x_o)^2}{a^2} + \frac{(z_q - z_o)^2}{b^2}} \tag{4.55}$$

   where x is the along-track distance, z is the elevation value, and a and b are the major and minor axes of the elliptical search area, respectively.
2. Calculating the core distance (CD) [Equation (4.56)] and reachability distance (RD) [Equation (4.57)] of each photon point based on the input parameters ε and MinPts. Figure 4.27 illustrates the core and reachability distances for the center point "o."

$$\text{CD}_{\varepsilon,\text{MinPts}}(o) = \begin{cases} \text{undefined}, & N_\varepsilon(o) < \text{MinPts} \\ \varepsilon', & N_\varepsilon(o) \geq \text{MinPts} \end{cases} \tag{4.56}$$

$$\text{RD}_{\varepsilon,\text{MinPts}}(o) = \begin{cases} \text{undefined}, & N_\varepsilon(o) < \text{MinPts} \\ \max\left(\text{CD}_{\varepsilon,\text{MinPts}}(o), \text{dist}(q,o)\right), & N_\varepsilon(o) \geq \text{MinPts} \end{cases} \tag{4.57}$$

   where ε is the neighborhood radius, MinPts is the minimum number of photons in the ε neighborhood, and $N_\varepsilon(o)$ is the photon number of the center point "o" in the ε neighborhood.

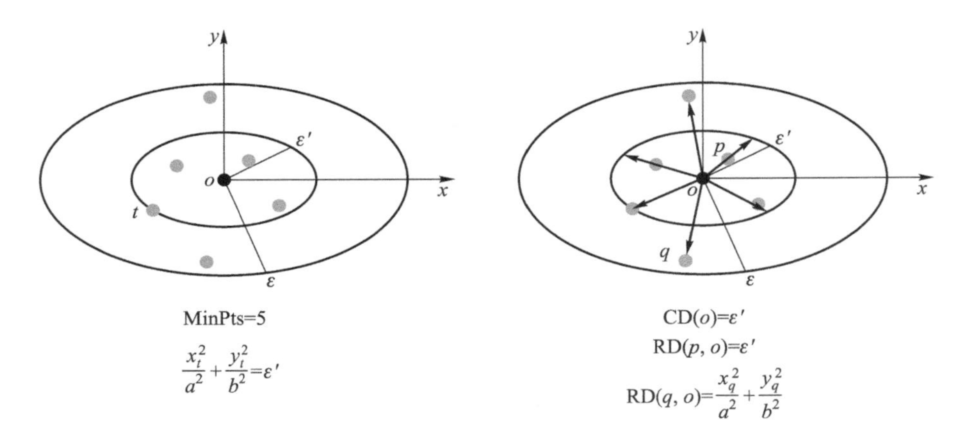

FIGURE 4.27 The core distance and the reachability distance of the center point "o" calculated in the process of an ellipse search.

3. Considering that a photon might be directly reachable by many core photons and have more than one RD value, the minimum RD value of each photon is set to the optimal RD value.
4. Signal photons are distributed in a more concentrated manner than noise photons. A photon with a high RD value indicates that it falls away from other photons, so this photon would be a noise photon. In contrast, a photon with a low RD value is regarded as a signal photon. For each window, all photons are first sorted by their RD values. Then, we derive the threshold of optimal RD values in each window by using the OTSU method. If the RD value of the photon point is greater than the threshold, the photon is marked as a noise photon; otherwise, it is marked as a signal photon. The basic principle of the OTSU method is described by Equation (4.58):

$$\begin{cases} w_0(t) = \dfrac{t}{N} \\ w_1(t) = \dfrac{N-t}{N} \\ \mu_0(t) = \dfrac{\sum_{i=1}^{t} \mathrm{RD}(i)}{t} \\ \mu_1(t) = \dfrac{\sum_{i=t+1}^{N} \mathrm{RD}(i)}{N-t} \\ \mu(t) = w_0(t) \times \mu_0(t) + w_1(t) \times \mu_1(t) \\ \sigma^2(t) = w_0(t) \times (\mu_0(t) - \mu(t))^2 + w_1(t) \times (\mu_1(t) - \mu(t))^2 \end{cases} \tag{4.58}$$

where N is the number of photons in each window, t is the serial number of the sorted photons, $w_0(t)$ is the weight of the first t photons to the total number of photons, $w_1(t)$ is the weight of the remaining $N-t$ photons to the total number of photons, $\mu_0(t)$ is the average reachability distance of the first t photons, $\mu_1(t)$ is the average reachability distance of the remaining $N-t$ photons, $\mu(t)$ is the average reachability distance of all photons, and $\sigma^2(t)$ is the variance of the reachability distance of photons.

Figure 4.26(c) shows the fine denoising results based on the improved OPTICS algorithm. It is observed that the algorithm effectively distinguishes the noise and signal photons. Table 4.9 shows the pseudo-code of the noise photon removal algorithm.

4.3.1.3 Denoising Based on Photon Elevation Distribution Features

Through the previous steps, most of the noise photons are removed, but some noise points remain under the ground, in the air, near the canopy surface, or near the ground. They can be removed by combining the characteristics of the elevation distribution. That is, the photon elevation in a certain size window conforms to the normal distribution. In this case, the method of interval estimation is used to eliminate the photon noise near the top of canopy and near the ground by setting the confidence interval. The final signal photons are shown in Figure 4.26(d).

4.3.2 Photon-Counting Classification

The photon denoising algorithm can extract the signal photons. However, to retrieve the surface structures such as canopy height in a forest, the signal photons need to be classified into ground and canopy photons. This section introduces a photon classification algorithm based on moving curve fitting (Zhu et al., 2018a). It classified the denoised photons into ground photons, canopy surface photons, canopy photons, and residual noise photons. The specific process is shown in Figure 4.28.

4.3.2.1 Extraction of Ground Photons

To improve the accuracy of ground photon extraction in areas with a complex terrain or dense vegetation coverage, the ground photon is extracted through four steps: initial ground photon extraction, pseudo ground photon removal, ground photon densification, and ground photon fitting. The specific process is shown in Figure 4.29.

1. Initial ground photon extraction. In order to reduce the influence of near-ground noise photons on the extraction of ground photons, an elevation frequency histogram is established based on the photons in the moving window. The peak position with the lowest elevation is determined. The elevation of the lowest peak is compared with the elevation of the lowest photon point. If their elevation difference is less than a certain threshold (the threshold is set adaptively according to the terrain), the peak is designated as the ground peak. The photon points with the highest density and within a certain elevation range of the ground peak are selected as the initial ground photons. Figure 4.30(a) shows the extraction results of the initial ground photons. It can be seen that the initial ground photons still contain "pseudo ground photons" which are actually vegetation canopy photons and near-ground photons.

TABLE 4.9
Pseudo-Code of the Noise Photon Removal Algorithm

Input: ***P***, original photons; ***s_along***, grid size in the along-track distance; ***b_elevation***, Elevation buffer of signal photons; ***eps***, Ellipse search radius, and the long and short half-axis values of the ellipse search area are ***a*** and ***b***, respectively; ***min_pts***, the minimum number of photons in an elliptical neighborhood;
Output: ***P_Signal***, signal photons; ***P_Noise***, noise photons;
Algorithm:

```
import original photons P;
construct grid G of point set P using grid size s_along;
for g_j in G
        constructing elevation frequency distribution histogram of g_j;
        obtain the peak elevation P_elevation of the frequency histogram;
        # Obtain initial signal photons
        for p_i in g_j
            if (p_i.z >= P_elevation – b_elevation and p_i.z <= P_elevation + b_elevation)
                    place p_i to initial signal photon set O;
            else
                    save p_i to P_Noise;
            end if;
        endfor;
        # Calculate the core distance (CD) and reachability distance (RD) of each photon by the OPTICS algorithm
        for o_i in g_j
            find all photons Q within the elliptical neighborhood of o_i;
            if Q.size < min_pts
                    CD(o_i) = UNDEFINED;   RD(o_i) = UNDEFINED;
            else
                    CD(o_i) = eps′; RD(o_j) = max(CD(o_i), dist(q, o_i));
            endif;
         endfor;
         # Calculate the RD_threshold of grid g_j using the OTSU algorithm;
         RD_threshold = OTSU(RD);
         for o_i in g_j
            if (RD(o_i) > RD_threshold)
                    save o_i to P_Noise;
            else
                    save o_i to P_Signal;
            endif;
         endfor;
endfor;
return P_Noise, P_Signal;
```

2. Pseudo–ground photon removal. In the process of initial ground photon extraction, near-ground photons, vegetation canopy photons, and noise photons might be misclassified as initial ground photons. These pseudo ground photons are further removed by the EMD method (Kopsinis &

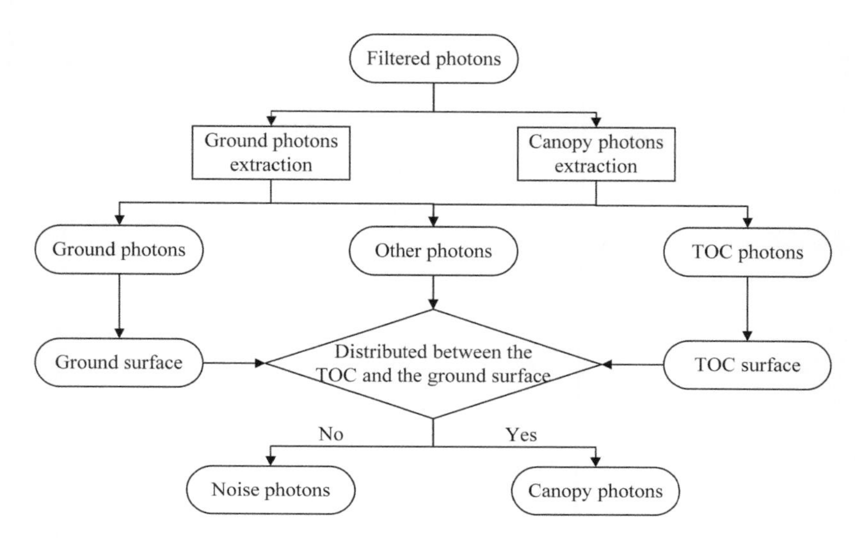

FIGURE 4.28 Photon-counting classification flowchart.

McLaughlin, 2009). First, the initial ground photons are regarded as the original signal. The EMD method decomposes the original signal into several intrinsic mode components and residual components with different frequencies [Equation (4.59)]. Then the OTSU method is used to determine the *k* value of Equation (4.61) to further divide the intrinsic mode components into low-order intrinsic mode components (high-frequency noise) and high-order intrinsic mode components (low-frequency signals). Next, the low-order intrinsic mode components are processed by the set threshold [Equation (4.60)]. Finally, the high-order and processed low-order intrinsic mode components are used for signal reconstruction [Equation (4.61)].

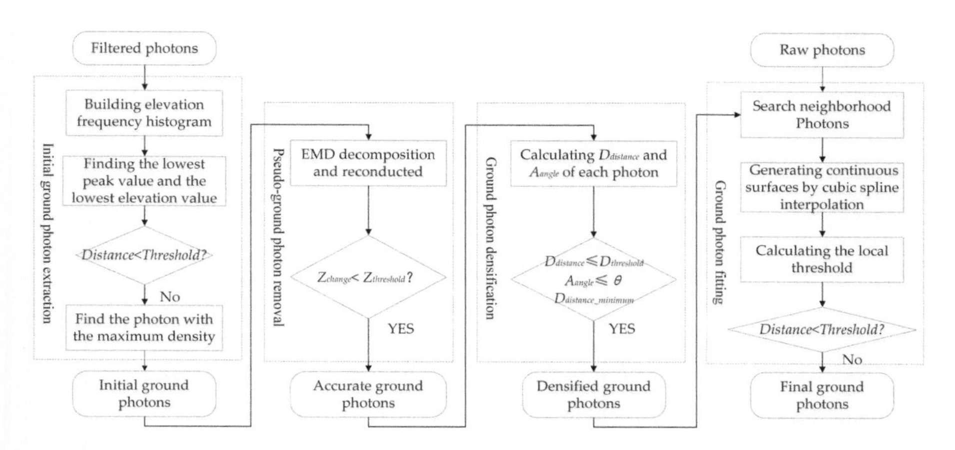

FIGURE 4.29 Ground photon extraction flowchart.

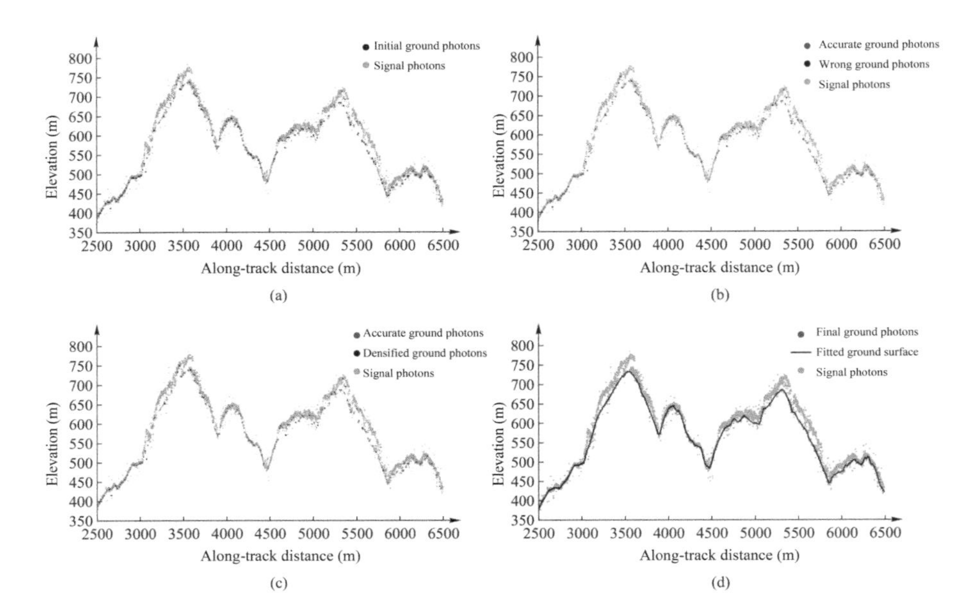

FIGURE 4.30 Extraction of ground photons. (a) Initial ground photon extraction. (b) Fine ground photon extraction. (c) Ground photon densification. (d) Ground fitting result.

The pseudo ground photons are removed by comparing the elevation values of the original and reconstructed signals. The final result of ground photon extraction is shown in Figure 4.30.

$$z(x)=\sum_{i=1}^{n}\mathrm{IMF}_i(x)+r_n(x) \tag{4.59}$$

$$\begin{cases} \mathrm{IMF}_i'(j)=\begin{cases}\mathrm{IMF}_i(j) & \mathrm{IMF}_i(j)\geq \mathrm{th}_i \\ 0 & \mathrm{IMF}_i(j)<\mathrm{th}_i\end{cases} \\ \mathrm{th}_i=\sigma_i\sqrt{2\ln N}=\dfrac{\mathrm{median}(\mathrm{abs}(\mathrm{IMF}_i))}{0.6745}*\sqrt{2\ln N} \end{cases} \tag{4.60}$$

$$z'(x)=\sum_{i=1}^{k}\mathrm{IMF}_i'(x)+\sum_{i=k+1}^{n}\mathrm{IMF}_i(x)+r_n(x) \tag{4.61}$$

where $\mathrm{IMF}_i(x)$ is the intrinsic mode component, $r_n(x)$ is the residual component, $z(x)$ is the original signal, $z'(x)$ is the reconstructed signal, th_i is the intrinsic mode component threshold, n is the number of intrinsic mode components, and N is the photon number of the original signal.

3. Ground photon densification. The progressive triangle densification method is used to increase the number of ground photons to improve the photon

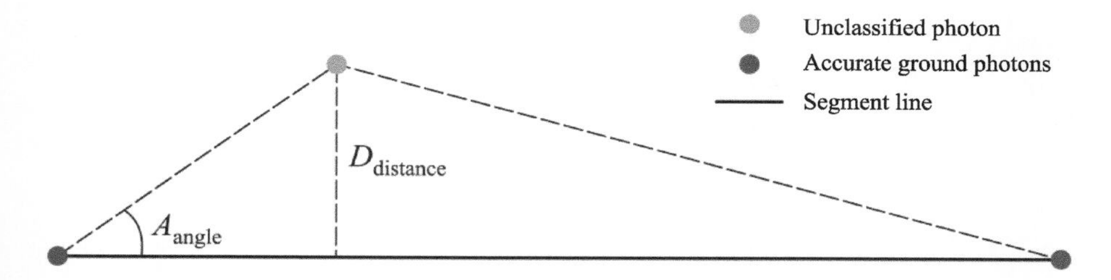

FIGURE 4.31 Ground photon densification.

classification accuracy in areas with complex terrain or dense vegetation coverage. Specifically, two parameters of the target photon (i.e., photon to be classified) are calculated to achieve ground photon densification. The one parameter is $D_{distance}$, the distance from the target photon to the corresponding ground segment line (Figure 4.31). The other parameter is A_{angle}, the angle between the ground segment line and the segment line comprising the target photon and the closest accurate ground photon (Figure 4.31). If the $D_{distance}$ of the target photon is less than the distance threshold d, the A_{angle} is less than the angle threshold θ, and its $D_{distance}$ is minimum, the target photon is marked as the ground photon. The result of ground photon densification is shown in Figure 4.30(c).

4. Ground photon fitting. To obtain the complete ground photons and generate the ground surface, the ground photons are fitted with cubic splines. First, a proximity search is conducted for each ground photon to obtain a limited number of ground photons that are closest to it. Then, the ground photons are fitted with a local cubic spline curve. The cubic spline curve coefficients are solved based on the least squares algorithm. Finally, a certain height difference threshold is set to achieve the separation of ground photons and non-ground photons. The final result of ground fitting is shown in Figure 4.30(d).

4.3.2.2 Extraction of Canopy Photons

Canopy surface identification is a prerequisite for canopy photon extraction. Through the steps of near-canopy surface noise photon removal, canopy surface photon extraction, and canopy surface fitting, an accurate canopy surface is generated.

1. Near-canopy surface noise photon removal. After denoising, the remaining photons might contain near-canopy surface noise photons, which affect the accurate extraction of canopy surface photons. In order to remove noise photons near the canopy surface, the denoised photons are first divided into windows, and then the photon elevation quantiles are calculated based on all the denoised photons in the window. Previous studies (Popescu et al., 2018) demonstrate that the photons corresponding to the photon elevation quantile intervals [0.96, 1] and [0.99, 1] during the day and night are very likely to be noise photons. Hence, the photons in this quantile interval are

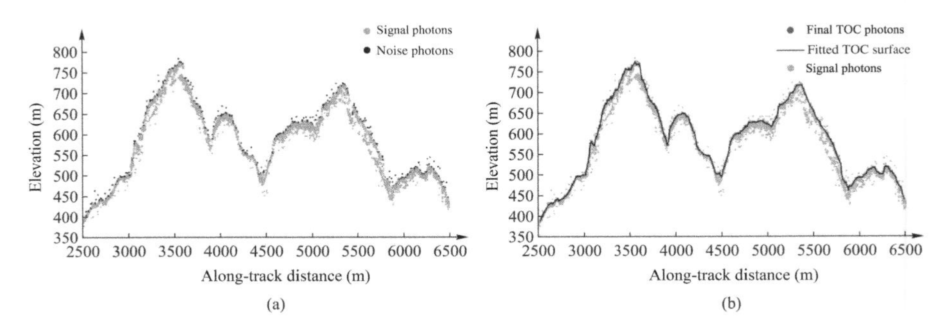

FIGURE 4.32 Canopy photon extraction. (a) Near-canopy surface noise photon removal. (b) Result of TOC surface fitting.

eliminated. The noise removal result of near-canopy surface is shown in Figure 4.32(a).

2. Canopy surface photon extraction. Since the denoised photons might contain isolated noise photons, the elevation quantiles are recalculated based on the denoised signal photon. The photons corresponding to the photon elevation quantile interval [0.96, 0.99] are extracted and regarded as the canopy surface photons.
3. Canopy surface fitting. The surface of the top of canopy (TOC) is derived by fitting the extracted canopy surface photons with a cubic spline curve. The fitting result of the TOC surface is shown in Figure 4.32(b). Table 4.10 shows the pseudo-code of the signal photon classification algorithm.

TABLE 4.10
Pseudo-Code of Signal Photon Classification Algorithm

Input: ***P***, denoised photons; ***s_along***, grid size in the along-track direction; ***initial_threshold***, initial ground photons judgment threshold; ***EMD_threshold***, EMD algorithm reconstruction threshold; ***d***, distance threshold; ***θ***, angle threshold; ***ground_threshold***, ground photons judgment threshold; ***toc_threshold***, top of canopy photons judgment threshold;

Output: ***P_ground***, ground photons; ***P_toc***, top of canopy photons; ***P_canopy***, canopy photons;

Algorithm:

import denoised photons ***P***;

construct grid *G* of point set *P* using grid size ***s_along***;

Obtain initial ground photons and top of canopy photons

for g_j in *G*

construct elevation frequency distribution histogram of g_j;

obtain the lowest peak of the frequency histogram, *Peak_min*;

obtain the lowest elevation of photons in g_j, *z_min*;

obtain photons with elevation percentiles ranging from 0.96 to 0.99 and place them in the top of canopy photons set TOC;

if (*Peak_min*.z - *z_min*) < ***initial_threshold***

find the photon with the maximum photon density around the *Peak_min* and place it as the initial ground photonsset IG;

(Continued)

TABLE 4.10 *(Continued)*
Pseudo-Code of Signal Photon Classification Algorithm

endif;

endfor;

Obtain accurate ground photons

develop EMD decomposition and reconstruction using all initial ground photons IG;

for ig_i in IG

if abs(ig_i.z-ig_i.z')< ***EMD_threshold***

place this photon as the accurate ground photons set AG;

endif;

endfor;

construct grid R of point set P utilizing the along-track distance of AG;

for r_j in R

for p_i in r_j

calculate the $D_{distance}$ and A_{angle} between p_j and adjacent accurate ground photon;

if $D_{distance}$<=***d*** and A_{angle}<=$\boldsymbol{\theta}$

place the photon with the smallest $D_{distance}$ to AG;

endif;

endfor;

endfor;

Obtain *ground_surface* and *toc_surface* using Cubic Spline Curve Fitting algorithm.

ground_surface = *cubicsplinefitting* (AG);

toc_surface = *cubicsplinefitting* (TOC);

for p_i in P

if (dist (p_i.z, *ground_surface*)<=***ground_threshold***)

save p_i to ***P_ground***;

elseif (dist (p_i.z, *toc_surface*)<=***toc_threshold***)

save p_i to ***P_toc***;

elseif (p_j.z > *ground_surface* and p_j.z < *toc_surface*)

save p_i to ***P_canopy***;

endif;

endfor;

return *P_ground*, *P_toc*, *P_canopy*;

4.4 SUMMARY

LiDAR data processing is the basis for subsequent applications. The processing accuracy directly affects the quantitative application effects. This chapter introduces in detail the key processing steps of three kinds of LiDAR data (i.e., discrete point cloud, full waveform, and photon counting) and discusses point cloud denoising, point cloud filtering, point cloud classification, waveform denoising, waveform decomposition, waveform deconvolution, waveform feature parameter extraction, photon denoising, and photon classification.

EXERCISES

1. Briefly introduce the similarities and differences between typical point cloud noise and atypical point cloud noise and explain their causes.
2. Present three algorithms for LiDAR point cloud filtering and introduce their advantages, disadvantages, and adaptability to complicated terrain.
3. Introduce the core idea and the algorithm steps of LiDAR waveform decomposition.
4. List at least ten LiDAR point cloud features, write down their calculation equations, and briefly explain the meaning of these features.
5. List at least ten LiDAR waveform features, write down their calculation equations, and briefly explain the meaning of these features.
6. Explain the similarities and differences between machine learning and deep learning in LiDAR point cloud classification.
7. Briefly describe the process of LiDAR photon counting denoising and the significance of each step in the process.
8. Briefly describe the process of LiDAR photon counting classification and the significance of each step in the process.

REFERENCES

Ankerst, M., Breunig, M. M., Kriegel, H. P., & Sander, J. (1999). OPTICS: Ordering points to identify the clustering structure. *ACM SIGMOD Record*, *28*(2), 49–60.

Bartels, M., & Wei, H. (2010). Threshold-free object and ground point separation in LiDAR data. *Pattern Recognition Letters*, *31*(10), 1089–1099.

Bassier, M., Bonduel, M., Genechten, B. V., & Vergauwen, M. (2017). Segmentation of large unstructured point clouds using octree-based region growing and conditional random fields. ISPRS - International Archives of the Photogrammetry, Remote Sensing and Spatial Information Sciences, XLII-2/W8, 25–30.

Boudraa, A. O., Cexus, J. C., & Saidi, Z. (2013). EMD-based signal noise reduction. *Proceedings of World Academy of Science Engineering and Technology*, *1*(1), 33–37.

Breiman, L. (2001). Random forests. *Machine Learning*, *45*(1), 5–32.

Chen, C., Li, Y., Li, W., & Dai, H. (2013). A multiresolution hierarchical classification algorithm for filtering airborne LiDAR data. *ISPRS Journal of Photogrammetry and Remote Sensing*, *82*, 1–9.

Comaniciu, D., & Meer, P. (2002). Mean shift: A robust approach toward feature space analysis. *IEEE Transactions on Pattern Analysis and Machine Intelligence*, *24*(5), 603–619.

Cortes, C., & Vapnik, V. (1995). Support-vector networks. *Machine Learning*, *20*(3), 273–297.

Ding, S., Liu, R., Cai, Y., & Wang, P. (2019). A point cloud adaptive slope filtering method considering terrain (in Chinese). *Remote Sensing Information*, *34*(4), 108–113.

Drake, J. B., Dubayah, R. O., Clark, D. B., Knox, R. G., Blair, J. B., Hofton, M. A., Chazdon, R. L., Weishampel, J. F., & Prince, S. (2002). Estimation of tropical forest structural characteristics using large-footprint LiDAR. *Remote Sensing of Environment*, *79*(2-3), 305–319.

Duda, R. O., & Hart, P. E. (1972). Use of the Hough transformation to detect lines and curves in pictures. *Communications of the ACM*, *15*(1), 11–15.

Dutta, S., Banerjee, S., Biswas, P. K., & Bhowmick, P. (2014). Mesh denoising using multi-scale curvature-based saliency. 12th Asian Conference on Computer Vision (ACCV), Singapore, Springer International Publishing, Part II 12, pp. 507–516.

Ester, M., Kriegel, H.-P., Sander, J., & Xu, X. (1996). A density-based algorithm for discovering clusters in large spatial databases with noise. Association for the Advancement of Artificial Intelligence Conference on Artificial Intelligence, AAAI Press, vol. 96, No. 24, pp. 226–231.

Fischler, M. A., & Bolles, R. C. (1981). Random sample consensus: A paradigm for model fitting with applications to image analysis and automated cartography. *Communications of the ACM*, *24*(6), 381–395.

Han, W. J., & Zuo, Z. Q. (2012). Noise removing algorithm of LiDAR point clouds based on TIN smoothing rules (in Chinese). *Science of Surveying and Mapping*, *37*(6), 153–154. https://doi.org/10.16251/j.cnki.1009-2307.2012.06.015

Harding, D. J., & Carabajal, C. C. (2005). ICESat waveform measurements of within footprint topographic relief and vegetation vertical structure. *Geophysical Research Letters*, 32(21), L21S10.

Hartigan, J. A., & Wong, M. A. (1979). Algorithm AS 136: A K-means clustering algorithm. *Journal of the Royal Statistical Society*, *28*(1), 100–108.

Hofton, M. A., Minster, J.-B., & Blair, J. B. (2000). Decomposition of laser altimeter waveforms. *IEEE Transactions on Geoscience and Remote Sensing*, *38*(4), 1989–1996.

Huang, N. E., Shen, Z., Long, S. R., Wu, M. C., Shih, H. H., Zheng, Q., Yen, N. C., Chi, C. T., & Liu, H. H. (1998). The empirical mode decomposition and the Hilbert spectrum for nonlinear and non-stationary time series analysis. *Proceedings of the Royal Society A: Mathematical Physical and Engineering Sciences*, *454*(1971), 903–995.

Jutzi, B., & Stilla, U. (2006). Range determination with waveform recording laser systems using a Wiener Filter. *ISPRS Journal of Photogrammetry and Remote Sensing*, *61*(2), 95–107.

Kamousi, P., Lazard, S., Maheshwari, A., & Wuhrer, S. (2016). Analysis of farthest point sampling for approximating geodesics in a graph. *Computational Geometry*, *57*, 1–7.

Kilian, J., Haala, N., & Englich, M. (1996). Capture and evaluation of airborne laser scanner data. *International Archives of Photogrammetry and Remote Sensing*, *31*, 383–388.

Kopsinis, Y., & McLaughlin, S. (2009). Development of EMD-based denoising methods inspired by wavelet thresholding. *IEEE Transactions on Signal Processing*, *57*(4), 1351–1362.

LeCun, Y., Bengio, Y., & Hinton, G. (2015). Deep learning. *Nature*, *521*, 436–444.

Lefsky, M. A., Keller, M., Pang, Y., De Camargo, P. B., & Hunter, M. O. (2007). Revised method for forest canopy height estimation from Geoscience Laser Altimeter System waveforms. *Journal of Applied Remote Sensing*, *1*(1), 013537.

Liu, Y., Xi, X., Wang, C., Nie, C., & Wang, P. (2020). An improved processing TIN densification filtering algorithm for point clouds (in Chinese). *Science of Surveying and Mapping*, *45*(5), 106–111. https://doi.org/10.16251/j.cnki.1009-2307.2020.05.016

Mallet, C., & Bretar, F. (2009). Full-waveform topographic LiDAR: State-of-the-art. *ISPRS Journal of Photogrammetry and Remote Sensing*, *64*(1), 1–16.

Marquardt, D. W. (1963). An algorithm for least-squares estimation of nonlinear parameters. *Journal of the Society for Industrial and Applied Mathematics*, *11*(2), 431–441.

Mongus, D., & Zalik, B. (2012). Parameter-free ground filtering of LiDAR data for automatic DTM generation. *ISPRS Journal of Photogrammetry and Remote Sensing*, *67*, 1–12.

Ng, A. Y., Jordan, M. I., & Weiss, Y. (2002). On spectral clustering: Analysis and an algorithm. Conference and Workshop on Neural Information Processing Systems. MIT Press, 849–856.

Peng, S., Xi, X., Wang, C., Dong, P., & Nie, S. (2019). Systematic comparison of power corridor classification methods from ALS point clouds. *Remote Sensing*, *11*(17), 1961.

Popescu, S. C., Zhou, T., Nelson, R., Neuenschwander, A., Sheridan, R., Narine, L., & Walsh, K. M. (2018). Photon counting LiDAR: An adaptive ground and canopy height retrieval algorithm for ICESat-2 data. *Remote Sensing of Environment*, *208*, 154–170.

Qi, C. R., Su, H., Mo, K. C., & Guibas, L. J. (2017a). Pointnet: Deep learning on point sets for 3d classification and segmentation. IEEE Conference on Computer Vision and Pattern Recognition (CVPR), Honolulu, IEEE, 77–85.

Qi, C. R., Yi, L., Su, H., & Guibas, L. J. (2017b). PointNet++: Deep hierarchical feature learning on point sets in a metric space. 31st Annual Conference on Neural Information Processing Systems (NIPS), Long Beach, Advances in Neural Information Processing Systems, volume 30.

Somekawa, T., Galvez, M. C. D., Fujita, M., Vallar, E. A., & Yamanaka, C. (2013). Noise reduction in white light LiDAR signal using a one-dim and two-dim Daubechies wavelet shrinkage method. *Advances in Remote Sensing*, *2*(1), 10–15.

Sun, G., Ranson, K. J., Kimes, D. S., Blair, J. B., & Kovacs, K. (2008). Forest vertical structure from GLAS: An evaluation using LVIS and SRTM data. *Remote Sensing of Environment*, *112*(1), 107–117.

Suveg, I., & Vosselman, G. (2002). Automatic 3D building reconstruction. *Proceedings of SPIE - The International Society for Optical Engineering*, *4661*, 59–69.

Wagner, W., Ullrich, A., Ducic, V., Melzer, T., & Studnicka, N. (2006). Gaussian decomposition and calibration of a novel small-footprint full-waveform digitising airborne laser scanner. *ISPRS Journal of Photogrammetry and Remote Sensing*, *60*(2), 100–112.

Wang, C., Tang, F., Li, L., Li, G., Cheng, F., & Xi, X. (2013). Wavelet analysis for ICESat/GLAS waveform decomposition and its application in average tree height estimation. *IEEE Geoscience and Remote Sensing Letters*, *10*(1), 115–119.

Wang, P., Xi, X., Wang, C., & Xia, S. (2017). Study on power line fast extraction based on airborne LiDAR data (in Chinese). *Science of Surveying and Mapping*, *42*(2), 154–158.

Wu, J., Van Aardt, J. A. N., & Asner, G. P. (2011). A comparison of signal deconvolution algorithms based on small-footprint LiDAR waveform simulation. *IEEE Transactions on Geoscience and Remote Sensing*, *49*(6), 2402–2414.

Wu, X., & Huang, Y. (2017). Smoothing and resampling of point cloud based on moving least squares (in Chinese). *Microcomputer and Its Applications*, *36*(11), 47-49.

Yang, B., Liang, F., & Huang, R. (2017). Processing, challenges and perspectives of 3D LiDAR point cloud processing (in Chinese). *Acta Geodaetica et Cartographica Sinica*, *46*(10), 1509–1516. https://doi.org/10.11947/j.AGCS.2017.20170351

Zhang, X., & Liu, J. (2004). Airborne laser scanning altimetry data filtering. *Science of Surveying and Mapping*, *29*(6), 50–53.

Zhang, W., Qi, J., Wan, P., Wang, H., Xie, D., Wang, X., & Yan, G. (2016). An easy-to-use airborne LiDAR data filtering method based on cloth simulation. *Remote Sensing*, *8*(6), 501.

Zhu, X., Nie, S., Wang, C., Xi, X., & Hu, Z. (2018a). A ground elevation and vegetation height retrieval algorithm using micro-pulse photon-counting LiDAR data. *Remote Sensing*, *10*(12), 1962.

Zhu, X., Nie, S., Wang, C., Xi, X., & Zhou, H. (2020). A noise removal algorithm based on OPTICS for photon-counting LiDAR data. *IEEE Geoscience and Remote Sensing Letters*, *18*(8), 1471–1475.

Zhu, X., Wang, C., Xi, X., Wang, P., Tian, X., & Yang, X. (2018b). Hierarchical threshold adaptive for point cloud filter algorithm of moving surface fitting (in Chinese). *Acta Geodaetica et Cartographica Sinica*, *47*(2), 153–160. https://doi.org/10.11947/j.AGCS.2018.20170491

5 LiDAR Remote Sensing Applications

5.1 TOPOGRAPHIC MAPPING

Topographic mapping is the process of measuring the horizontal projection position and elevation of the Earth's surface and presenting them with topographic maps, with symbols and notes. Traditional topographic mapping is time-consuming, labor-intensive, and difficult to guarantee high measurement accuracy in complex terrains. LiDAR can quickly, directly, and accurately acquire high-density three-dimensional (3D) spatial information of Earth's surface and has been widely used in topographic mapping. Compared with aerial photogrammetry, using LiDAR point clouds to produce digital terrain products improves work efficiency and product accuracy, greatly reducing labor costs.

LiDAR technology is used to rapidly produce high-precision digital elevation models (DEMs), digital surface model (DSMs), contour lines, etc. The production process is shown in Figure 5.1. A DEM is a ground model that represents the ground elevation with a set of ordered numerical arrays, which is derived by interpolating the ground points after point cloud filtering with the aid of feature lines. A DSM is a model that contains static objects such as buildings, bridges, and trees, which can be generated by interpolating ground points with permanent object points with the aid of feature lines. Contour lines are vector data with elevation attributes, which can be generated by interpolating the extracted ground points from ground point cloud thinning or interpolating the extracted feature points from the DEM.

The following is an example of an airborne LiDAR point cloud and its applications in terrain mapping. The data coverage is shown in Figure 5.2, with an area of about 18,930 m^2. The surface coverage is dominated by buildings, vegetation, and water. A total of 1,971,891 laser points were collected. The LiDAR topographic mapping usually includes three steps: point cloud data processing, surface model construction, and contour line generation, which are described next.

5.1.1 LiDAR Point Cloud Processing

The LiDAR point cloud processing mainly includes point cloud registration, point cloud denoising, and point cloud filtering.

5.1.1.1 Point Cloud Registration

The survey area is large and usually cannot be completely covered by one measurement, so the original point clouds from multiple measurements need to be registered and mosaicked together. Point cloud registration is achieved by rotating and

DOI: 10.1201/9781032671512-5

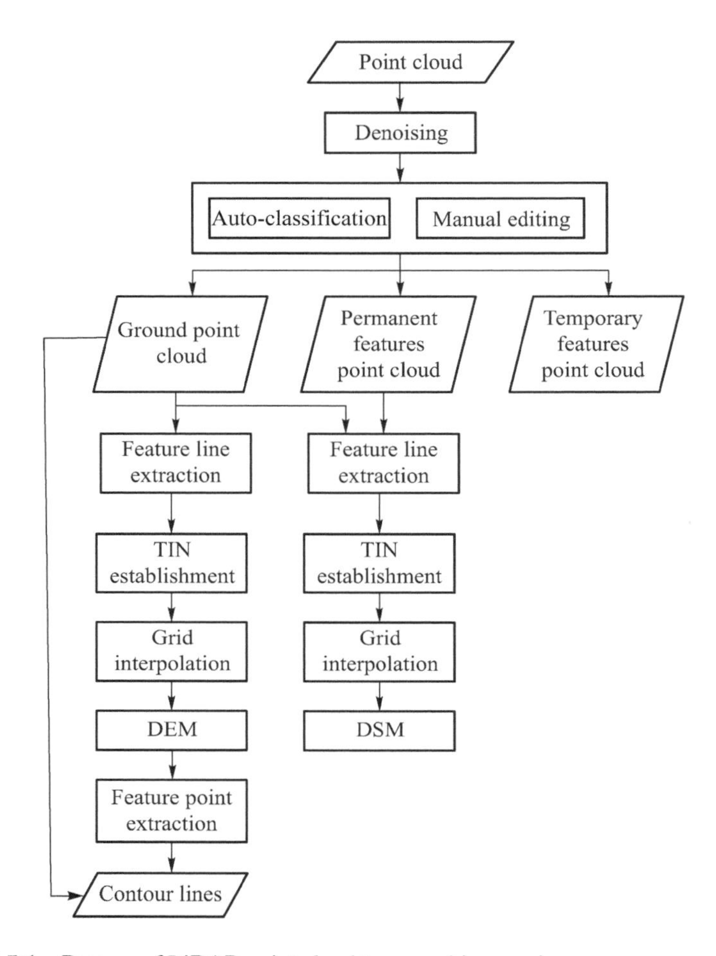

FIGURE 5.1 Process of LiDAR point cloud topographic mapping.

translation matrices. It mainly includes two steps: coarse registration and fine registration. The coarse registration is generally used for the registration of two point cloud sets with unknown initial relative positions. This step provides reasonable initial transformation matrices for the subsequent fine registration. The fine registration optimizes the initial transformation matrix by multiple iterations to obtain the global optimal transformation matrix. The details of point cloud registration are described in Section 3.1.3.

5.1.1.2 Point Cloud Denoising

Raw LiDAR point clouds inevitably have noise. The point cloud denoising algorithm is detailed in Section 4.1 of this book. Figure 5.2 shows the denoising results of the point cloud in a study area after manual editing and processing with the statistical denoising algorithm. It is observed that the noises in the boxes are removed.

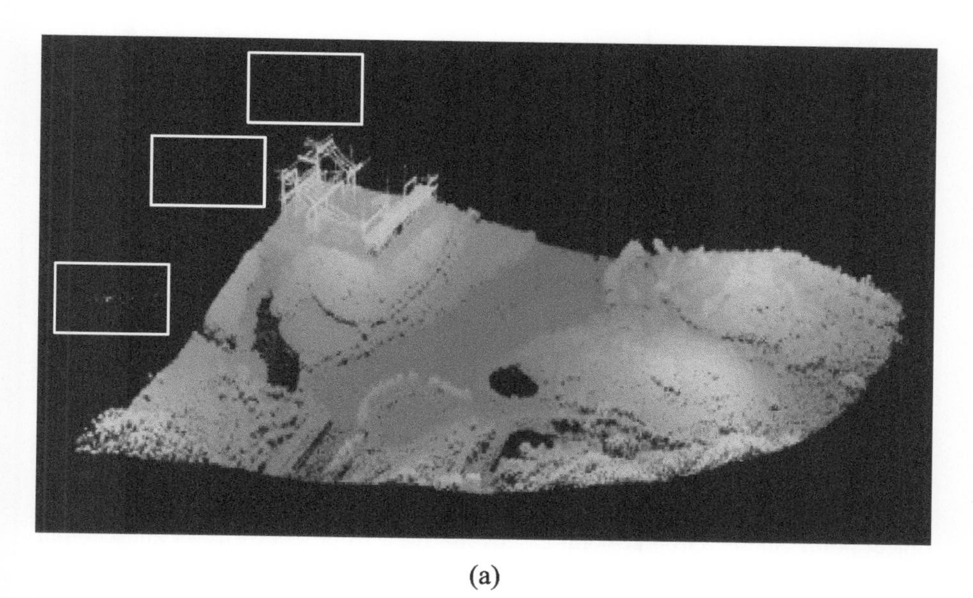

(a)

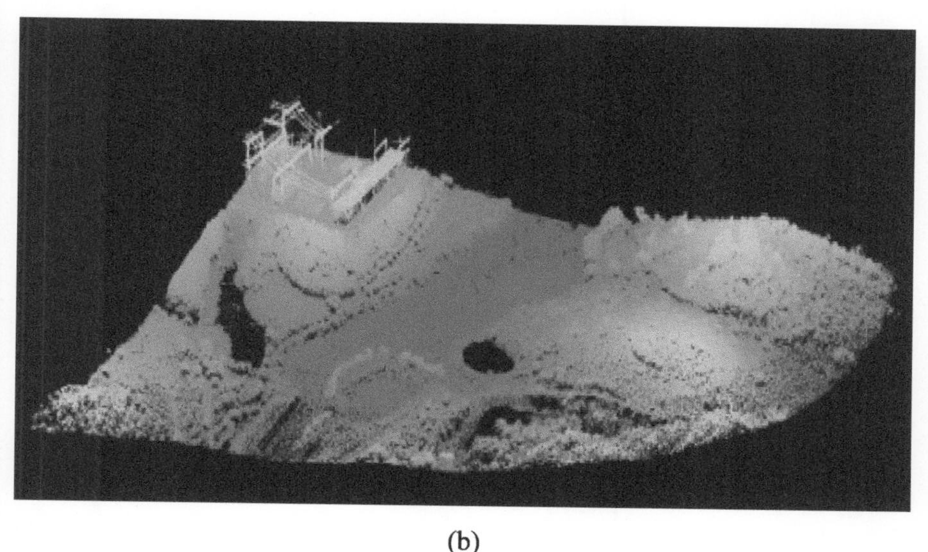

(b)

FIGURE 5.2 Point cloud denoising. (a) Before denoising; (b) after denoising.

5.1.1.3 Point Cloud Filtering

According to China's industry standards such as "CH/T 8023—2011 Specification for data processing of airborne LiDAR" and "CH/T 9022—2014 Digital products of fundamental geographic information—1:500 1:1000 1:2000 1:5000 1:10000 digital surface models," point clouds collected by airborne LiDAR can be divided into three categories: ground point clouds, permanent object point clouds, and temporary

FIGURE 5.3 Point cloud filtering result with cloth simulation filter (CSF) algorithm.

object point clouds. Ground point clouds are used to produce DEMs and contour lines. Permanent object point clouds include buildings, vegetation, etc., which can be combined with ground point clouds to produce DSM products. Temporary object point clouds include stationary and moving temporary objects (such as vehicles, pedestrians, and animals). Ground point clouds are obtained by point cloud filtering. The Point Cloud Magic (PCM) software program integrates various point cloud filtering algorithms, such as cloud simulation filtering and moving surface filtering, which are described in Section 4.1.2. The point cloud filtering result of a study area is shown in Figure 5.3.

5.1.2 DEM and DSM Constructions

There are two main expression forms of the surface model, i.e., triangulated irregular network (TIN) surface model and regular grid surface model. Before constructing the model, the feature lines are usually extracted to improve the model construction accuracy, and then the DSM and DEM are generated by spatial interpolation algorithms.

5.1.2.1 Extraction of Feature Lines

Due to the variability of terrain features and absorption of laser by water, there might be some "voids" in the survey area, which causes errors in the production of DEMs and DSMs. This problem can be avoided by constructing feature lines and setting constraints during interpolation.

There are soft feature lines, hard feature lines, and fault feature lines. Soft feature lines describe the discontinuity of a slope and make the surface model more realistic in areas such as ridges and rivers. Hard feature lines maintain the sharpness of the surface and are mainly used for sudden changes in elevation such as coastlines and dams. Fault feature lines are used for surface fractures such as geological faults, steep cliffs, cliff walls, and building edges. Each location on the surface has both high and low elevation values, so adding feature lines at the top and bottom of the fault can correctly describe the real condition of the ground surface in the area. The feature lines are usually obtained in three ways: photogrammetry, automatic extraction from the point cloud, and manual acquisition. The photogrammetry method is highly accurate but requires strict registration with the point cloud. The automatic extraction method from the point cloud does not require registration, but the accuracy is affected by the point cloud density. The manual acquisition is highly accurate, but time-consuming and costly.

5.1.2.2 DEM and DSM Constructions Based on Spatial Interpolation

Spatial interpolation is the process of estimating the unknown point values based on the properties and spatial distribution of the known points, with the assumption that the unknown points are influenced by the known points. The classical spatial interpolation algorithms include TIN interpolation, inverse distance weighted (IDW), natural neighborhood interpolation, and Kriging interpolation. The latter three methods derive regular grid products, while the TIN derives vector products.

1. TIN interpolation algorithm. The three nearest known points are connected to form a network of triangles covering the whole area. The slope and direction of each triangle are fixed. Then the interpolation is performed according to the angle and distance between the predicted point and the nearest triangular surface. Its core function is to reduce the sample points from multiple proximity points to the three closest points. Further interpolation can be performed by linear interpolation based on TIN or natural neighborhood interpolation based on area weights. In this example, the filtered ground point clouds and all point clouds without temporary features are connected into continuous triangular surfaces under the feature line constraints to generate the TIN-based DEM and DSM (Figure 5.4). The constraints of feature lines on the TIN mainly include (1) not intersecting with triangles, (2) always being the sides of triangles, and (3) limiting the slope of triangular surfaces (Tang et al., 2010).
2. Natural neighborhood interpolation algorithm. Find the subset of input samples closest to the query point and interpolate them with weights based

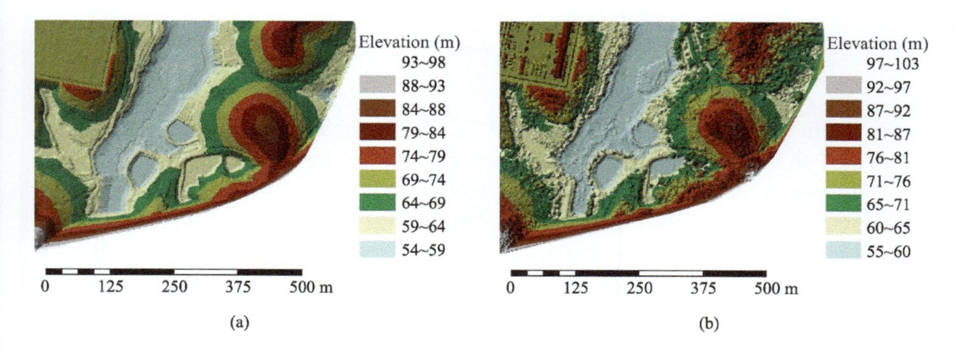

FIGURE 5.4 TIN-based DEM and DSM.

on the overlap area size of the samples. This method can be locally adjusted according to the structure of the input data without a user-defined search radius, sample count, or shape-related parameters. Also, it has the same effect for both regularly and irregularly distributed point clouds. However, the method interpolates only with a subset of samples around the query point, which is local in nature. Additionally, the interpolated elevation range is smaller than the elevation range of the input point cloud.

3. IDW interpolation algorithm. This algorithm determines the weights based on the distance between known and predicted points. The closer the distance, the greater the influence; conversely, the greater the distance, the smaller the influence, as expressed in Equation (5.1).

$$Z = \sum_{i=1}^{n} \frac{1}{(D_i)^p} Z_i \bigg/ \sum_{i=1}^{n} \frac{1}{(D_i)^p} \tag{5.1}$$

where Z is the estimated value, Z_i is the elevation of the i-th known point, D_i is the distance between the i-th known point and the estimated point, and p is the power. When the value of p is 1, the interpolation weight is linear decay. The larger the value of p, the more the predicted point is influenced by the near point and the finer the interpolation surface, but the smoothing effect is lost. The smaller the value of p, the more the predicted point is influenced by the farther point, the smoother the interpolation surface, and the detail will be lost. Generally, a p value of 0.5–3 is a reasonable choice.

4. Kriging interpolation algorithm. This algorithm is commonly used in geostatistics. Similar to the IDW, this algorithm needs to calculate the weight of the influence of known points on the predicted points. However, unlike IDW, which calculates the weight based on distance only, Kriging interpolation quantifies the spatial autocorrelation by considering the distance, the overall location arrangement of predicted points, and known points; see Equation (5.2). The distribution of spatial points is usually expressed in a linear, Gaussian, spherical, exponential, or circular model.

$$\hat{Z}(S_0) = \sum_{i=1}^{N} \lambda_i Z(S_i) \tag{5.2}$$

where S_i is the i-th known point, $Z(S_i)$ is the measured elevation value at the i-th known point, λ_i is the unknown weight at the i-th known point, S_0 is the predicted point, $\hat{Z}(S_0)$ is the predicted elevation value of the predicted point, and N denotes the number of measured known points. The Kriging algorithm considers the variance distribution of elevation in space, determines the distance range that has influence on the point to be interpolated, and then uses the sampling points within this range to estimate the elevation value of the point to be interpolated. Reliability is higher when there are more sampling points.

TABLE 5.1
DEM Grid Spacing for Different Scales

Scale	DEM Grid Spacing (m)
1: 500	0.5
1: 1000	1.0
1: 2000	2.0
1: 5000	2.5
1: 10,000	5.0

When using the regular grid interpolation algorithms to produce DEM and DSM products, the equal spacing grid is usually used, i.e., the vertical and horizontal spacings of the grid are the same. According to the relevant standards, the standards of grid spacing for different scales are listed in Table 5.1.

5.1.3 Contour Line Generation

A contour line product is an intuitive representation form of terrain, which expresses the terrain with different arrangements and patterns. It is a common method of topographic mapping. For example, the contour lines of steep terrain are closely spaced, and the contour lines bend toward the upstream of rivers. Contour lines connect points with equal elevation values. The vertical distance between adjacent contour lines is called contour distance. According to the requirements of scale and contour distance, contour line products can be generated by the regular grid surface model or TIN surface model.

1. Contour lines generated from the regular grid surface model
   Usually, it is divided into high-order surface interpolation and grid linear interpolation, among which grid linear interpolation is more widely used. The grid linear interpolation first determines the starting point of the contours to be interpolated in the grid and then traverses the interpolated contour points of the grid from the starting point to generate contour lines. The grid spacing affects the generated contour lines. A larger grid spacing leads to uneven and rough contour lines. Hence, a suitable grid spacing should be selected in conjunction with the scale. In addition, the contour lines generated directly from the original regular grid surface model might be a square shape or uneven, making the contour lines unsmooth. Hence, the smoothing process should be performed on the original regular grid surface model, so that the contour lines do not pass through the center of the pixel of the grid. Focal statistics is a commonly used smoothing method. It applies a custom weighting template to calculate the weighted average of the eight neighborhoods around each pixel as the smoothed value and then uses a segmented linear interpolation to generate the contour lines [Figure 5.5(a)].

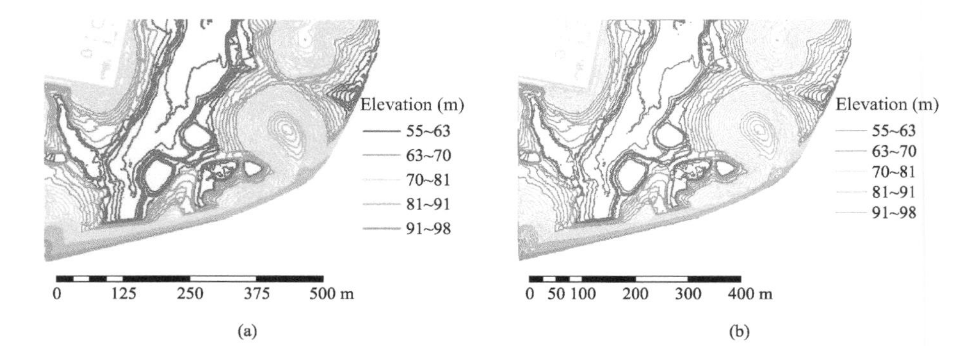

FIGURE 5.5 Contour lines generated from (a) regular grid surface model and (b) TIN surface model.

2. Contour lines generated from the TIN surface model

   Automatically generating contour lines from the TIN includes two steps: (1) checking contour lines and triangles and (2) contour line drawing. First, check whether each triangle is passed by contour lines. If so, interpolate the two nodes of the contour line linearly along the edge of the triangle to determine the position of contour line nodes. Then connect all contour nodes to generate initial contour lines. Finally use the spline function fitting algorithm to smooth the initial contour lines to generate contour line products [Figure 5.5(b)].

5.2 FOREST INVESTIGATION

Forests are the core components of terrestrial ecosystems and play an irreplaceable role in maintaining the ecological balance. Traditional methods for forest investigation mostly rely on field measurements to obtain information about tree species, tree height, crown width, diameter at breast height (DBH), etc. These methods are time-consuming, labor-intensive, and inefficient. Optical remote sensing provides horizontal structure information of the forest canopy, but it is difficult to directly acquire vertical structure information. In contrast, LiDAR provides complete 3D structural information of the forest canopy, thus showing unique advantages in forest investigation (Li et al., 2016). This section mainly introduces the applications of LiDAR technology in forest structural parameter retrievals (canopy height, gap fraction, vegetation coverage, biomass, etc.) and tree species fine classification.

5.2.1 Individual Tree Parameter Retrievals

5.2.1.1 Individual Tree Segmentation

Individual tree segmentation identifys the point cloud of each single tree through related algorithms, aiming to obtain the position and structural parameters of each individual tree. The individual tree segmentation methods based on airborne LiDAR data are mainly divided into two categories: one is the canopy height model

(CHM)–based segmentation methods, including local maximum, watershed segmentation, region growth, and region-based hierarchical cross-sectional analysis; the other one is point cloud–based segmentation methods, including *K*-means clustering, adaptive distance clustering, mean-moving clustering, and graph cut. Here, we take the airborne point cloud data in the Blue Ridge area as an example to introduce the individual tree segmentation algorithm based on graph cut (Wang et al., 2019). The specific steps are as follows, and its pseudo-code is as in Table 5.2.

First, the original point cloud is filtered and interpolated to create a DEM and DSM (Figure 5.6). The CHM is generated based on DSM and DEM. Then the local maximum in the CHM is calculated to determine the treetop, which is used as the prior knowledge of the individual tree position. Based on the prior individual tree positions, the iterative segmentation is conducted to segment each individual tree.

TABLE 5.2
Pseudo-Code of Individual Tree Segmentation Algorithm Based on Graph Cut

```
Input: P, point cloud data
Output: all_trees, segmented trees
Algorithms:
import point cloud data P;
# Step 1: filtering and interpolation, generate DEM and DSM
DEM, DSM = apply_filters_and_interpolation(P);
# Step 2: generate Canopy Height Model (CHM)
CHM = calculate_difference(DSM, DEM);
# Step 3: calculate local maxima of the CHM
tree_tops = calculate_local_maximal(CHM);
# Step 4: iterative segmentation
segmented_trees = [];
for tree_top in tree_tops:
   single_tree = segment_single_tree(CHM, tree_top);
   segmented_trees.append(single_tree);
endfor;
# Step 5: consider canopy occlusion and structural stratification, handle potential missed detections
potential_trees = detect_potential_trees(segmented_trees, CHM);
# Step 6: iterative detection until no new trees are segmented
all_trees = segmented_trees + potential_trees;
while True:
   new_trees = []
   for tree in all_trees:
      more_trees = detect_more_trees(CHM, tree);
      if more_trees:
         new_trees.append(more_trees);
      endif;
   endfor;
   all_trees.append(new_trees);
endwhile;
return all_trees;
```

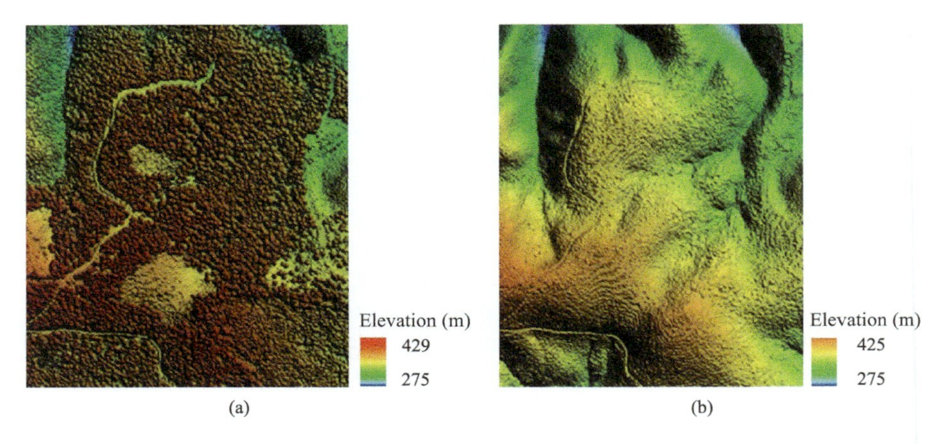

FIGURE 5.6 DSM and DEM constructed from point cloud data.

In addition, canopy occlusion and structural stratification might lead to missing detection of individual trees in the middle and lower layers of the canopy. To address this problem, for unclassified point clouds, the global maximum is used instead of the local height maximum as the prior condition of the individual tree segmentation. The possible individual trees in the unclassified point cloud are iteratively detected from high to low by considering the shape and the minimum point number of the tree. Figure 5.7 shows the result of individual tree segmentation in a study area.

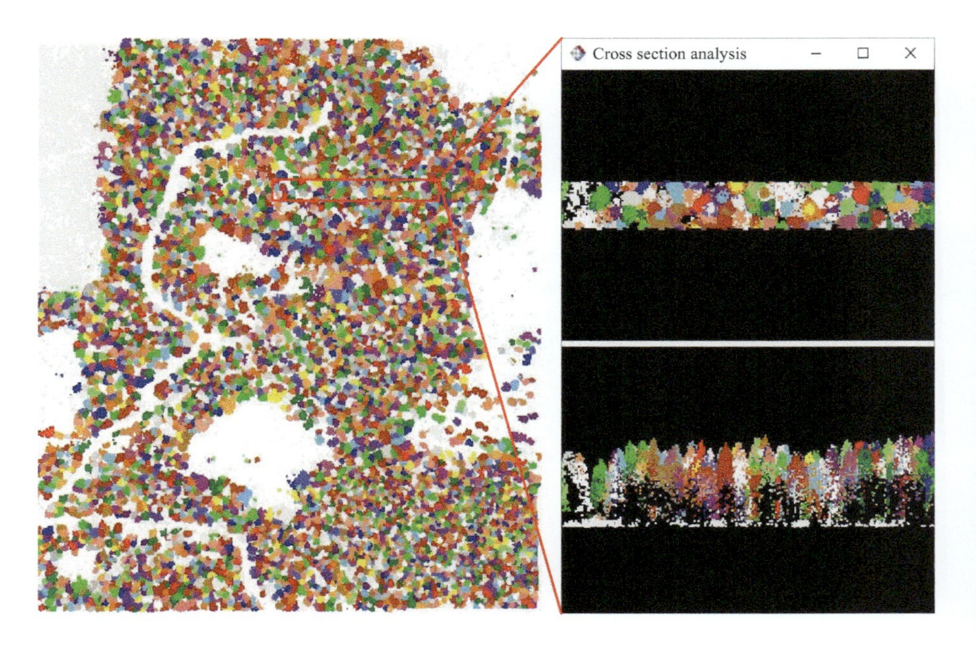

FIGURE 5.7 Individual tree segmentation result (each color represents an individual tree).

5.2.1.2 Calculation of Individual Tree Parameters

The segmentation result is used to extract the structural parameters at the individual tree scale, including tree height, crown width, etc.

1. Tree height. Tree height is one of the basic parameters for the assessment of forest growth, biomass, and volume, and is also an important variable in forest resource investigation. For the LiDAR point cloud, tree height is usually considered to be the height difference between the highest point of an individual tree and the ground point. Using the individual tree height extraction tool in the PCM software, the information of all the individual trees in the study area was extracted. The results are shown in Table 5.3. The output information includes tree number, tree position, and tree height.
2. Crown area. The crown area is extracted by projecting the point cloud belonging to an individual tree onto the horizontal plane and calculating the area of the convex hull formed by the projected point cloud. There are three steps in the calculation of crown area. First, the point cloud is segmented to derive a point cloud of each individual tree. Second, the segmented point cloud is projected onto the horizontal plane and then used to generate the convex polygon. Third, calculate the area of each convex polygon, that is, the crown area.

5.2.2 Canopy Parameter Retrievals

5.2.2.1 Canopy Height Model

A CHM is the main data source for the retrieval of forest structural parameters at the stand scale. It is also used for individual tree segmentation and individual tree structural parameter retrievals. There are two main ways to generate CHM based on a LiDAR point cloud. The first one is to make a difference between DSM and DEM.

TABLE 5.3
Position and Height of Individual Trees

Tree ID	*X*	*Y*	Tree Height(m)
1	187450	4989881	59.38
2	187454	4989856	58.89
3	187966	4989965	57.38
4	187864	4989695	57
5	187964	4989954	56.07
6	187994	4989936	56
7	187858	4989693	56
8	187904	4989705	56
…	…	…	…
15402	188004	4989500	2

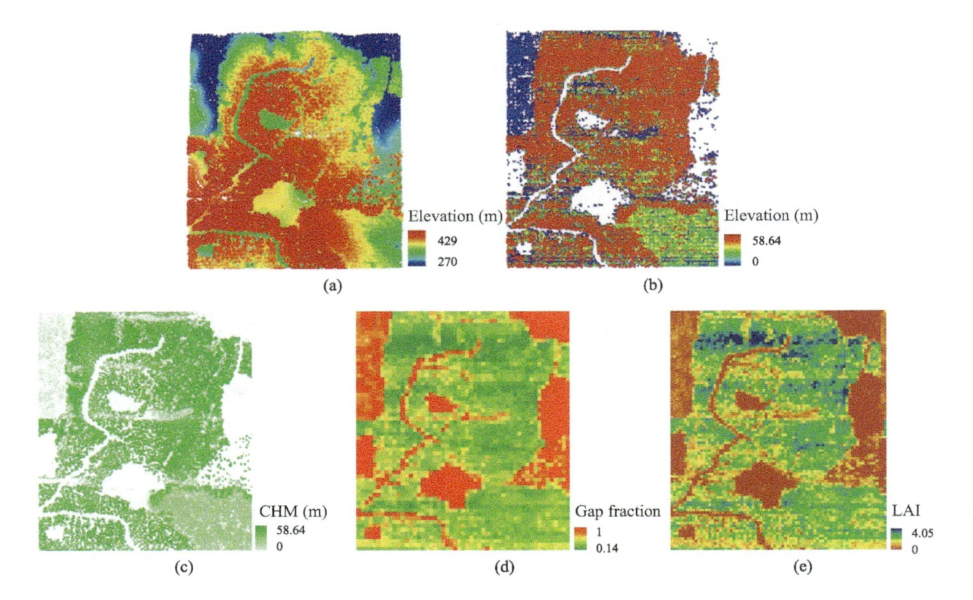

FIGURE 5.8 Forest structural parameter retrievals based on point clouds. (a) Original point cloud. (b) Normalized point cloud. (c) CHM. (d) Gap fraction. (d) Leaf Area Index (LAI).

This method requires two times of spatial interpolation, resulting in a loss of canopy information. Also, it fails to describe the complexity of the forest canopy structure. The second one is to use the DEM to normalize the elevation of the LiDAR point cloud and then interpolate the normalized point cloud to generate the CHM. The second method generates the CHM directly from the point cloud, which theoretically has less information loss than the first method. Here, we use the second method to generate the CHM in the following example.

The study area shown in Figure 5.8(a) is about 147 ha. The point cloud density is 4.86 pts/m^2. The acquisition time is in spring 1999. The forest type is dominated by plantation. The PCM software was first used to normalize the LiDAR point cloud, and the result is shown in Figure 5.8(b). Then the normalized point cloud is spatially interpolated to generate the CHM. The commonly used interpolation algorithms are introduced in Section 5.1. Figure 5.8(c) shows the CHM generated by using the IDW interpolation algorithm.

5.2.2.2 Calculation of Canopy Structural Parameters

1. Gap fraction. Gap fraction refers to the probability that light/photons pass through the forest canopy to reach the ground, and its value ranges from 0 to 1 (from complete vegetation coverage to no vegetation coverage) (Li & Strahler, 1988). Gap fraction is closely related to the incident angle of light, canopy structure (LAI and leaf angle distribution, etc.). It is calculated by the Beer-Lambert law. Both LiDAR point cloud and waveform data have been successfully applied to forest gap fraction retrievals. The point

cloud–derived gap fraction P_{gap} is expressed as the ratio of the number of ground points to the number of all points, as shown in Equation (5.3).

$$P_{gap} = \frac{n_{ground}}{n} \tag{5.3}$$

where n_{ground} is the number of ground points and n is the number of all points. For full-waveform LiDAR data, the gap fraction can be replaced by the laser penetration index (LPI), which fully considers the energy transfer of the laser within the forest canopy. Hence, the gap fraction is calculated as the ratio of the ground return intensity to the total return intensity, as in Equation (5.4).

$$P_{gap} = \text{LPI} = \frac{\sum I_{\text{Ground}}}{\sum I_{\text{Ground}} + \sum I_{\text{Vegetation}}} \tag{5.4}$$

where LPI is calculated from the LiDAR intensity, $\sum I_{\text{Ground}}$ is the total intensity of ground returns, and $\sum I_{\text{Vegetation}}$ is the total intensity of canopy returns.

Figure 5.8(d) shows the gap fraction distribution map obtained by using the gap fraction module of the PCM software. To derive reliable results, the size of the spatial grid should be larger than the crown width, and the elevation threshold should be able to separate vegetation points from ground points. Here the spatial grid and the elevation threshold are set as 15 × 15 m and 2 m, respectively.

2. LAI. LAI is defined as half of the surface area of all leaves projected on a unit surface area (Chen & Black, 1992). Traditional optical remote sensing is used for large-scale LAI estimation, but it suffers from the problem of signal saturation. The LiDAR is capable of obtaining the canopy vertical structure due to the strong penetration ability of a laser pulse; thus, it has been widely used in forest LAI estimation. A commonly used forest LAI inversion method is introduced in detail next, and its pseudo-code is as in Table 5.4.

   The attenuation of laser energy in the canopy is closely related to the LAI, which is expressed by the Beer-Lambert law:

$$I = I_0 e^{-k \cdot \text{LAI}} \tag{5.5}$$

where I is the LiDAR intensity below the canopy, I_0 is the LiDAR intensity above the canopy, and k is the extinction coefficient, which depends on leaf inclination and beam direction. Equation (5.5) shows that LAI can be calculated based on gap fraction, which is replaced by LPI; thus, the LAI is calculated based on the extinction coefficient k and LPI:

$$\text{LAI} = -\frac{1}{k} \ln\left(\frac{I}{I_0}\right) = -\frac{1}{k} \ln \text{LPI} \tag{5.6}$$

Since the return intensity is affected by various factors such as laser emission energy, surface reflectivity, atmospheric attenuation, and transmission

TABLE 5.4
Pseudo-Code of Forest LAI Estimation Based on LiDAR Point Cloud

Input: ***P***, forest point cloud; $\boldsymbol{\theta}$, average leaf angle; ***s***, spatial resolution.
Output: **LAI**, forest LAI.
Algorithm:
mark the point cloud ***P*** as vegetation points P_v and ground points P_g by filtering.
generate DEM from P_g.
generate 2D grids *G* with resolution ***s***.
for each grid g_i in *G*:
    search all points located in g_i.
    initialize the ground/vegetation return intensity I_g=0, I_v=0.
    for each ground point P_{g_i} located in g_i:
        $I_g = I_g + P_{g\_i}.intensity$;
    endfor;
    for each vegetation point P_{v_i} located in g_i:
        $I_v = I_v + P_{v\_i}.intensity$;
    endfor;
    $\mathbf{LAI}_i = -1/\cos(\boldsymbol{\theta}) * \ln(I_v / I_g)$.
endfor;
return LAI;

distance, intensity correction is necessary to reduce the influence of the aforementioned factors and improve the estimation accuracy of forest structural parameters. The specific details of intensity correction can be found in Section 3.1. Figure 5.8(e) is the LAI distribution map generated by the PCM software. The spatial resolution, elevation threshold, and leaf inclination angle distribution are set as 15 m, 2 m, and 0.5, respectively.

3. Other canopy structural parameters. In addition to gap fraction and LAI, LiDAR point clouds are used to retrieve other canopy structural parameters, such as canopy coverage, canopy closure, leaf inclination distribution, and biomass.

    Both canopy coverage and canopy closure are important parameters reflecting the forest structure. Canopy coverage refers to the percentage of the vertical projected area of the vegetation canopy on the horizonal plane to the total area. Canopy closure refers to the proportion of the sphere of the sky that is obscured by tree branches and leaves as viewed from a point on the woodland. Korhonen et al. (2011) used the ratio of the number of first returns to the total number of returns to estimate the forest canopy coverage. Riaño et al. (2003) used the ratio of the canopy point number to all point numbers to estimate the canopy closure.

    Leaf angle distribution (LAD) refers to the distribution of the included angles between the each leaf and the vertical direction. The LiDAR provides a new technique for LAD estimation. Riaño et al. (2003) used voxels to simulate the process that laser beams penetrate the canopy and are intercepted, estimated the proportion of the interaction area between the blades of each layer and the laser, and achieved the LAD curve retrievals based on gap probability theory.

The forest biomass is defined as the total amount of living organic matter that exists in a unit area. It is used to describe the function and productivity of forest ecosystems and is an important input for forest carbon sink estimation. To estimate forest biomass from LiDAR data, a series of methods, such as multiple linear regression, machine learning, and deep learning, are used to establish a biomass estimation model based on LiDAR feature parameters or forest parameters (such as tree height, crown width, DBH, canopy coverage, and LAI) (Luo et al., 2019b; Yang et al., 2019). Specifically, spaceborne LiDAR has been used for forest biomass estimation at global and regional scales, while airborne LiDAR point cloud or waveform data have been used for forest biomass estimation at the stand scale.

In addition, feature parameters derived from the point cloud or waveform are used to extract other vegetation structure parameters. For example, the 80%–90% percentile height or the maximum height of the first returns are used to estimate the average tree height; the 20% or 30% percentile height has a strong correlation with the volume of timber, average DBH, and number of trees. The PCM software provides automatic extraction of 56 point cloud feature parameters (such as cumulative height percentage, average absolute deviation, crown amplitude fluctuation ratio, and coefficient of variation). Based on these feature parameters, the aforementioned forest structural parameters can be estimated.

5.2.3 Forest Parameter Mapping at the Regional Scale

Spaceborne LiDAR covers a wide range and has been widely used in large-area forest height inversion and mapping. Yang et al. (2015) and Wang et al. (2016) used Geoscience Laser Altimeter System (GLAS) data and other remote sensing data to produce forest canopy height distribution maps with 500-m resolution in China and around the world, respectively. The follow-on spaceborne LiDAR systems (ICESat-2/ATLAS and GEDI) have the greater footprint densities (see Section 3.3), providing more reliable data sources for high-resolution forest height mapping over large areas. This section uses the ICESat-2/ATLAS and GEDI data and other data to map 30-m resolution forest canopy heights by implementing a list of steps, including data processing, footprint-scale forest height inversion and consistency analysis, and forest height mapping. The process is shown in Figure 5.9.

5.2.3.1 Spaceborne LiDAR Data Processing

A list of LiDAR waveform processing operations, including waveform denoising, waveform decomposition, and waveform feature parameter extraction, are conducted to extract the forest canopy parameters from the GEDI waveform. The specific algorithms are introduced in Section 4.2. Additionally, photon denoising and classification are conducted on ICESat-2/ATLAS photon-counting data. Details about the relevant algorithms are introduced in Section 4.3.

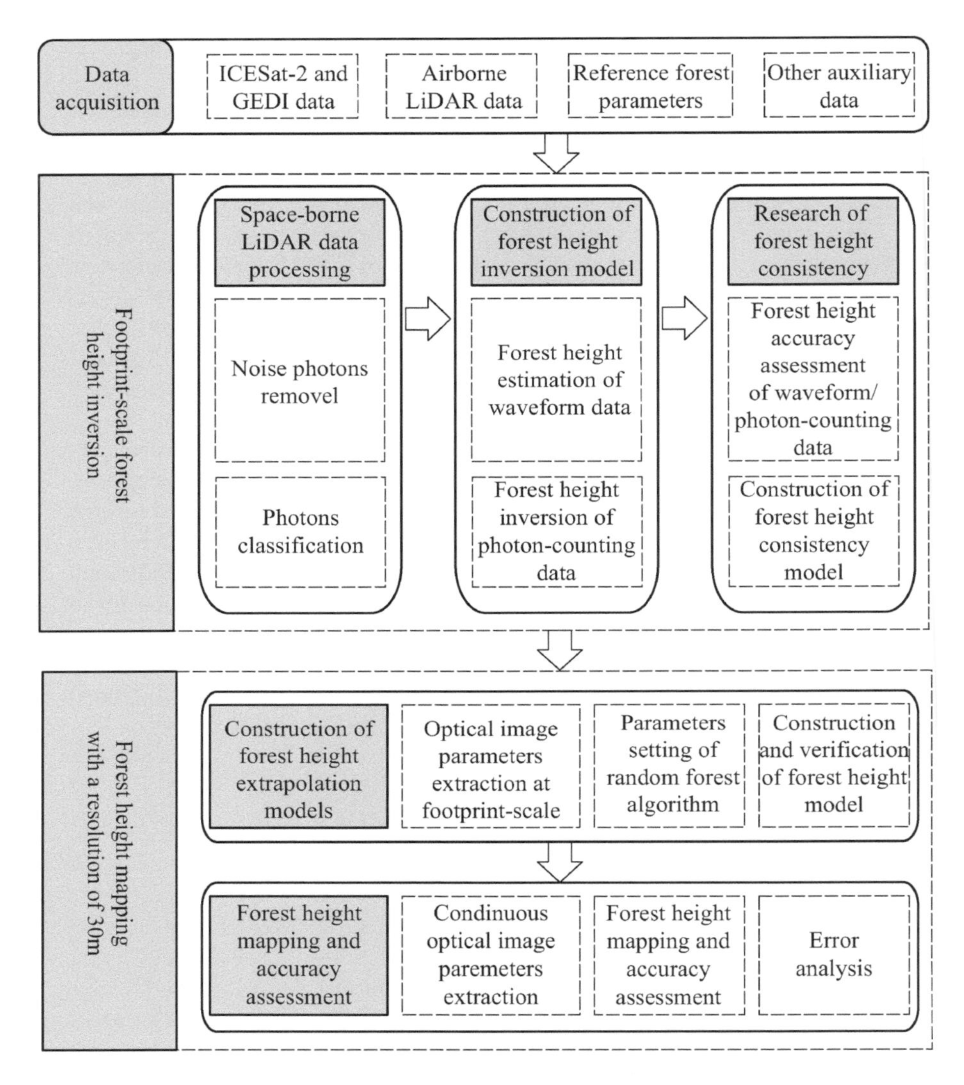

FIGURE 5.9 Flowchart of forest height mapping based on spaceborne ICESat-2 and GEDI LiDAR data.

5.2.3.2 Footprint-Scale Forest Height Inversion and Consistency Analysis

The ICESat-2 forest canopy height is calculated based on the canopy surface and ground surface extracted from ICESat-2 data. The GEDI forest canopy height is calculated based on the waveform feature parameters derived from GEDI data. There are obvious differences between ICESat-2 and GEDI in terms of forest detection mechanisms and data spatial distributions. Hence, differences exist in the forest heights derived from ICESat-2 and GEDI. To ensure the consistency and reliability of forest height retrievals between the two, the stepwise linear regression algorithm is used to establish a forest height consistency model linking GEDI forest height with ICESat-2 feature parameters (Zhu et al., 2022). Finally, the forest height samples with high consistency are obtained.

5.2.3.3 Forest Height Mapping

Taking the footprint-scale forest height extracted from spaceborne LiDAR as the main data source and combing with optical remote sensing imagery, topographic data, and climate data, the random forest (RF) regression algorithm is used to establish forest height extrapolation models in areas with different ecologies and forest types. Based on forest height extrapolation models, a forest height distribution map with 30-m resolution is produced.

5.2.4 Tree Species Classification

LiDAR point cloud or waveform data are used to extract structural information for tree species classification. Chen et al. (2019b) extracted structural feature parameters, texture feature parameters, and crown shape feature parameters from high-resolution point cloud data, and then achieved tree species classification based on these features, with a classification accuracy of 85%. Koenig and Höfle (2016) achieved tree species classification by extracting various features from LiDAR full-waveform data, such as waveform amplitude, canopy length-width ratio, and canopy volume. Overall, LiDAR data provide rich vertical structural information; however, it fails to provide the spectral information. Therefore, the fusion of LiDAR and hyperspectral data can effectively improve the classification accuracy of tree species. Existing methods usually use hyperspectral imagery to extract spectral features (e.g., spectral reflectance), vegetation index features (e.g., normalized vegetation index), and texture features (e.g., gray level co-occurrence matrix) and use LiDAR data to extract vertical distribution features (e.g., CHM, tree height percentile) and finally combine these features and adopt machine learning or deep learning methods for tree species classification. The fusion of LiDAR and hyperspectral features is generally divided into two categories. One is direct stacking, which directly takes the features extracted from the two kinds of data as the input of the classifier. The other mixes the features of the two kinds of data through component substitution, guided filtering, wavelet transform, and other methods and then inputs the classifier. Here is an example of using the hyperspectral data and LiDAR data of the Guangxi Gaofeng Forest Farm for fine tree species classification. Figure 5.10 shows the classification results of tree species based on a combination of spectral-vertical-spatial features, with a classification accuracy of 96.1% (Feng et al., 2020).

5.3 POWER LINE INSPECTION

The scale of high-voltage or ultra-high-voltage transmission corridors is growing rapidly, which puts forth higher demands on the efficient management and safe operation of transmission corridors. Power line inspection is the process of accurately and quickly obtaining spatial information and dynamic changes of transmission corridors and surrounding environments. The means for power line inspection have

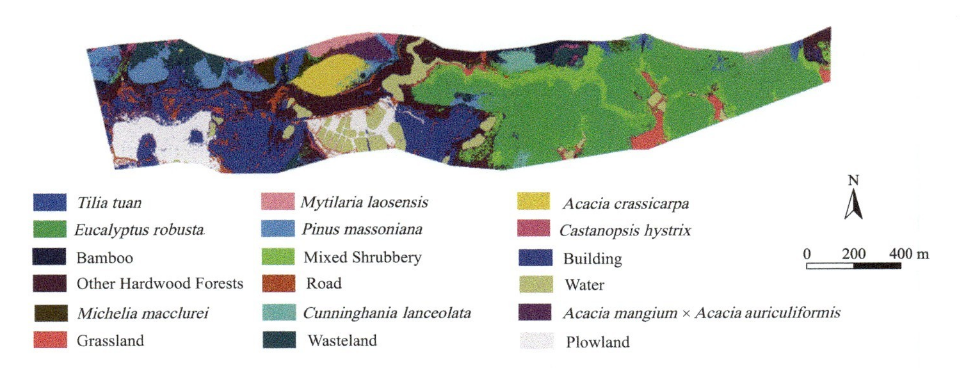

FIGURE 5.10 Tree species classification result of Guangxi Gaofeng Forest Farm.

developed from the initial manual inspections, to the remote sensing and photogrammetry inspections, to airborne LiDAR inspections. Although traditional remote sensing and photogrammetry technologies can reduce operation and maintenance costs to a certain extent, there are problems such as low measurement accuracy and large inspection errors. Airborne LiDAR systems can directly acquire high-density, high-precision 3D spatial data, providing a new technical means for the acquisition of spatial information of transmission corridors and safety inspection. The main process of airborne LiDAR transmission corridor inspection is shown in Figure 5.11, which includes transmission corridor point cloud classification, 3D modeling, and safety analysis. The following is an introduction to the airborne LiDAR power corridor inspection process, taking an ultra-high-voltage transmission corridor as an example. The corridor point cloud is collected by the unmanned aerial vehicle (UAV) with the Riegl VUX1 system and is about 52 km long and 100 m wide, with a point cloud density of about 68 pts/m^2. The main objects include vegetation, power towers, power lines, and buildings. The area is densely vegetated and has a large undulating terrain.

5.3.1 Transmission Corridor Point Cloud Classification

The commonly used LiDAR point cloud classification methods for transmission corridors are divided into two categories, i.e., hierarchical classification for key elements and machine learning classification, which are described below.

5.3.1.1 Hierarchical Classification for Key Elements

Hierarchical classification for key elements separates the target point clouds from the overall point cloud data one by one by analyzing the spatial geometric features of the objects and the other point clouds. Figure 5.12 shows the flowchart of the hierarchical classification for key elements. The specific steps are as follows.

1. Power line point cloud extraction

   Power line point cloud extraction methods are generally divided into four categories: (1) methods based on elevation distribution features;

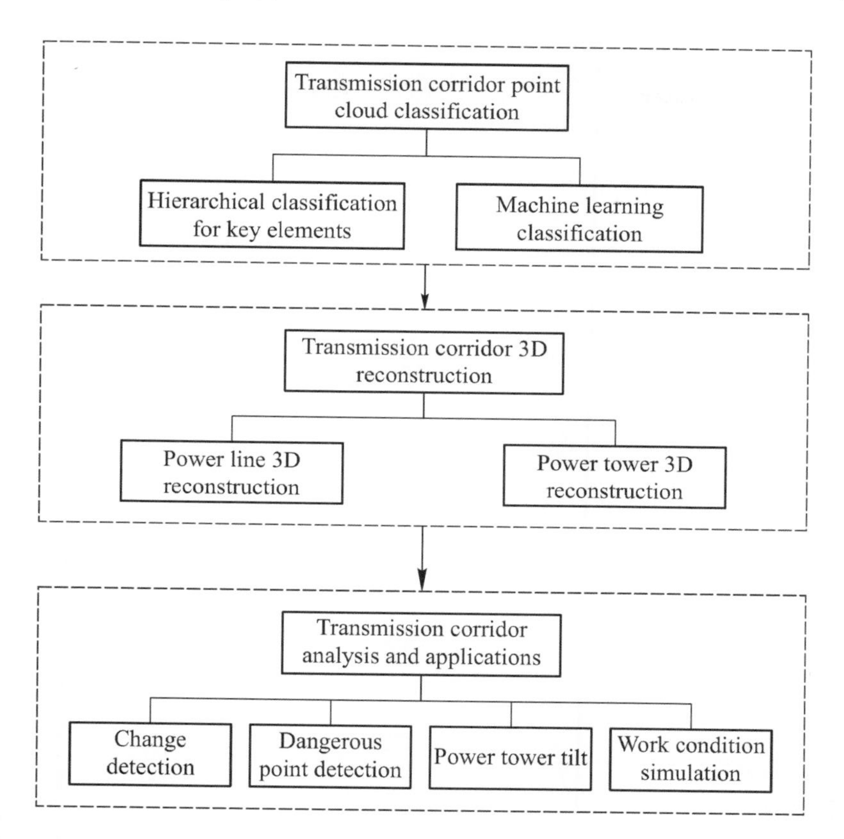

FIGURE 5.11 Flowchart of airborne LiDAR transmission corridor inspection.

(2) methods based on connectivity analysis and region growing; (3) straight or parabolic detection methods based on the Hough transform or random sampling consensus (RANSAC); and (4) methods based on power line point cloud density, echo information, and eigenvalues. In practice, due to variable arrangements of power lines and sparse or missing point cloud data, the features related to the power lines weaken or disappear, which makes the power line point cloud difficult to identify. Therefore, the combination of multiple methods is a commonly used approach for improving the classification accuracy, but it increases the complexity of the algorithm. Figure 5.13 shows the power line point cloud extracted by the RANSAC algorithm (shown in red).

2. Power tower point cloud extraction

   Fast and accurate extraction of power tower point clouds from airborne LiDAR data is the basis for 3D digital modeling. When extracting the tower point cloud, it is necessary to fully analyze the spatial structure characteristics of the tower point cloud. The commonly used extraction methods are divided into the image processing method, region growing method, and hybrid extraction method.

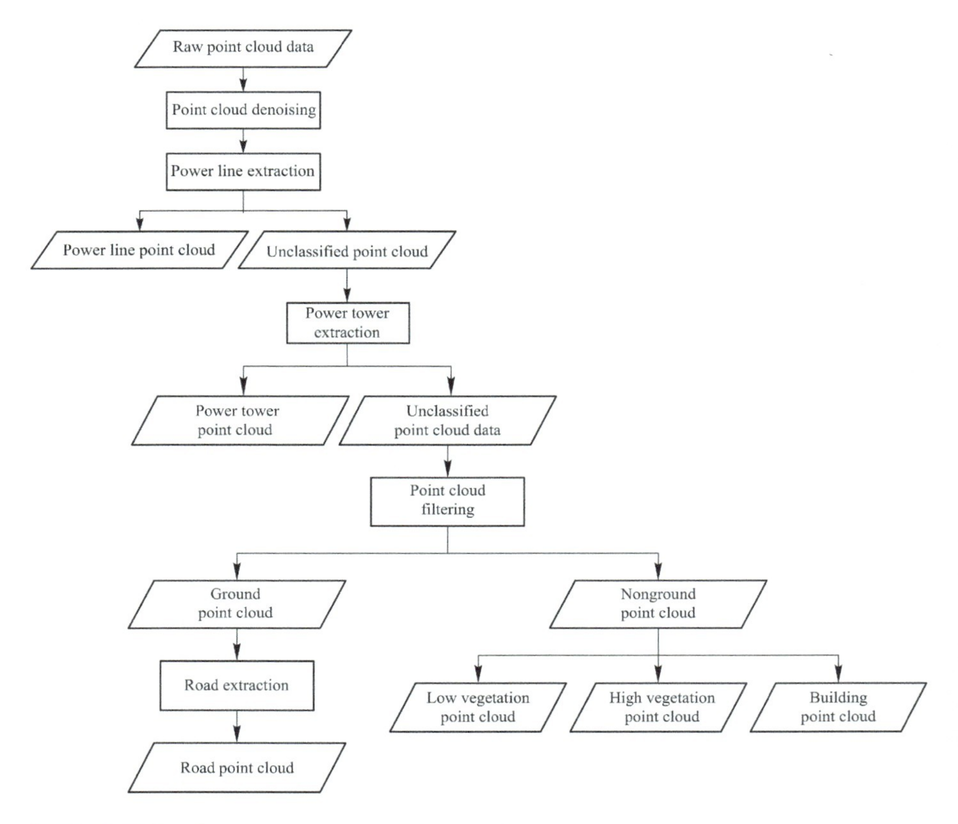

FIGURE 5.12 Flowchart of hierarchical classification for key elements.

The image processing method refers to the idea of tracking the linear structure of a 3D power tower using binary image contour tracking, thereby achieving the accurate extraction of the power tower point cloud. The region growing method is more commonly used. The method first projects the point cloud onto the horizontal plane and grids it. Then the gridded data containing the power towers are clustered by using a region growing algorithm. Some constraints (e.g., vertical projection area of the tower point cloud at a certain height and the tower head length threshold) are set to filter out the point cloud data that do not meet the requirements, to segment an independent tower point cloud. However, these two methods

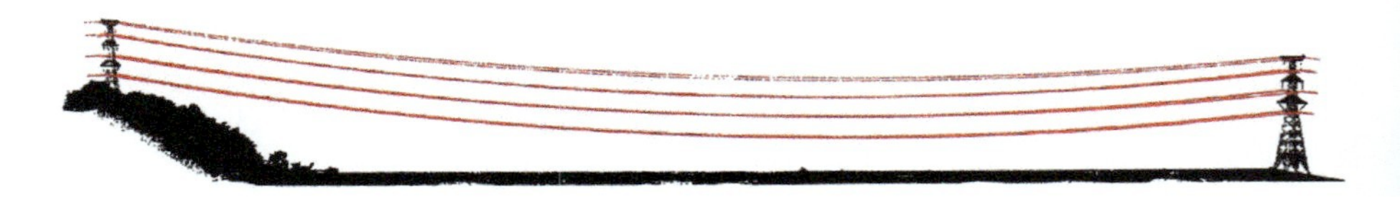

FIGURE 5.13 Power line point cloud of a single transmission corridor.

only coarsely extract the power towers. The results contain a great deal of vegetation, power lines, and other miscellaneous points, which need to be further eliminated manually. Considering the variety of tower types and complex structures, it is difficult to directly extract the complete tower point cloud using a single method. Hence, a combination of multiple extraction methods, i.e., hybrid extraction method, is usually considered. For example, first, the point cloud of the towers is coarsely extracted using the planar grid clustering method and preprocessed by Euclidean cluster denoising and spatial grid region growing. Then the point cloud of the tower's main body and edges is extracted based on the RANSAC spatial linear fitting method, and the surrounding noise is removed using the region-growing method. Finally, the bottom noise is removed for different types of towers. The hybrid extraction method can use the appropriate extraction algorithm according to the point cloud distribution features in different parts of the tower and usually obtains better extraction results. Figure 5.14 shows the extraction results of the T-shape and O-shape tower point clouds using hybrid extraction method.

3. Extraction of point clouds of other objects

Once the power tower point cloud and power line point cloud are extracted, other objects are further classified.

Road point cloud extraction. The road point cloud is extracted based on point cloud spatial geometric features, such as connectivity, elevation differences, slope, and shape, by using algorithms such as region growing.

Building point cloud extraction. Building roofs are generally smooth. The normal vectors of point clouds of flat building roofs have similarity, so the normal vector-based region growing algorithm is used to extract a mathematically significant flat point cloud, which is considered as the building point cloud.

Vegetation point cloud extraction. The vegetation point cloud is classified according to the height from the ground, i.e., height difference threshold. The points with a height difference greater than the threshold are classified

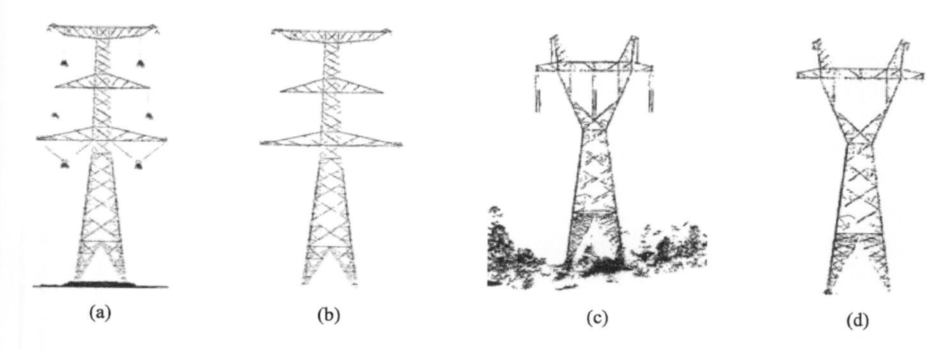

FIGURE 5.14 Power tower point cloud extraction. (a) Original T-shape tower point cloud; (b) extracted T-shape tower point cloud; (c) original O-shape tower point cloud; (d) extracted O-shape tower point cloud.

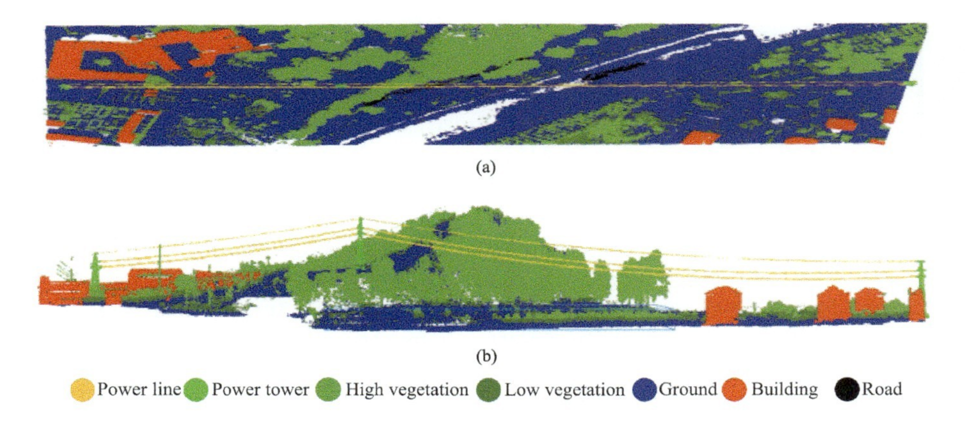

FIGURE 5.15 Results of the point cloud classification by hierarchical classification for key elements in the transmission corridor. (a) Top view; (b) side view.

as high vegetation points, while the points below the threshold are classified as low vegetation points.

Figure 5.15 shows the results of the point cloud classification by hierarchical classification for key elements in the transmission corridor. In general, this method is less hardware dependent and has high classification accuracy and classification efficiency. Also, it basically meets the demand for real-time analysis and processing in the in situ fieldwork. However, the types of features it identifies are limited.

5.3.1.2 Machine Learning Classification

Machine learning classification uses the training set to train the corresponding model, and through the model, classifies the test point cloud. First, the transmission corridor point cloud is divided into a test set and a training set according to a certain ratio. Also, the point cloud is denoised. The typical objects in the transmission corridor scene are specified as power towers, power lines, ground, vegetation, buildings, etc. Then the spatial geometric features of each object are analyzed. By setting appropriate neighborhood dimensions and based on nearest neighbor relationships, point primitive features are extracted and input to the classifier for training. Finally, the test point cloud set is input to the classifier model for test. The accuracy of final classification results is evaluated.

This section uses four commonly used machine learning classifiers to classify transmission corridor point clouds, i.e., K-nearest neighbors (KNN), logistic regression (LR), RF, and gradient boosting decision tree (GBDT). The point cloud features used are listed in Table 4.3 of Section 4.1.3. The classification results are shown in Figure 5.16, with some misclassified areas in the black boxes.

Comparing the classification results of the four classifiers, the results of the three classifiers, KNN, RF and GBDT, are more stable, and the RF and GBDT have similar classification results. The LR algorithm has the worst classification accuracy and is most affected by the data source. It obviously misclassifies the power towers

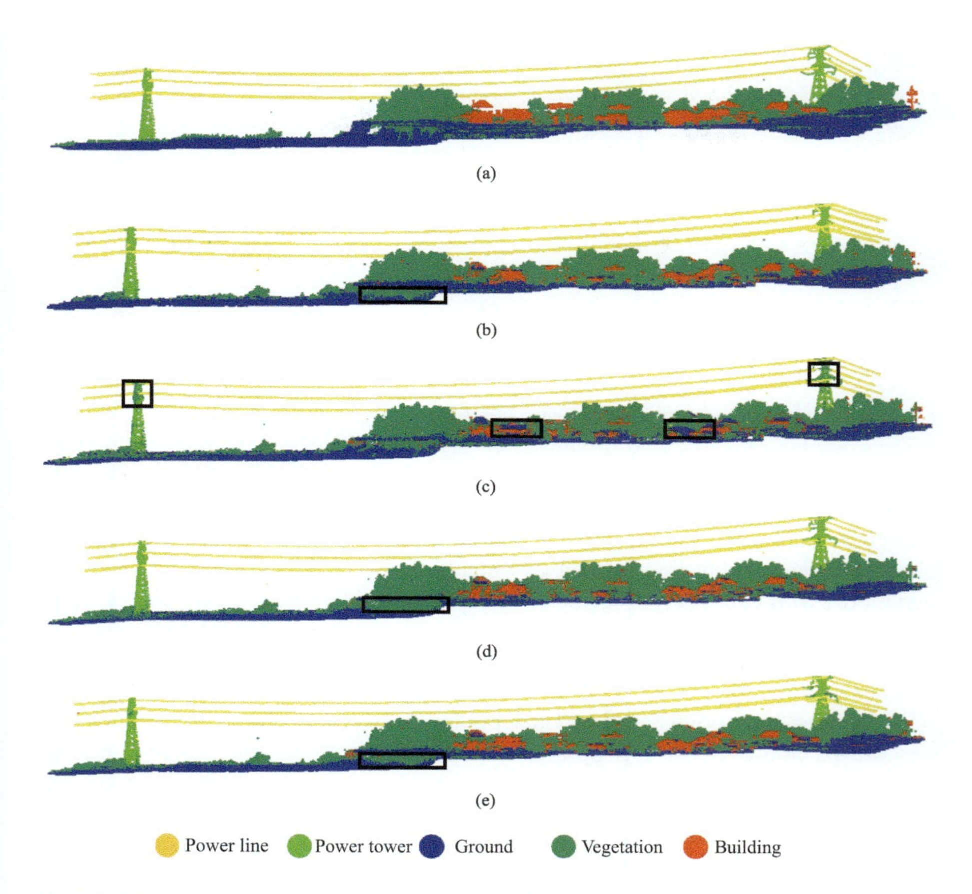

FIGURE 5.16 Results of machine learning classification. (a) Reference classification data; (b) KNN; (c) LR; (d) RF; (e) GBDT.

and power lines. Overall, the machine learning–based transmission corridor classification methods are highly accurate, with a classification accuracy of over 95% for key power elements, which meets the analysis requirements of a wide range of applications. However, they have high computational requirements and poor timeliness. In specific engineering applications, we can choose suitable automatic classification algorithms or combinations of algorithms to improve the efficiency of data processing.

5.3.2 3D Modeling of the Transmission Corridor

5.3.2.1 3D Modeling of Power Lines

Potential danger detection and structural stability analysis are the main components of transmission corridor safety analysis. The point cloud–based power line 3D modeling is the basis for applications, such as 3D visualization, working condition

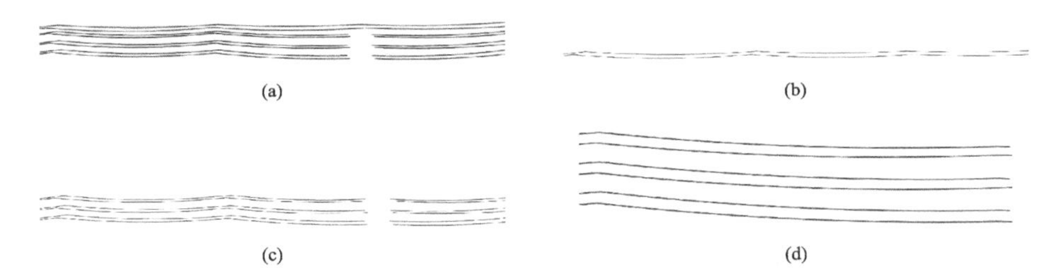

FIGURE 5.17 Power line 3D modeling. (a) Original power line point cloud. (b) Earth wire model. (c) Wire model. (d) Conductor bundle model.

simulation, and early warning analysis. The fitting of power lines is usually described using a combination of 2 two-dimensional (2D) models in the XOY horizontal plane and the XOZ vertical plane or directly using a 3D model. These models are usually classified into five types: models combining straight lines and suspended chain lines, models combining straight lines and a quadratic polynomial with one variable, models combining straight lines and binary quadratic polynomials, polynomial model, and model with multiple line segments (including straight line segment model, parabolic line segment model, and suspended chain line segment models). The modeling process involves the solution of three types of equations: straight line, parabola, and suspended chain line.

Power line hang points are the contact points between the power line and insulators. Based on the power line point cloud and the 3D model, the power line hang points are extracted by analyzing the point cloud local elevation change. Finally, the 3D model of power lines is reconstructed, as shown in Figure 5.17.

5.3.2.2 3D Modeling of Power Tower

Point cloud–based power tower 3D modeling is the basis for real-time and accurately monitoring power tower conditions. There are four main methods for power tower model construction, i.e., manual or semi-automatic modeling using modeling software, data-driven modeling, model-driven modeling, and hybrid-driven modeling (Zhou et al., 2017). Here we introduce the hybrid-driven modeling method for the power tower 3D construction based on the airborne point cloud (Chen et al., 2019a). First, based on the similarity of the power tower structure, the reoriented power tower point cloud is divided into three parts: complex structure, quadrangular frustum pyramid structure, and inverted triangular pyramid structure, corresponding to the point cloud parts in Figure 5.18.

For the upper complex structure, the 3D coordinates of the connection points and the topological correlation between them are determined to obtain the contour corner points. For the lower inverted triangular pyramid structure, find its partition plane and extract the fit points and then perform a linear fit to the points using the RANSAC algorithm. For the central quadrilateral pyramid structure, the external tower profile is extracted to be fitted by a 2D linear equation. The interior of the power tower is divided into four types, i.e., XX, XV, VX, and VV (Figure 5.19). The coordinates of the intermediate intersection points are calculated. The points are divided into

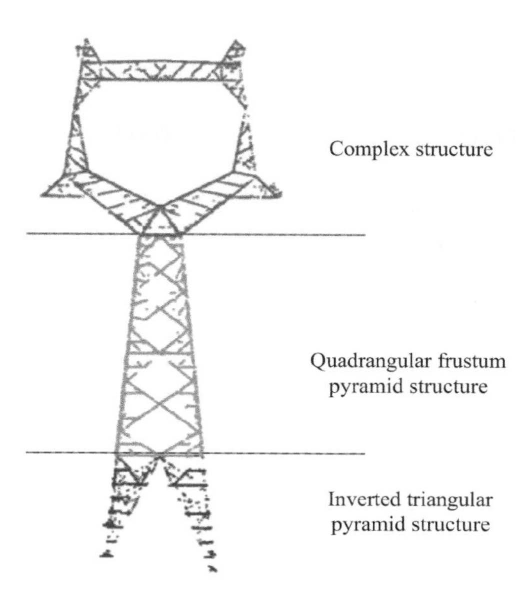

FIGURE 5.18 Power tower segmentation.

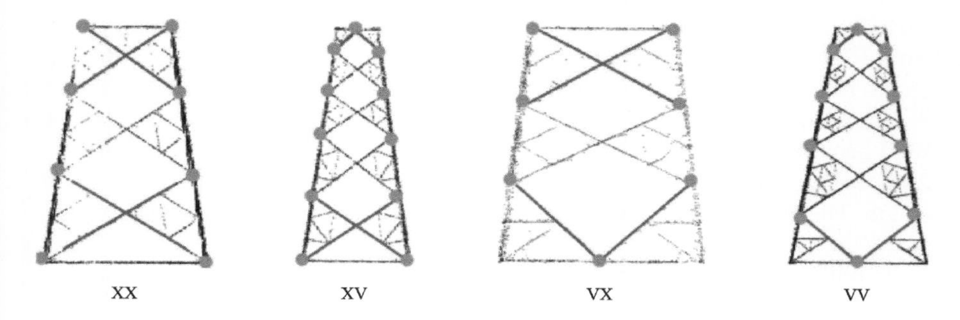

FIGURE 5.19 Types of internal structures of towers.

clusters based on the DBSCAN algorithm and then are fitted to the internal straight lines. The least squares iterations are used to eliminate crossover errors and determine the type of internal structure. Finally, the upper complex structure, the lower inverted triangular pyramid structure, and the central quadrilateral pyramid structure vectors are connected to derive a complete 3D model of the tower. Figure 5.20 shows the results of the 3D modeling of a variety of power towers, with an average modeling error of 0.32 m and an average calculation time of 0.8 s per tower.

5.3.3 Transmission Corridor Safety Analysis

Based on the results of transmission corridor point cloud classification, it is possible to accurately calculate the safety distance between the ontology elements and other objects

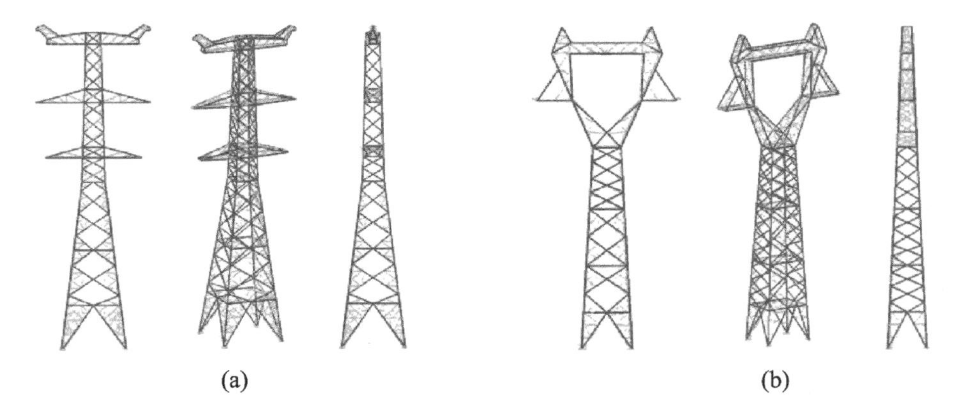

FIGURE 5.20 3D modeling results for two types of power towers. (a) T-shape tower. (b) O-shape tower.

in the corridor, accurately simulate the working conditions of the corridor, make danger predictions, and quantitatively analyze the change patterns of the objects therein.

5.3.3.1 Change Detection

Compared with change detection based on 2D images, the multi-temporal airborne LiDAR data not only allow for the extraction of changed areas and changed objects from 3D space but also provides a more accurate understanding of their change properties. Here we use the multi-temporal point cloud provided by State Grid General Aviation Company in the 800-kV transmission corridor for experiments. The data were acquired by the UAV LiDAR system Riegl VUX-1LR in October 2016 and 2018. Figure 5.21 shows the point cloud of one transmission corridor, with a 505-m length. The data of the two phases are preprocessed, filtered, and classified.

1. Building change detection. The building point cloud data of the two phases are normalized in elevation separately. The changed areas are analyzed and extracted. As shown in Figure 5.22, the new buildings are marked in blue and boxed with red dashed lines. Two new buildings were added to the area over a 2-year period.

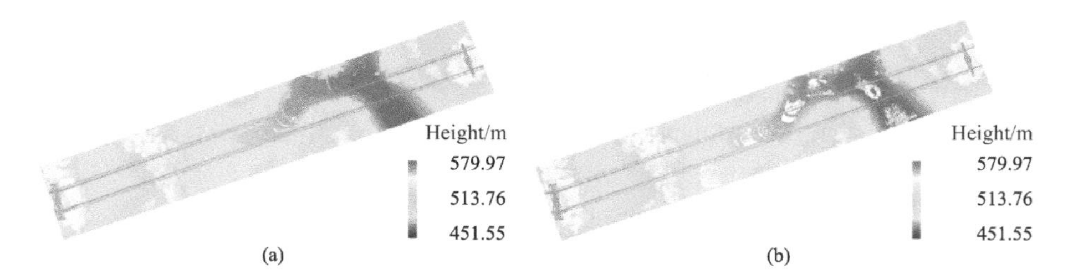

FIGURE 5.21 Top view of transmission corridor point cloud data acquired in (a) 2016 and (b) 2018.

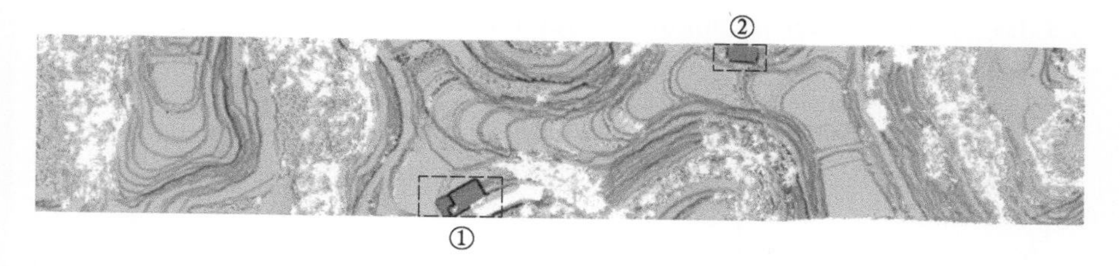

FIGURE 5.22 Areas of building change.

2. Vegetation change detection. Vegetation height change is one of the important factors of the transmission corridor safety inspection. The elevation threshold is set to extract the changed point cloud in the two phases of point cloud data and mark them to get the changed areas.

5.3.3.2 Danger Detection

Dangers might be harmful to the safe operation of power transmission corridors. Transmission corridors are located in urban and rural areas, with long and extensive lines exposed in the fields. For example, unauthorized tree planting under high-voltage lines might cause wires to discharge against trees. Hence, it is of great importance to monitor dangers and conduct a safety warning for transmission corridors based on airborne LiDAR data. To intuitively view the dangers, the specifications used for the danger analysis, the results of the analysis, and the details of the dangers can be automatically output in the form of a report. Figure 5.23 shows the details of a tree between power towers No. 15 and No. 16 and 6.855 m away from tower No. 15, where the distance to the power line is less than the safety threshold and is determined as a danger. This tree needs to be felled.

No.	Power tower intervals	Distance from fore-tower/m	Distance from aft-tower/m	Coordinates	Measured distance/m		
					Horizontal	Vertical	Headroom
1	#15-#16 lower left phase	6.855	392.793	107°19'4.34"E, 29°43'42.22"N	1.02	4.32	4.44

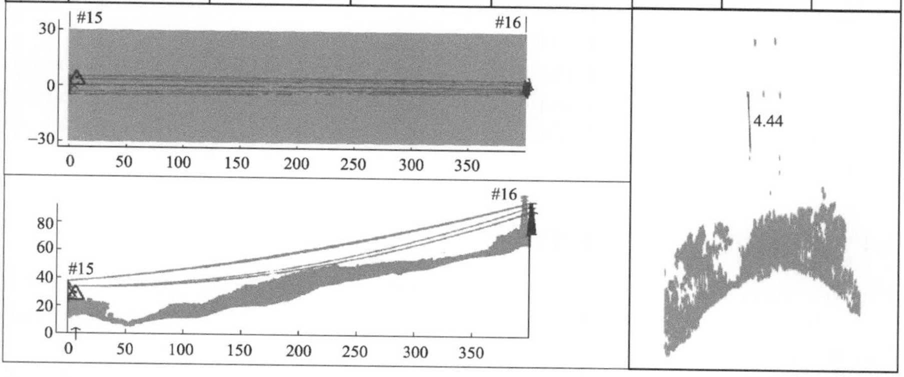

FIGURE 5.23 Example of safety analysis report.

5.3.3.3 Power Tower Inclination Detection

When the transmission corridors pass through areas such as coal mining areas, soft soil areas, and riverbed areas, towers are prone to tilting, which causes unbalanced forces on the tower earth wires, resulting in changes in tower forces and affecting the normal operation of the line. Hence, the tilting degree of the tower needs to be checked regularly, and their changes need to be predicted. The horizontal section method is a common method for calculating tower inclination (Cai et al., 2018). It assumes that the tower is tilted in a single direction and there is no distortion or deformation in the central axis of the tower. The main steps include (1) stratifying the tower point cloud vertically; (2) extracting the coordinates of the center point of each tower layer; and (3) calculating the coefficients of the spatial linear equations of each tower layer center using least squares fitting and calculating the inclination gradient and inclination angle of the tower.

5.3.3.4 Simulation of Working Conditions

Simulation of working conditions achieves a safety warning by simulating the transmission corridor working conditions. For example, the wires in the transmission corridor are simulated by constructing the wire state equation to achieve wire safety detection. The wire state equation is the condition equation in which the wire changes from one state to another, as in Equation (5.7). In terms of vegetation growth prediction, the safe distance between trees and wires is monitored by constructing the tree growth model to predict their growth dynamics. In terms of ground change, the warning of ground change is carried out by simulating changes in the horizontal and vertical structure of the ground under the same transmission corridor at different times.

$$\sigma_n - \frac{\gamma_n^2 l^2 E\cos^3\beta}{24\sigma_n^2} = \sigma_m - \frac{\gamma_m^2 l^2 E\cos^3\beta}{24\sigma_m^2} - \alpha E\cos\beta(t_n - t_m) \tag{5.7}$$

where σ_n, σ_m are the horizontal pressures of wire sag in the two states, respectively; γ_n, γ_m are the wire-specific loads in the two states, respectively; t_n, t_m are the wire temperature in the two states, respectively; l, β are the span and height difference angle, respectively; and α, E are the temperature expansion coefficient and elasticity coefficient of the wire, respectively.

Figure 5.24 shows an example of transmission corridor working condition simulation. By simulating changes in power lines under conditions of high temperature, high wind, and heavy ice, possible danger points are predicted to reduce the likelihood of power accidents.

5.4 BUILDING 3D MODELING

The planning, construction, and management of urbanization have an increasingly strong demand for a 3D digital building model (DBM). As an important part of geospatial information data, DBM has a wide range of applications in smart cities, autonomous driving, and energy demand assessment (Yang & Lee, 2019). The DBM construction based on point clouds is mainly divided into two

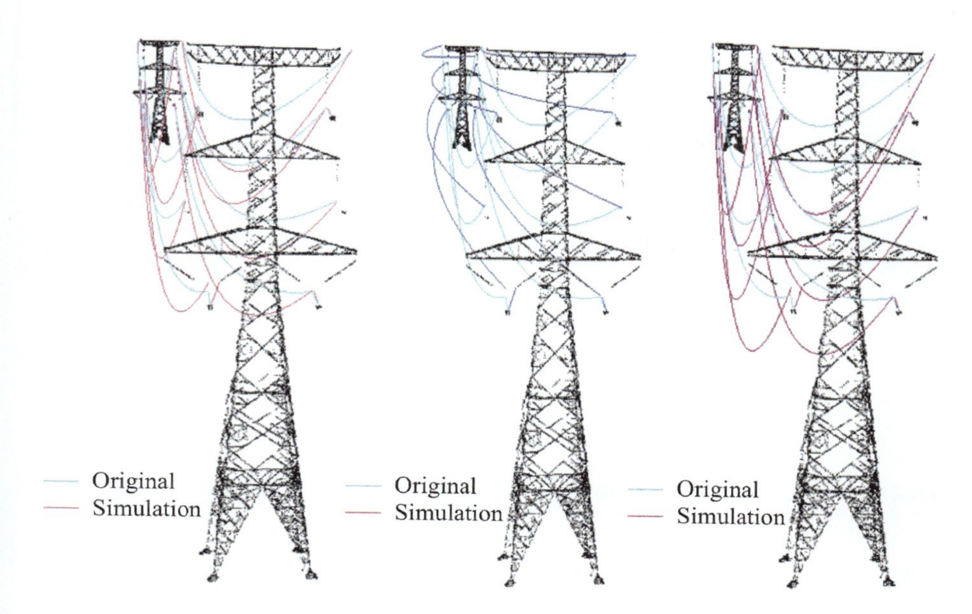

FIGURE 5.24 Example of the transmission corridor working condition simulation in conditions of (a) high temperature; (b) high wind; and (c) heavy ice.

categories: model-driven modeling methods and data-driven modeling methods (Maas & Vosselman, 1999).

The model-driven modeling methods usually assume that real buildings such as houses and high-rise buildings are composed of basic geometries. By pre-establishing a model library with simple geometries, the methods find the best model in the database that matches the point cloud based on prior knowledge such as building topology by using the algorithms, e.g., Bayesian inference algorithm (Morris, 1983) or reversible jump Markov Chain Monte Carlo (RJMCMC) algorithm (Green, 1995), and finally build the DBM model. The model-driven modeling methods have strong robustness, low data quality requirements, and can model well even when some point cloud is missing. The models have a high degree of regularization, do not require individual processing of the building point cloud, and retain a lot of semantic information. However, this kind of method needs to establish a building model library in advance and is not suitable for modeling complex buildings.

The data-driven modeling methods can achieve relatively complex building modeling without any prior knowledge and have obvious advantages in modeling scale and speed. They assume that all the building outlines are closed polygons. First, building points are extracted from the input point cloud. Then a series of individual processing steps are performed, such as roof segmentation and boundary extraction. Specifically, the feature points and feature lines of the outer boundary of the building are extracted by using clustering and other methods. Finally, these feature points are connected according to topology rules to construct the DBM. The data-driven methods have strong applicability and can describe more detailed information of the building structures. The disadvantage is high requirements for data quality.

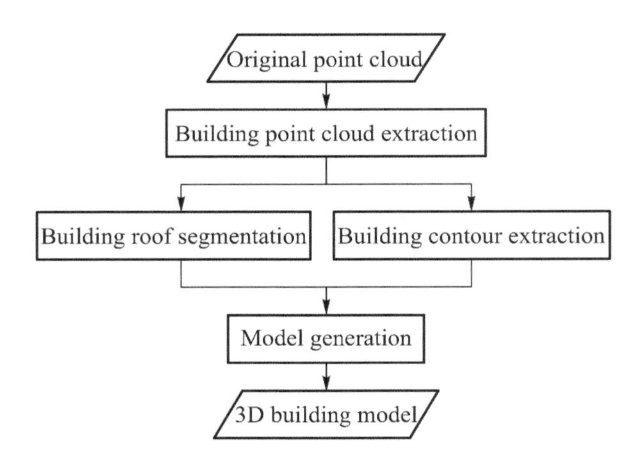

FIGURE 5.25 Flowchart of building 3D modeling.

Also, when many noise points exist or the point cloud is seriously missing, modeling is difficult, and semantic information is largely missing.

Overall, the two modeling methods have their own advantages and disadvantages in building modeling. The model-driven method is not sensitive to data quality and can quickly reconstruct building models with strong regularization, the correct topology, and a beautiful appearance. However, its accuracy is limited by the richness of the model library, and it cannot achieve 3D modeling of complex buildings. The data-driven method can reconstruct building models with any shape, and the reconstructed models include more roof details, but the data processing is complicated.

This section introduces a typical data-driven building modeling method and demonstrates a practical case. First, the building point cloud is detected and extracted from the original point cloud. Then the roof patch is extracted using the segmentation algorithm. Finally, the 3D building model is constructed by combining the roof patch information and building boundary information. The key steps include building point cloud extraction, building roof segmentation, building contour extraction, and building 3D model generation (Figure 5.25). The data used in the example are the Area 3 data from the Vaihingen dataset released by the German Association of Photogrammetry and Remote Sensing. The dataset was acquired by the Leica ALS50 system in August 2008. The field of view range is ±45°. The average flight altitude is 500 m. The average point cloud density is 4 pts/m^2. Modeling accuracy is assessed by the Organizing Committee of the ISPRS Working Group (WG) II/4.

5.4.1 Building Point Cloud Extraction

The process of extracting the building point cloud from the original point cloud is shown in Figure 5.26. First, remove the obvious wall points and vegetation points with the normal vector feature and multi-echo feature. Then, extract the ground points with the normal vector–based region growing algorithm; determine whether the point cloud region grows according to the normal vector and distance constraints; and mark

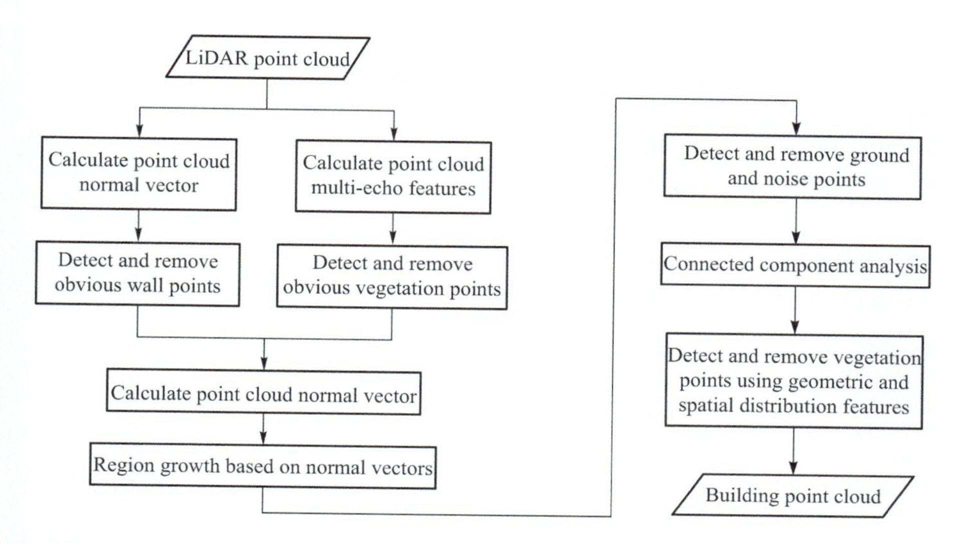

FIGURE 5.26 Flowchart of building point cloud extraction.

the point sets with a small point number or large elevation fluctuations as noise points. Finally, the connected component analysis method is used to perform 3D Euclidean space clustering. Some geometric features (such as area, length-width ratio, and average point cloud density) are used to distinguish buildings and vegetation points.

5.4.2 Building Roof Segmentation

Building roof segmentation separates the roof patches that do not belong to the same plane, which is one of the key steps in building 3D modeling from point clouds. The separated roof patches reflect the topology of the roof. The intersection lines and points between the roof patches help to construct the building models.

Here we introduce a common roof segmentation algorithm, i.e., the normal vector region growing roof segmentation algorithm. Taking the Area 3 data of the Vaihingen dataset as an example [Figure 5.27(a)], the covariance eigenvalues of the point cloud are first used to detect non-planar points [Figure 5.27(b)]. Then the normal vector–based

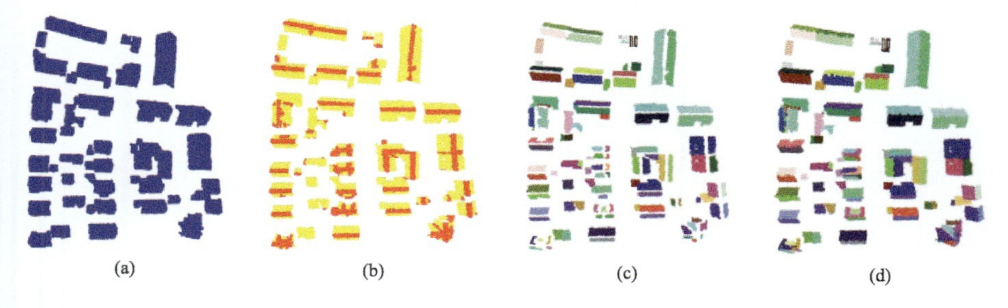

FIGURE 5.27 Building roof segmentation of Area 3. (a) Roof point cloud; (b) extracted non-planar points (red); (c) plane point segmentation result; (d) final segmentation result.

region growing algorithm is conducted to detect roof planes [Figure 5.27(c)]. In this process, the distance from the point to the plane and the distance from the point to other points in the plane are taken into account. Finally, the building roof is obtained by assigning the non-planar point cloud into the roof planes [Figure 5.27(d)]. The non-planar point here generally means that the point cloud in the neighborhood of

TABLE 5.5
Accuracy Evaluation of Roof Segmentation of Area 3

Data	Compl.Roof* [%]	Corr.Roof ** [%]	Compl.Roof 10 [%]	Corr.Roof 10 [%]
Area 3	73.2	100	83.1	100

* Compl.Roof (10) refers to the recall rate of all roof segmentation results (or roofs with area greater than 10 m^2).
** Corr.Roof (10) refers to the accuracy rate of all roof segmentation results (or roofs with area greater than 10 m^2).

TABLE 5.6
Pseudo-Code of the Normal Vector–Based Region Growing Algorithm for Roof Point Cloud Segmentation

Input: ***P***, roof point cloud; ***thres_nor***, threshold of normal difference; ***thres_cur***, curvature threshold.
Output: {*C_list*}, list of classes.
Algorithm:
initialize the unvisited points $\boldsymbol{P}_{un} = \boldsymbol{P}$;
initialize the class list {***C_list***};
while $\boldsymbol{P}_{un}$!= NULL
    initialize a new class ***C_list***.
    initialize a seed point set $\{P_S\}$.
    select the point with minimum curvature P_{min} from $\boldsymbol{P}_{un}$.
    add P_{min} into the seed point set $\{P_S\}$.
    add P_{min} into the class ***C_list*** and remove it from $\boldsymbol{P}_{un}$.
    for each P_i in $\{P_S\}$:
        find the neighboring points $\{P_{\text{neigh}}\}$ of point P_i.
        for each P_j in $\{P_{\text{neigh}}\}$:
            if abs(P_j.normal –P_j.normal) < *thres_nor*
                add P_j into the class ***C_list*** and remote it from $\boldsymbol{P}_{un}$.
                if abs(P_j.curvature –P_i.curvature) < *thres_cur*)
                    add P_j into $\{P_S\}$.
                endif;
            endif;
        endfor;
    endfor;
    add the class ***C_list*** to the classlist {***C_list***}.
endwhile;
return {*C_list*}.

the point does not belong to the same plane, while the plane point cloud means that the point cloud (including this point) in the neighborhood of the point does belong to the same plane. The evaluation results of model accuracy show that the roof segmentation based on region growing extracts the building roof plane correctly, and the accuracy of roof segmentation in Area 3 is 100% for roofs with area larger than 10 m^2 (Table 5.5). The pseudo-code of the normal vector–based region growing algorithm for roof point cloud segmentation is given in Table 5.6.

5.4.3 Building Contour Extraction

There are two kinds of methods for extracting building contours (or building boundaries). One is to convert the point cloud into an image and obtain the contours by using an edge detection algorithm of image processing; the other is to directly process the point cloud and determine the inflection point according to the adjacency relationship between points, then connect each inflection point to generate the architectural contour, such as the α-shape algorithm (Edelsbrunner et al., 1983). These two methods only derive the general outline of the building, with different degrees of the "sawtooth" phenomenon, which is seriously inconsistent with the real building. In practice, the building contours are parallel or vertical to each other. Therefore, it is necessary to regularize the boundary line after boundary extraction to make it more realistic. Boundary regularization is a key step of contour extraction. Usually, the key points of the boundary are extracted first. The new contour is then formed by connecting these key points. Finally, the adjacent boundary is changed to vertical relationship by forcing orthogonality.

Generally, the steps of building contour extraction are as follows. First, resample the building point cloud data to obtain the raster point cloud. Then use the 2D α-shape algorithm to obtain the building boundary. Next use the RANSAC algorithm to extract the line segments on the building edge. Finally connect these line segments and adjust them to be parallel or vertical to achieve the boundary regularization.

5.4.4 Building Model Generation

The key to building model generation is to obtain the correct intersection lines and intersection points between roofs. Generally, building model generation includes the following steps: determine the mathematical intersection line with the roof surface equations; calculate the optimal intersection line based on the intersection line between roofs and topological relations between roofs; obtain the intersection points between roofs by the topological map; and generate the 3D building model by combining the intersection points of the building boundary and roof intersection lines.

Roof intersection line determination algorithms usually assume that the optimal roof intersection line separates roof facets absolutely and is located near the mathematical intersection line. Therefore, according to the mathematical roof intersection line l_m, change the equation of the intersection line in a certain range (adjacent points between adjacent roofs), and score the intersection line. The intersection line

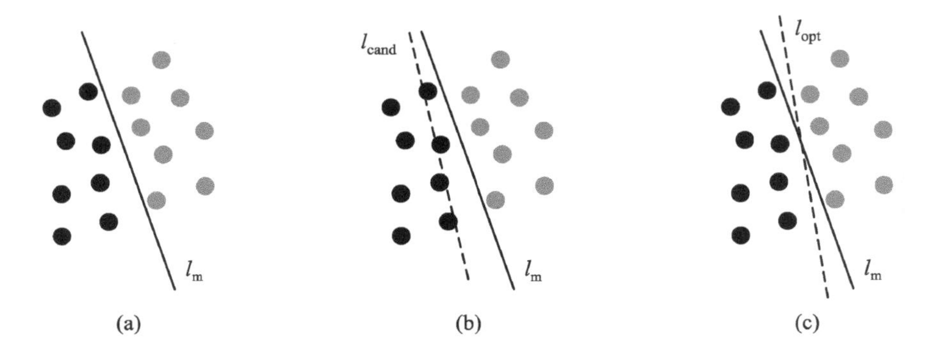

FIGURE 5.28 Determination of optimal roof intersection line. (a) Determination of mathematical roof intersection line. (b) Determination of candidate roof intersection line. (c) Determination of optimal roof intersection line.

with the highest score is selected as the best roof intersection line. The specific steps are as follows:

1. According to the roof segmentation results, the intersection line of the plane equations is calculated to determine the mathematical intersection line of the roof l_m [Figure 5.28(a)].
2. According to the specified range, determine the possible region of the intersection line. Change the slope of the center point position of the variable l_m, and obtain the candidate roof intersection lines l_{cand} [Figure 5.28(b)].
3. Score l_{cand} according to the relationship between the point cloud of adjacent roofs and l_{cand}.
4. According to the scoring results, the line with the maximum score is selected as the best roof intersection line l_{opt} [Figure 5.28(c)].

After determining the roof intersection line, the corresponding undirected graph is constructed by the topological relation between roofs. Then the minimum closed loop in the graph is detected and the intersection point between roofs is determined. According to the topological relation (plane intersection relation) in Figure 5.29(a),

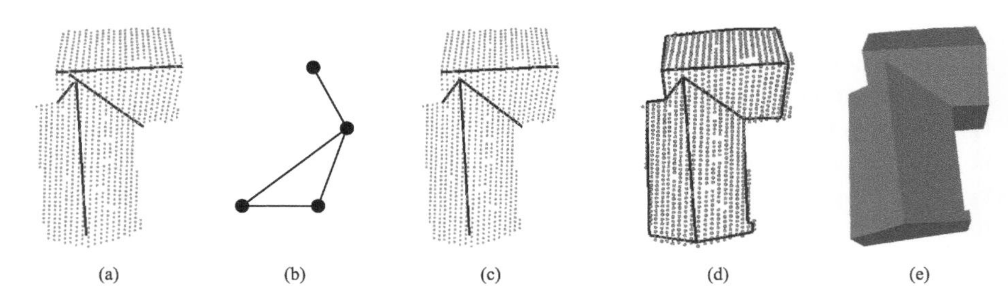

FIGURE 5.29 Building 3D model generation based on topological graph. (a) Roof intersection line; (b) roof topological relationship; (c) roof intersection points; (d) roof wireframe model; (e) building 3D model.

an undirected graph of a roof topological relation is constructed, as in Figure 5.29(b), in which each endpoint represents a roof plane, and the lines between planes indicate that there are intersection lines between roofs. By detecting the minimum closed loop in the undirected graph, the planes intersecting at a point are determined. Then the intersection points of roof intersection lines are obtained by the intersection constraints of these planes. The results are shown in Figure 5.29(c).

Finally, the intersection nodes between the roof intersection lines and the regular boundary and other key nodes are derived based on the roof regular boundary and the intersection points and lines between roofs. The final 3D model is generated based on these nodes. The final building wireframe model is shown in Figure 5.29(d), and its 3D model is obtained by extending the building boundary to the ground surface, as shown in Figure 5.29(e). Figure 5.30 shows the final results of the building models in Area 3.

Two indicators, root mean square (RMS) and root mean square of Z value (RMSZ), are used to quantitatively evaluate the modeling accuracy of Area 3. The result is RMS = 0.8 m, RMSZ = 0.1 m. The RMS value represents the 2D plane precision of the model. By projecting the model onto the XY plane, the precision of the building boundary is compared with that of the reference model boundary. The RMSZ represents the geometric accuracy of the elevation of the building model, which is obtained by comparing the DSM differences between the modeling results and the reference data. The result shows the boundary regularization method based on the point cloud can derive accurate building boundaries, and the building model generation method in the example can accurately extract intersection lines and intersection points with high modeling accuracy.

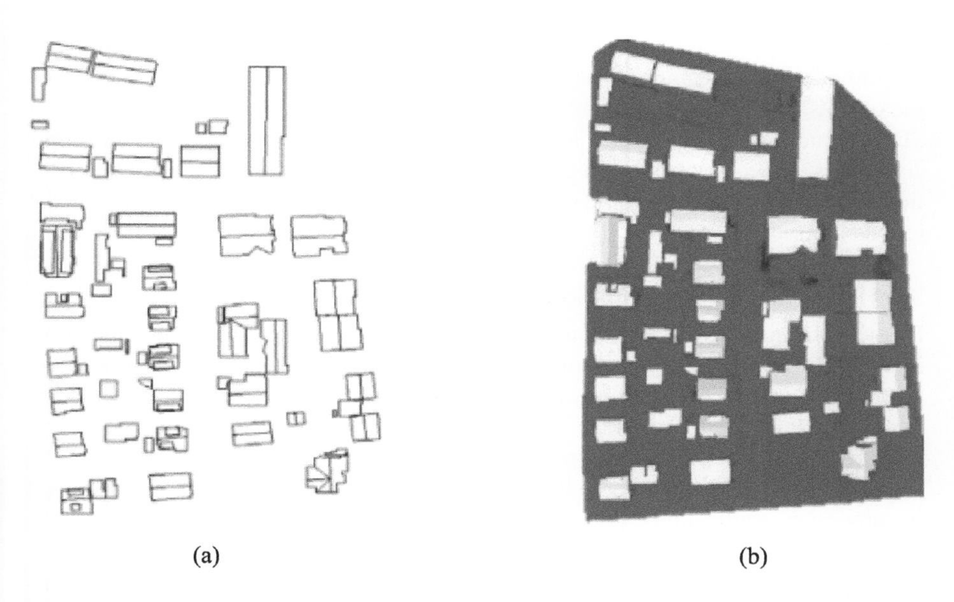

FIGURE 5.30 Building 3D model of Area 3. (a) Roof wireframe model; (b) building 3D model.

5.5 AUTOMATIC DRIVING

The vehicles equipped with an autonomous driving system (ADS) have the capability of self-driving and effectively avoiding traffic accidents. It is beneficial to reduce casualties and property losses. In general, the ADS consists of environmental perception, autonomous positioning, planning decision, and control module. Environmental perception enables an unmanned vehicle to accurately understand the road environment, and autonomous positioning provides a spatial location for the vehicle. Both are the basis for the safe driving of the unmanned vehicle (Yurtsever et al., 2020). In the 21st century, LiDAR is used for environmental perception and autonomous positioning, which has promoted the development of unmanned driving technology. First, the LiDAR point cloud provides the high-accuracy position of a road environment. The environmental perception based on the LiDAR point cloud determines the classification, position, and direction of objects at the same time, guiding the driving of the unmanned vehicle. In addition, the a priori map composed of the LiDAR point cloud has been widely used for autonomous positioning of unmanned vehicles. This section describes the application of LiDAR in environmental perception and autonomous positioning.

5.5.1 Environmental Perception

The unmanned vehicle sensor continuously scans the surrounding environment and extracts important information in the environment, such as the classification, location, and driving direction of vehicles, pedestrians, and other objects. Therefore, the unmanned vehicle can fully understand the road environment and accelerate, decelerate, change lanes, and stop, ensuring the driving safety. Specifically, semantic segmentation and object detection are important components of environment perception. The former subdivides the complete road scene into multiple sub-parts with the same internal attributes. The latter determines the classification and location of the objects. Compared with images, the LiDAR point cloud provides high-precision 3D information of objects and is not easily affected by illumination and weather conditions. Here we introduce the application of LiDAR in the environment perception from two aspects: semantic segmentation and object detection, and describe the related algorithms.

5.5.1.1 Semantic Segmentation

Image information extraction is very important for image-based applications. However, useful or interesting information is often only located in some specific areas of the image. For example, image-based vehicle detection needs to obtain the position of the vehicle, while the vehicle is usually located on the lane in the image. Semantic segmentation segments a complete image into several sub-parts, and the pixels in each sub-part belong to the same classification, narrowing the search scope of subsequent applications. Like the semantic segmentation of an image, the semantic segmentation of a point cloud divides the input point cloud data into a series of clusters, and all points in the same cluster have the same classification label. For automatic driving, these labels include vehicles, pedestrians, buildings, ground, trees, road signs, traffic lights, etc.

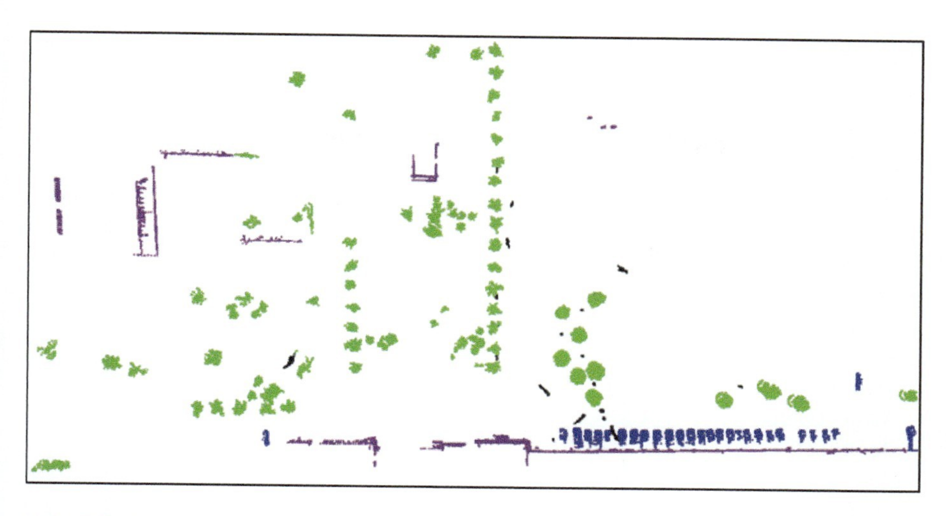

FIGURE 5.31 Result of point cloud semantic segmentation.

The semantic segmentation of a point cloud is defined as follows. There is a group of laser points $X = \{x_1, x_2, \ldots, x_n\}$ and category labels $Y = \{y_1, y_2, \ldots, y_k\}$ in 3D space. Each laser point x_i $(i = 1,2,\ldots,n)$ is assigned a label $y_i (j = 1,2,\ldots,k)$. Figure 5.31 shows the semantic segmentation results of point cloud data. In the figure, blue points are vehicles, green points are trees, purple points are buildings, and black points are other objects.

Traditional point cloud semantic segmentation is achieved by extracting features (such as distance and edge), estimating geometry, supervised/unsupervised learning, and so on. In recent years, the application of deep learning has been expanded from an image to a 3D point cloud. The deep learning model for images requires that the input data have a regular structure. Thus, disordered and unstructured point clouds cannot be directly input into the deep learning model used for images. The early deep learning–based point cloud semantic segmentation needs to convert the point cloud into data with a regular structure (such as image and voxel). In contrast, the subsequent PointNet and PointNet++ network architectures solve this problem of disordered and unstructured data, realizing the direct processing of point clouds.

At present, a variety of stable and efficient deep learning models have been used for segmenting the point cloud to meet the requirements of real-time environment perception. According to the format difference of input data, models are divided into image-based model, voxel-based model, and point cloud–based model (Li et al., 2020). The image-based models require that virtual images are set at multiple angles, and the point cloud is projected into the virtual image to generate multi-view images. After projection, the unordered point cloud is transformed into an image. Then, a suitable deep learning model is further constructed based on the mature image network framework to segment the road point cloud. For the voxel-based models, the point cloud is first voxelized and then input into the deep learning model to achieve semantic segmentation (Zhou & Tuzel, 2018). The point cloud–based models use the deep learning model for the point cloud to extract local and global features.

For example, PointNet can extract features from a point cloud for a subsequent module of learning and finally achieve the semantic segmentation.

5.5.1.2 Object Detection

An image can clearly record the road scene; thus, image-based object detection is widely used in the environmental perception of unmanned vehicles. Object detection includes two steps: classification determination and spatial location calculation. Although the location of the object in the road scene can be estimated by using a single camera or multi-camera system, the feature matching process increases the time cost. In addition, the accuracy of the location estimation still needs to be improved. The LiDAR point cloud contains high-precision position coordinates. The object detection of a point cloud can directly obtain the object position.

Object detection of a point cloud is usually expressed as a 3D bounding box with a classification label. The bounding box is the smallest box containing the object, which approximately represents the position, classification, and direction of the object. The 3D bounding box is defined as (x,y,z,h,w,l,θ,c) (Li et al., 2020). (x,y,z) is the center of this box representing the object's location. l, w, h is the box's length, width, and height, respectively. c is the object class (same as the class of point cloud). θ is the object direction. Figure 5.32 (a) and (b) shows the initial and output bounding boxes during the object detection, respectively. θ is the rotation angle required to obtain the output bounding box from the initial one with the rotation center of (x,y,z).

The point cloud data of the road scene contains many road points, which are usually connected with other objects and interfere with the detection. Therefore, the road points can be removed, which not only reduces the amount of data but also avoids the interference to detection to a certain extent. For example, the edge line recognition method is usually used to extract road points. The specific steps are as follows. First, part of the non-road point cloud is eliminated by setting an average elevation threshold. Then, the road is divided into several sub-sections along the cross-section direction of the road to reduce the impact of the road fluctuation. Finally, using the

FIGURE 5.32 Object detection results of the point cloud of a road scene.

high-intensity characteristic of the road edge point, the edge points of each subsection are extracted and fitted into a smooth edge line. The points between the edge lines are exactly the road point cloud.

The traditional object detection algorithms of a point cloud, such as Euclidean clustering and region growth, usually cluster the point cloud and determine the category by analyzing the characteristics of the cluster. For example, traffic signs are detected according to the height, intensity, flatness, and size. In addition, machine learning can be used for traffic sign detection. For instance, the vehicles, streetlight poles, traffic signs, and other targets can be detected from a point cloud by combining super-voxel segmentation with Hough Forest.

A deep learning model of a point cloud is widely used in real-time detection of the road environment. Like semantic segmentation, the deep learning models include the image-based model, voxel-based model, and point cloud–based model. Generally, the image-based model first projects the point cloud into an image (such as the top view), then uses the deep learning model to detect the object, and finally projects the object back to the point cloud to achieve the boundary box generation of the object. The voxel-based model first converts a point cloud into 3D grids for detection and finally outputs the target bounding box. The point cloud–based model usually involves three steps. (1) The point cloud is divided into multiple clusters to obtain the rough position of the object. (2) Features are extracted from the region where the object is located. (3) The object bounding box is calculated based on the extracted feature. In recent years, point cloud instance segmentation, which integrates semantic segmentation and object detection, is gradually receiving attention in point cloud processing. It aims to combine semantic information with the results of object detection.

5.5.2 Autonomous Positioning

Accurate autonomous positioning is the guarantee for the safe driving of unmanned vehicles. The integration of the global navigation satellite system (GNSS) and inertial measurement unit (IMU) can achieve the autonomous positioning of unmanned vehicles. However, the GNSS has a low positioning accuracy in densely built areas or tunnels or is even unable to position. Simultaneous localization and mapping (SLAM) can be used for accurate positioning. However, stability and computing costs are still the challenges faced by SLAM when used for automatic driving. At present, the positioning method based on an a priori map is widely used. First, the laser point cloud is used to construct an a priori map (such as a high-precision map). Then the point cloud collected by the LiDAR sensor equipped on the vehicle is matched with the a priori map to obtain the vehicle position. This section introduces the application of LiDAR in the positioning of unmanned vehicles by taking the example of the a priori map–based method. First, construction of the a priori map, i.e., high-precision map, is briefly explained, then the relevant positioning methods are introduced.

5.5.2.1 High-Precision Map Production

As a prior map, a high-precision map contains important information, such as the location of lane lines, the location and type of guidance signs, and the location of traffic lights. Such information provides a location reference for unmanned vehicles. The laser point cloud collected by LiDAR is an important data source for high-precision map production. It can generate a point cloud map and an annotated map. The former provides the location information of features, and the latter provides the classification and the topological relationship of objects, such as lane lines and sidewalks.

The production of high-precision maps involves three steps: field data collection, back-end data processing, and manual verification and release. To ensure the accuracy of annotation, the combination of manual and semi-automatic annotation of a point cloud is still the main production mode of high-precision maps.

5.5.2.2 Prior Map–Based Unmanned Vehicle Positioning

The prior map constructed by the LiDAR point cloud records the location information of the road scene. In most cases, the ground objects near the lane (such as buildings and streetlights) have little changes. Hence, the unmanned vehicle can achieve positioning by finding the best match between the collected point cloud and the prior map. The coordinate system where the prior map is located is the global coordinate system. The point cloud collected by the unmanned vehicle is in the local coordinate system with the LiDAR as the origin. In case of the best match, the coordinate of the origin of the local coordinate system in the global coordinate system is the position of the vehicle.

Landmarks of road scenes can be used for the a priori map–based positioning. First, the landmarks, such as curbstones, streetlamps, and traffic lights, are extracted from the point cloud collected by the unmanned vehicle. Then the point cloud of the landmarks is compared with the prior map to obtain the best match. The priori map–based positioning highly depends on the landmarks. When there are few landmarks, the performance of this method needs to be improved.

A common method of a priori map–based positioning uses a combination of matching and GNSS. First, the rough position of the unmanned vehicle in the prior map is obtained by GNSS and dead reckoning. Then, the point cloud collected by the unmanned vehicle is iteratively matched with the prior map near the rough position to find the best matching position, achieving the accurate positioning of the unmanned vehicle.

5.6 CROP MONITORING

The applications of LiDAR technology in crop monitoring include estimation of plant height and leaf length, retrieval of crop structural parameters, e.g., LAI, leaf area volume density (LAVD), LAD, fraction of absorbed photosynthetically active radiation (FPAR), and aboveground biomass (Luo et al., 2019a; Qin et al., 2017; Su et al., 2018). In terms of data sources, some studies used airborne or terrestrial LiDAR waveform or point cloud data directly, while others combined the LiDAR data with hyperspectral/multispectral remote sensing images. In terms of retrieving methods, some studies used the multivariate statistical regression to establish the correlation model between the LiDAR feature parameters and the crop structural parameters, while other studies establish the physically based retrieval models of crop structural parameters based on the LiDAR feature parameters. Here, we take an example of corn to introduce the application of LiDAR technology in crop monitoring.

5.6.1 Identification of Corn Planted Area

The corn planted area should be identified before retrieving corn structural parameters. Traditionally, the crop planted area is identified from the hyperspectral or multispectral remote sensing images by using supervised classification methods. With the rapid development of LiDAR technology, more studies use LiDAR data or combine LiDAR data and spectral images to identify crop planted areas (Li et al., 2015; Qin et al., 2017). Specifically, the features of LiDAR data and optical imagery (e.g., waveform features, point cloud feature, reflectance features) are extracted and then input into the classifier for classifying or identifying the crop planted area. Taking the airborne LiDAR full-waveform data as an example, there are three steps during the identification of corn planted areas. First, the LiDAR waveform is decomposed. Second, a variety of waveform feature parameters are extracted, such as amplitude and width of each waveform component, mixed ratio parameter, height, symmetric parameter, and vertical distribution. Finally, these extracted feature parameters are input into the classifier for identifying the corn planted area.

5.6.2 Identification of Stem and Leaf Points

The identification of stem and leaf points is necessary for estimating the corn structural parameters accurately. The difference of normal (DoN) method is a commonly used method to identify the corn stems and leaf points. This method considers the obvious structure differences between corn leaves and corn stems to identify the LiDAR points of corn stems and leaves. Specifically, the structural difference between stem and leaf can be quantified using the DoNs in two different searching radiuses. For a given random point p, the DoN Δn is computed as:

$$\Delta n(p, r_1, r_2) = [n(p, r_1) - n(p, r_2)]/2 \tag{5.8}$$

where $\Delta n(p, r_1, r_2)$ is the directional DoN in two scales, r_1 and r_2 are two searching radiuses, and $r_1 < r_2$.

Figure 5.33 shows an example of identification of stems and leaf points. Figure 5.33(a) is the raw corn plant points including leaves and stem, and Figure 5.33(b) is the identified corn leaf points using the DoN method with the search radiuses $r_1 = 0.1$ m and

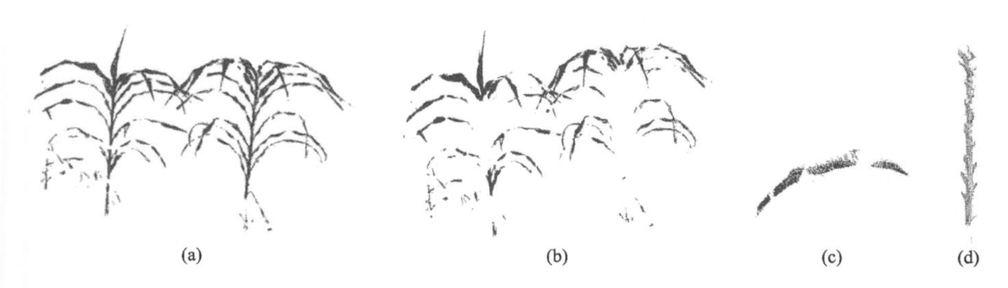

FIGURE 5.33 Identification of corn stem and leaf points: (a) raw LiDAR point cloud; (b) identified corn leaf points; (c) zoomed corn leaf points; (d) zoomed corn stem points.

$r_2 = 1$ m. Figure 5.33(c) and (d) show the zoomed views of the point cloud of identified corn leaf and stem. It is seen that most leaf points and stem points are identified accurately. Unfortunately, there are still some stem and leaf points whose normals are similar and are not identified correctly.

5.6.3 Estimation of Leaf Angle Distribution

LAD is an important canopy structure parameter. The crop canopy with a different LAD has different weakening capacity for interception of solar radiation. Therefore, estimating crop LAD accurately is vital in agriculture. There are five common LADs: (1) spherical distribution: continuous random distribution of leaf inclination angle and random azimuth angle; (2) conical distribution: a fixed leaf inclination angle within [0°, 90°] range and random azimuth; (3) horizontal distribution: leaf inclination angle of 0° and random azimuth; (4) vertical distribution: leaf inclination angle of 90° and random azimuth; and (5) ellipsoidal distribution: continuous ellipsoidal distribution of leaf inclination angle and random azimuth. The Campbell ellipsoidal distribution is a common crop LAD, which is computed as (Campbell, 1990):

$$g(\alpha) = \frac{2\chi^3 \sin\alpha}{A\left(\cos^2\alpha + \chi^2 \sin^2\alpha\right)^2} \tag{5.9}$$

where α is the leaf inclination angle with the value ranges in $0 \le \alpha \le \frac{\pi}{2}$; $g(\alpha)$ is the probability density of α; and χ is the ellipsoidal distribution parameter, which is calculated as the ratio of the horizontal and vertical semi-axes of the ellipsoid. A is the parameter related to χ and is calculated as follows:

$$A = \begin{cases} \chi + \dfrac{\sin^{-1}\varepsilon}{\varepsilon} & \chi < 1, \varepsilon = \left(1-\chi^2\right)^{\frac{1}{2}} \\ 2 & \chi = 1 \\ \chi + \dfrac{\ln[(1+\varepsilon)/(1-\varepsilon)]}{2\varepsilon\chi} & \chi > 1, \varepsilon = \left(1-\chi^{-2}\right)^{\frac{1}{2}} \end{cases} \tag{5.10}$$

In Campbell ellipsoidal distribution, χ is an uncertain parameter. The χ value of corn LAD in different growing seasons can be computed using the LiDAR point cloud:

$$\chi^2 = \frac{1}{3\sin^2\alpha_{\max}} + 1 \tag{5.11}$$

where $\alpha_{\max}$ is the leaf inclination angle with the highest frequency.

Here we take an example of LAD estimation based on the LiDAR point cloud acquired on four corn growing stages in four towns (Zhuozhou, Gaobeidian, Dingxing, and Rongcheng) of Baoding City, Hebei Province, China. The LiDAR point cloud scanning was conducted on stem elongation stage, heading stage,

TABLE 5.7
Proportion of the Leaf Angles (in 10° Intervals) on the Four Growing Stages

		Proportion of Leaf Inclination Angle (%)								
Date	χ	**0°–10°**	**10°–20°**	**20°–30°**	**30°–40°**	**40°–50°**	**50°–60°**	**60°–70°**	**70°–80°**	**80°–90°**
July 10 (stem elongation stage)	1.223	7.83	11.33	14.2	16.1	16.68*	15.05	10.63	5.68	2.5
July 10 (heading stage)	1.206	6.49	8.98	10.73	12.42	14.18	15.41*	14.67	11.1	6.02
August 19 (flowering stage)	1.214	7.1	10.11	11.3	11.89	13.4	14.79*	14.02	10.85	6.54
September 4 (grain-filling stage)	1.195	6.17	8.76	10.02	10.73	12.19	14.54	15.79*	13.53	8.27

Note: * marks the interval with the highest frequency.

flowering stage, and grain-filling stage (Su et al., 2019). The LADs of four key growing seasons are acquired by computing the normal of identified leaf points. Then the χ value is computed by Equation (5.11), as listed in Table 5.7. It is seen that the maximum leaf angle probability is 40°–50° on July 10 (stem elongation stage), 50°–60° on July 26 (heading stage) and on August 19 (flowering stage), and 60°–70° on September 4 (grain-filling stage).

5.6.4 Estimation of Leaf Area Volume Density

LAI is defined as the total one-sided leaf area per unit ground area. The LAVD is defined as the one-sided leaf area per unit height and per unit volume. The relationship between LAI and LAVD is $\text{LAI} = \int_0^z \text{LAVD}(z)dz$. $\text{LAVD}(z)$ describes the variation of leaf area in the vertical direction, which is affected by vegetation species, growing stage, and surrounding environment. It is an important input for ecological process models.

Taking the estimation of LAVD using terrestrial LiDAR as an example, there are three steps for LAVD estimation. (1) Transform the LiDAR point cloud into 3D voxels. (2) Assign the ray-tracing status for each voxel (also called voxel status) by tracking the LiDAR returning position (Figure 5.34). If the laser beam is intercepted by plant leaves or stems within the voxel, the voxel state is marked as 1. If the laser beam passes through the voxel, the voxel state is marked as 2, and the state of this voxel is assigned after the farthest voxel where the intercepted laser beam is determined. If the voxel is not reached by any laser beam, the voxel state is marked as 3. (3) Accordingly, the LAVD of the corn plant is calculated based on the voxel state, as in Equation (5.12).

$$\text{LAVD}(h,\Delta H) = \frac{\cos\theta}{G(\theta)} \cdot \frac{1}{\Delta H} \cdot \ln\left(\sum_{k=1}^{N} \frac{n_i(k)}{n_i(k) + n_p(k)} \right) \tag{5.12}$$

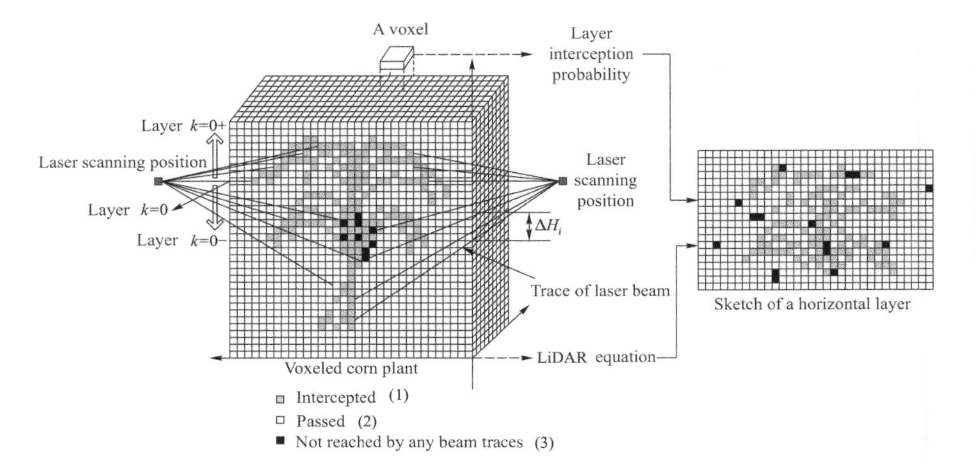

FIGURE 5.34 LAVD estimation of corn plants using a voxel-based method.

where LAVD is the leaf area volume density at height h, ΔH is the vertical stratification height interval, θ is the zenith angle of the incident laser beam, $G(\theta)$ is the projection of the unit leaf area in the plane perpendicular to the laser beam direction, N is the number of voxel layers of the point cloud, $n_i(k)$ is the number of voxels with state 1 in the kth layer, and $n_p(k)$ is the number of voxels with state 2 in the kth layer.

This method is used to calculate the corn LAVD based on the terrestrial laser scanning (TLS) point clouds acquired in the greenhouse at the China Agricultural University. Validated by the measured LAVD in the field campaign, the mean absolute percentage errors for the two corn plants are 24.5% and 15.7%, respectively. This example indicates the voxel-based method estimates the corn LAVD accurately from the TLS point cloud. The cumulative value of the LAVD profile is exactly the corn LAI (Su et al., 2018).

5.6.5 Estimation of Factional Absorbed Photosynthetically Active Radiation

Photosynthetically active radiation (PAR) refers to solar radiation with a spectrum of 400–700 nm that is used by green plants for photosynthesis. FPAR is the ratio of PAR absorbed by vegetation to the total incident PAR, which represents the capability of vegetation to absorb light energy. The vertical distribution of FPAR describes the variation of FPAR with the canopy height. This parameter is used to monitor the growth and health of vegetation (or crops). The FPAR vertical profile is calculated as the ratio of the canopy return energy profile to the total return energy. Taking airborne full-waveform LiDAR data to calculate the FPAR vertical profile as an example, the specific steps are as follows. First, the cumulative energy ratio above the height z is calculated from airborne full-waveform data by Equation (5.13). Then the single-layer energy ratio is calculated by Equation (5.14). Finally, the vertical profile of the energy ratio is generated by combining all the single-layer energy ratios (Qin et al., 2017).

$$E_r(z) = \frac{R_v(z)}{R_v(0)} \cdot \frac{1}{1 + \frac{\rho_v}{\rho_g} \frac{R_g}{R_v(0)}} \tag{5.13}$$

$$E_l(i) = E_r\left(z_{i,t}\right) - E_r\left(z_{i,b}\right) \tag{5.14}$$

where $E_r(z)$ is the cumulative energy ratio above the height z; R_g is the ground return energy; $R_v(z)$ is the cumulative return energy from the canopy top to the height z; $R_v(0)$ is the cumulative return energy from the canopy top to bottom; ρ_g and ρ_v are the ground reflectance and canopy reflectance, respectively; $E_l(i)$ is the energy ratio of i-th layer; and $z_{i,t}$ and $z_{i,b}$ are the top and bottom heights of i-th layer, respectively.

Here we collected airborne LiDAR full-waveform data of a maize area in Huailai Ecological Experimental Station, Chinese Academy of Sciences. Four maize plots were randomly selected, and their FPAR vertical distribution profiles were field measured. The aforementioned method is used to estimate maize FPAR vertical profiles based on full-waveform data. Figure 5.35 shows the filed-measured maize FPAR vertical profiles and the LiDAR-derived vertical energy ratio profile. It is seen that the measured FPAR profile and the estimated FPAR profile show the same variation trend. The value of the LiDAR-derived energy ratio is slightly smaller than the field-measured maize FPAR. In addition, the minimum values of the signal-layer measured and estimated FPARs mostly appear in the top layer, while the maximum values appear in the lowest two layers. This is because the leaves at the top of the maize canopy are relatively sparse, and the number of leaves increases rapidly at first and then remains stable or decreases slightly from the top to bottom layers.

The cumulative value of the FPAR vertical profile is regarded as the maize FPAR. Based on the estimated and measured FPAR, a statistical regression model is established to map the maize FPAR in the entire area, as in Figure 5.36. The spatial distribution difference of the maize FPAR is related to various factors, including fertilization, irrigation conditions, and planting dates.

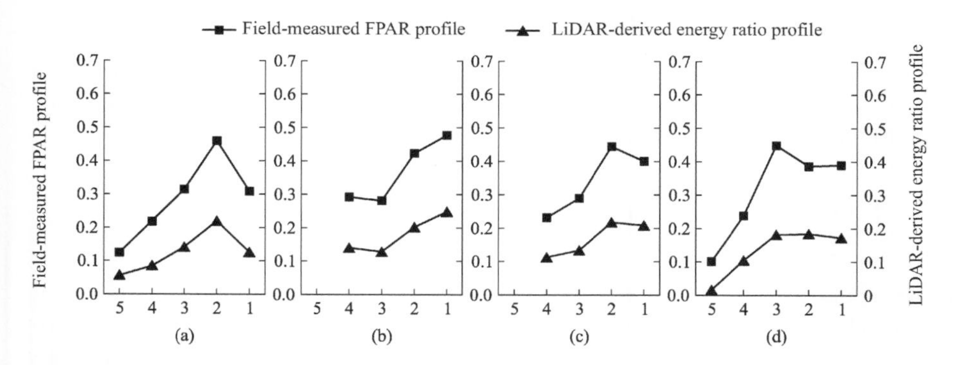

FIGURE 5.35 Comparison between field-measured vertical profile of maize FPAR and vertical profile of LiDAR energy ratio.

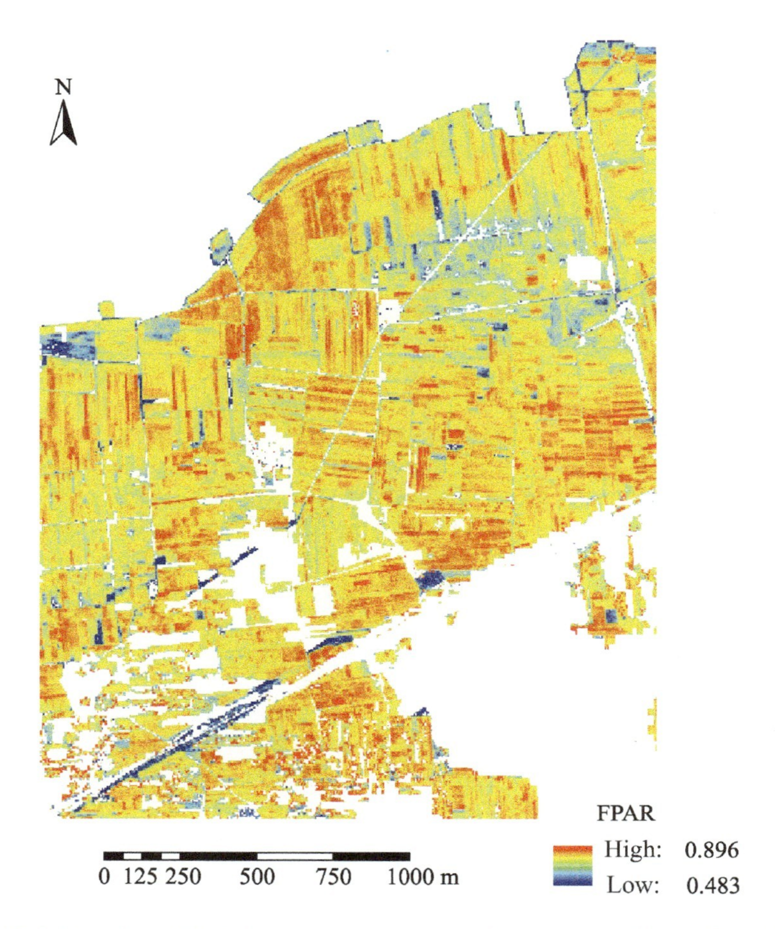

FIGURE 5.36 Maize FPAR distribution estimated using airborne full-waveform LiDAR in the study area.

5.7 CULTURAL HERITAGE CONSERVATION

The application of LiDAR technology in cultural heritage research mainly includes two aspects: digital conservation and archaeological discovery. Scholars have done many studies in the two aspects.

5.7.1 Cultural Heritage Conservation Based on Terrestrial Laser Scanning

The application of TLS in cultural heritage is mainly shown in four aspects as follows. (1) Original data archiving (mass point cloud) of sites and heritage for archaeological site digital recording and preservation, such as fast collection of high-density 3D point clouds of unearthed cultural relics, monuments, and ancient buildings. (2) Provide 3D model support for archaeological excavation. (3) 3D digital model reconstruction, rescue restoration, and digital management of heritage. (4) Change monitoring of heritage ontology based on multi-period data.

5.7.1.1 Archiving of Site/Heritage Original Materials

The TLS system can directly acquire the high-density, high-precision 3D point cloud of the heritage surface in a non-invasive way. Although the point cloud is very discrete, it contains multiple properties of the target surfaces, such as geometric topological relations, target reflected intensity, and color information. Even without post-processing, the detailed features, color information, and 3D spatial structure of a heritage site can be displayed in the computer in an all-round way and be accurately measured, especially suitable for objects with rich surface geometry and texture. As early as 2002, Stanford University used TLS technology to digitize Michelangelo's statue. Beijing University of Civil Engineering and Architecture scanned the Hall of Supreme Harmony of the Forbidden City in Beijing. The Aerospace Information Research Institute of the Chinese Academy of Sciences obtained a high-density, high-precision 3D point cloud of the Ta Keo Temple in Angkor Wat in Cambodia and the Huabiao (marble pillar) in Beijing. This is the most basic spatial information for digital archiving of heritage/sites, and TLS technology is the best means to achieve this goal.

5.7.1.2 Digital Recording and Preservation of Archaeological Sites

In the early days of archaeology, photographs were used to record and display the archaeological excavation process, which was often difficult to quantify and was incomplete. Some site information might be lost. In contrast, with a high-density point cloud of the archaeological site, TLS technology can reproduce the excavation site by drawing a high-precision planar graph, stratigraphic section map and a detailed exploration map of the excavation site. It can measure the volume of relics and analyze the erosion of the site surface and then achieve dynamic display, digital record, and change analysis of the archaeological process. TLS technology gives full play to the advantages of efficient, high-precision, all-round, and dynamic real-time data collection in archaeological excavation and digital construction, such as digital records of the Egyptian Khufu Pyramid and its surrounding archaeological sites, the No. 2 pit of the Qin Terracotta Army, No. 1 pit of Sanxingdui Site, and archaeological excavation and digital construction of the Shuntianmen site in Kaifeng.

5.7.1.3 3D Digital Model Reconstruction

Many ancient buildings, stone grottoes, and frescoes are facing the problems of disrepair, natural weathering, and human destruction. Their fine 3D digital models are considered the only way to "rescue" the cultural heritage sites on the verge of being damaged or destroyed. For example, after the Bamiyan Buddhas were destroyed in Afghanistan in 2007, an elaborate digital scan was used to reconstruct a virtual model. In 2013, the Chinese Academy of Cultural Heritage conducted fine scanning and rescue repair for the Thousand-handed Goddess severed finger in the Dazu Stone Carvings of Chongqing, China. In 2014, Yongle Bell and its 230,000 inscriptions in the Beijing Great Bell Temple Museum were digitalized and reconstructed. These 3D digital models were displayed and measured online in an all-round way, providing an immersive experience to readers. In addition, more than 40 precious cultural relics in the National

Museum of China were scanned in detail and their 3D models were displayed dynamically. Here, we take an example of the Ta Keo Temple at Angkor Wat in Cambodia and introduce the specific steps of the application of TLS technology in the reconstruction of 3D digital models of cultural heritage (Wan et al., 2014).

1. Data collection. Ta Keo covers a large area (more than 10,000 m^2) and has a complex building structure with four layers of a pyramidal structure. Under the premise of fully considering equipment performance and data integrity, scanning stations are laid scientifically and reasonably. In this case, a Riegl VZ-1000 scanner was used to set up 71 stations, and a 230-GB point cloud and image data were obtained.
2. Data processing, including point cloud registration and denoising. First, a rough registration is completed through measurement of the longitude and latitude of the site and orientation. Then fine registration is completed by multi-site matching. To avoid the transmission error accumulated through the registration of two consecutive sites, the data of all sites are divided into multiple blocks. The site data of each block is registered, ensuring minimal registration error within the block. The blocks are then merged until the overall joint registration of all sites is completed. Further point cloud denoising is conducted. Since Ta Keo is composed of numerous interconnected and intersecting facades, there is no unified filtering reference plane. The traditional point cloud denoising method cannot be fully used. Here we deleted the non-target point cloud around Ta Keo, such as trees and pedestrians, through human-computer interaction.
3. Model reconstruction. Ta Keo has many architectural components with rich textures, which is a complex curved surface architecture. The models of complex architectures are generally reconstructed by TIN. The Riegl RiScan software provides the function of a point cloud data model reconstruction based on TIN. The process is as follows. (1) According to the mapping relationship between the corrected texture image and the point cloud, the color information of the image is assigned to the corresponding point cloud [Figure 5.37(a)] to obtain the color point cloud [Figure 5.37(b)]. (2) "Convert to 3D" format conversion is carried out on the point cloud data of each station to generate the triangulated network data format defined by software. This process is a mapping operation, which maps the 3D point cloud to 2D data and constructs the triangulated network according to 2D neighborhood relations. (3) Mark the triangulated network data required to model, construct the TIN, and automatically render [Figure 5.37(c)]. (4) Resample the model according to the color point cloud to achieve the image mapping and generate the true 3D model [Figure 5.37(d)]. (5) Merge the 3D models constructed by the various stations and complete the 3D reconstruction of the overall model of Ta Keo.

5.7.1.4 Dynamic Change Monitoring of Heritage Ontology

The dynamic changes of cultural heritage are analyzed and predicted based on 3D digital models in different periods. For example, the University ät Hamburg in

(a) (b) (c) (d)

FIGURE 5.37 3D modeling of Ta Keo Temple. (a) Uncolored original point cloud; (b) point cloud after coloring; (c) uncolored 3D model; (d) color-rendering 3D model.

Germany obtained high-density point cloud data of hundreds of stone statues on Easter Island in the Pacific in different periods and analyzed the deformation of these stone statues over time by constructing 3D digital models (Kersten et al., 2009). Beijing University of Civil Engineering and Architecture acquired point cloud data of complex buildings such as the Jixiang Multigate tower in Baiju Temple, Xizang Autonomous Region, China, and obtained accurate deformation information of the tower through model construction, which provided the scientific basis for repair work (Wang et al., 2011).

TLS technology acquires 3D information of archaeological sites and relics with a low-cost, flexible method and produces high-precision, high-density data. More importantly, it can obtain complete information from the outside and inside of the object, which solves the contradiction between traditional mapping and non-contact measurement to the greatest extent. Therefore, it has been widely used in recording ancient buildings, fresco rock paintings, cultural relics, and large-scale archaeological site excavations.

5.7.2 Ancient Heritage Site Discovery Based on Airborne Laser Scanning

Ancient heritage site detection under forest areas is a challenging task for archaeologists due to the lack of high-resolution DEM, especially in complex dense forest areas. Airborne laser scanning (ALS) is the only technology that measures the topography of forest-covered areas with high precision, which provides the possibility for remote sensing archaeology in dense forests. The high-frequency laser pulses of the ALS system penetrate through the forest canopy gaps and acquire a high-density understory point cloud, which is used to generate a high-accuracy, high-resolution DEM. The DEM helps to discover the heritage sites by combining with other archaeological methods. Many scientists have carried out successful applications for ancient heritage site discovery using ALS, such as cases studies in England, Germany, the Netherlands, Greece, Ireland, Belgium, Austria, Italy, and the United States.

5.7.2.1 History of Forest Understory Archaeology with ALS

1. Early exploration period. In the early stage of LiDAR development, ALS records a point cloud with only one or several returned echoes. The low-density point cloud can only generate the low-resolution DEM, limiting the applications of LiDAR. Sittler (2004) used the ALS technique to generate a DEM with a resolution of 1 m and elevation accuracy of 50 cm and discovered some fossilized ridges and furrows in woodlands near the Rhine River, which were considered to be the results of medieval cultivation practices. Humme et al. (2006) used point cloud filtering and the Kriging interpolation method to produce a DEM and recognized Celtic remains in east Netherlands in the Bronze Age with 2500-year history. However, researchers could only extract limited remains, such as main roads, footpath, and ancient city walls, due to the low density of the point cloud.
2. Extensive application period. With the development of the ALS system, the pulse frequency is improved greatly and the cost of data acquisition declines. High-density point clouds are less expensive to be acquired, which improves the resolution of DEMs significantly and the applications in forest archaeology. The British scientists discovered an ancient Rome fort embedded in modern cultivated farmland, which caused the Heritage Management Bureau to carry out an archaeological project in an understory with ALS, and several ancient remains were revealed subsequently. The Germany Cultural Heritage Management Bureau arranged a 3-year plan for an archaeological project. They processed a part of the point cloud and developed a local relief model (LRM) for archaeology, which explains and depicts the characteristics of the local terrain elevation and small remains accurately and directly.
3. Intensive application period. The previous archaeological work is mainly based on a limited number of returns of ALS. The resolution and accuracy of the DEM derived from point clouds are obviously affected by structural complexities of research objects and surroundings. With the advent of the full-waveform ALS, scientists can acquire much more abundant and fine vertical information of the vegetation and understory in a forest, which greatly improves the resolution and accuracy of the DEM. The Austrian scientists generated a high-accuracy DEM from the full-waveform data of Riegl LMS-Q560 by removing the non-ground signal and then detected ancient remains of the Iron Age in a forest based on subtle terrain differences in the DEM (Doneus et al., 2008).

5.7.2.2 Typical Archaeological Discovery

There are many archaeological findings made using ALS technology. The most famous ones include LiDAR archaeology of the Mayan new ancient city by the University of Florida, United States, and LiDAR archaeology of Angkor Wat by the University of Sydney, Australia.

Funded by the National Aeronautical and Space Administration (NASA) in 2010, the University of Florida carried out a campaign to acquire 3D information of the dense rainforest in Caracol, Belize. They used ALS data to produce a fine 3D topographical map and discovered many unknown remains of ancient buildings, roads, and terraced fields in only 4 days. An ancient Mayan castle was rebuilt. Compared with the previous work of field archaeological investigation over 25 years, the ALS technology revealed ancient remains of nearly 200 km^2 within 1 month, which was 8 times the traditional field archaeological findings (Chase et al., 2010).

In 2012, the University of Sydney initiated an archaeological project in Angkor Wat, which was supported by archaeologists from Japan, France, Hungary, the United States, and Indonesia. The LiDAR system was mounted on a helicopter to acquire point clouds with a density of 4–5 pts/m^2 for the Angkor Wat site and surrounding forest area. They made high-precision 3D topographic maps, discovered Angkor Wat remains hidden in dense forest and paddy fields in the north of Phnom Kulen, and reconstructed a prosperous Angkor ancient city (Evans et al., 2013). These great findings not only extend the area of the Angkor ancient city from 9 km^2 to 35 km^2 but also push back the history of Angkor remains 350 years. With a fine 3D digital map of the understory, the archaeologists explained the spatial distribution pattern of the ancient road network, water system, farmland, and extent of the city. These findings were incredible and unreachable for the traditional field investigation and remote sensing methods.

5.7.2.3 Underwater Archaeology

ALS is extensively used in land archaeology. However, airborne laser bathymetry (ALB) used in nearshore underwater archaeology is very rare. The main reason is that the existing ALB systems are much fewer than land-survey LiDAR system, and they are relatively expensive. In practice, aquatic plants prevent the laser pulse from penetrating to the bottom of the water. Hence, the inclining scanning mode is used to lengthen the laser pulse range to increase the possibility of single laser pulse penetrating to the obstacles. Doneus et al. (2013) investigated the ancient remains in the Adriatic Sea and produced an underwater local relief model to enhance local microtopographic features. They found building remains submerged in water, including a platform that measured 80 m × 60 m. With the help of historical references, they deduced that the platform was an ancient wharf relic.

5.7.3 Archaeology by Combining Airborne and Terrestrial Laser Scanning

As mentioned earlier, there are advantages and disadvantages of TLS and ALS technologies. Therefore, people often combine the two in large archaeological site protection to achieve better effects. In 2007, the Surveying and Mapping Bureau of Hebei Province, China, scanned the ancient great wall of Shanhai Pass all around from the air and ground by using ALS and TLS, which was the first such study in China. Then, the high-precision, high-definition true 3D digital model of Shanhai Pass was reconstructed by combining the point cloud with the texture images acquired by high-resolution digital cameras integrated in the ALS and TLS systems. This work provided important raw data and model support for the site restoration (Cao, 2008). Wuhan University carried out a campaign in the Mogao Caves of Dunhuang City, China. They used TLS to acquire detailed information on murals and Buddha statues in the grottos and rebuilt their 3D digital models. Meanwhile, the ALS and close-range photogrammetry surveying and remote sensing technology are combined to fully scan the nine-story building of Mogao Caves and its surroundings. They achieved the goal of creating a "digital Dunhuang" and provided support for heritage management and restoration (Chang et al., 2011). Compared with traditional 3D modeling methods, the integrated technology of ALS with TLS not only greatly saves the human and economic costs of digitization and 3D modeling of cultural heritage in large area but also improves modeling efficiency and accuracy.

5.8 INDOOR 3D MODELING AND NAVIGATION

With the continuous advancement of the construction of smart cities and digital twin cities, there is an increasing demand for semantically rich, geometrically accurate, and topologically consistent building indoor models. The building indoor 3D models provide support for basic data in terms of emergency disaster relief, navigation and location services, intelligent buildings, and urban operations. The development of sensor (such as optical camera and LiDAR) and 3D computer vision technology provides reliable and rich ways for 3D data acquisition of indoor environments. In contrast to the outdoor counterparts, the constraints of the layout and occlusion among indoor objects brings serious challenges to the application of fine 3D modeling and navigation. 3D point clouds exhibit remarkable advantages in indoor fine 3D modeling and navigation applications due to the high-precision geometric structure and rich spatial context of objects.

5.8.1 Point Cloud–Based Indoor Modeling

The existing methods of 3D indoor modeling generally aim to satisfy the specific indoor application requirement by synthesizing three types of spatial information: geometry, semantics, and topology, which is used to vectorize the geometric structure of entity elements, adding semantic information (such as label and category) and reconstructing the topological relations such as connectivity and adjacency between different entity elements. Therefore, studies on 3D point cloud–based indoor modeling generally include three aspects: fine geometric model construction, rich semantic annotation, and complex spatial topology construction.

5.8.1.1 Fine Geometric Model Construction

Fine geometric model construction focuses on the modeling of geometric parameterization and texture mapping of the main building structures and the internal auxiliary facilities. Existing methods are mainly the robust estimation algorithms, such as normal estimation method, least squares estimation method, region growing method, RANSAC method, Bayesian sample consensus method, and deep learning. These methods first fit and parameterize the main structure of the building (e.g., walls, floors, ceilings) based on certain primitives (e.g., straight lines, planes). Then, geometric spatial constraints (such as orthogonality) are used to produce the polygonal-structured model of a building (Kang et al., 2016; Liu et al., 2018). In addition, the methods for automatic generation and dynamic updating of 3D indoor scene textures are constantly being refined and improved to provide high-fidelity indoor 3D geometric models. With the diversification and complication of the indoor space layout, researchers have put much emphasis on modeling indoor spaces with complex geometric structures (such as cylindrical walls, spherical ceilings, and other non-planar structures) (Wu et al., 2021).

5.8.1.2 Rich Semantic Annotation

Rich semantic annotation focuses on the description and analysis of semantic attributes (structural information and object functions) of the indoor space and internal auxiliary facilities of buildings by means of machine/deep learning–based algorithms (Khan et al., 2014). Most existing methods usually first subdivide point clouds into a set of homogeneous primitives according to the predefined rules. After that, the hand-crafted geometric, spectral, and texture features are extracted to describe the appearance and geometric attributes of primitives that make up the indoor elements, which are used as the input to machine learning–based semantic annotation models (such as decision trees, support vector machines, and random forests) for the classification tasks. Very often, using these low-level visual features alone cannot robustly represent the similarity of the same indoor elements due to occlusion and illumination variations, among other factors. Therefore, it is difficult to achieve effective semantic annotation tasks. In contrast, with the help of the hierarchical pyramid architecture, the end-to-end convolutional neural network can efficiently capture the spatiotemporal information of different scale contexts and effectively mine the high-level semantic features, which greatly improves the accuracy and efficiency of semantic annotation.

5.8.1.3 Complex Spatial Topology Construction

Complex spatial topology construction primarily focuses on the extraction, organization, management, and storage of various indoor elements and space information, as well as visualization of the indoor topology graph (such as TIN, node edge, and regular grid), which explicitly expresses and simplifies the spatial constraints between indoor elements in the indoor scene. In practice, the spatial topology of the indoor space is difficult to summarize as "adjacency, containment, disjoint" in Euclidean space due to the limitation of physical environment structures (such as rooms, doors, and windows). The representation of the topological structure between the indoor elements should establish the implicit spatial constraints (such as containment, connectivity, and adjacency). It can offer better indoor services such as indoor navigation, gas diffusion simulation, emergency rescue, and other location-based services. Taking point cloud data as the data source, the existing methods usually first identify and extract the indoor elements of interest (such as rooms, doors and windows, stairs, and corridors) and then define their attributes (such as whether they are navigable) according to the requirements of the specific indoor application. Next, the indoor elements of interest are represented and organized by one or more primitives (such as regular grids, irregular triangles, and nodes). Finally, the topological relationship between them is derived from the semantic attributes (such as openings and obstacles). In addition, the mobile mapping technology based on SLAM has gradually matured, which can easily capture the real-time context. The platform trajectory is generally used as auxiliary information to improve the accuracy and efficiency of the indoor element extraction and topology construction.

A real case of topological construction of the indoor environment is given in detail to demonstrate how to generate a topologically consistent indoor navigation network from the original point cloud (Yang et al., 2021). The flow is as shown in Figure 5.38. First, supervised learning tools, such as machine learning/deep learning, are used to interpret the original point cloud into the semantic point cloud, including doors, walls, the ground, etc., as in the upper subgraph of Figure 5.38. On that basis, according to the subspace definition with specific functions, the building architecture–guided room partitioning method based on the distance transformation and the watershed segmentation is developed to decompose the indoor space into a group of non-overlapping subspaces with specific functions, thereby adaptively tracking and locating the position of vertical elements (such as walls), as in the left subgraph in Figure 5.38.

In addition to the non-overlapping subspace partitioning, the topological relationship between them needs to be reconstructed to determine their spatial relationship. As the common opening, the door is usually considered a transition space connecting adjacent subspaces and acts as a bridge in the analysis of the connectivity relationship of subspaces. Therefore, here we adapted the door-guided topology construction method based on the semantic point clouds, as shown in the right subgraph in Figure 5.38. In our implementation, the door-guided topology construction process is classified into two categories: physical door-guided construction and virtual door-guided construction.

With regard to the physical door-guided construction, the state of the physical doors is divided into three types: open, semi-open, and closed. Their uncertain state might cause an incorrect topological relationship to be developed. For example, in the case of an open door, two doors will be detected instead of one: one door leaf in the arbitrary position, and a second one representing the open hole to move through. Figure 5.39(a) and (b) illustrate a conceptual model of an open door and its associated connectivity relationship. From Room #2 inside, there are two detected doors: one (i.e., closed) seemingly allows users to enter Room #1; the other (i.e., open) lets users go through to the hallway. The former is obviously unreasonable because a door is navigable if and only if it can be detected (or seen) from both Room # 1 and Room #2. Thus, a door model is designed to deduce the true door position for correcting the space transition. Figure 5.39(c) shows a conceptual example of the designed door model for space transition correction. In this case, the true positions of the door leaves are recovered and updated, and duplicated ones are filtered.

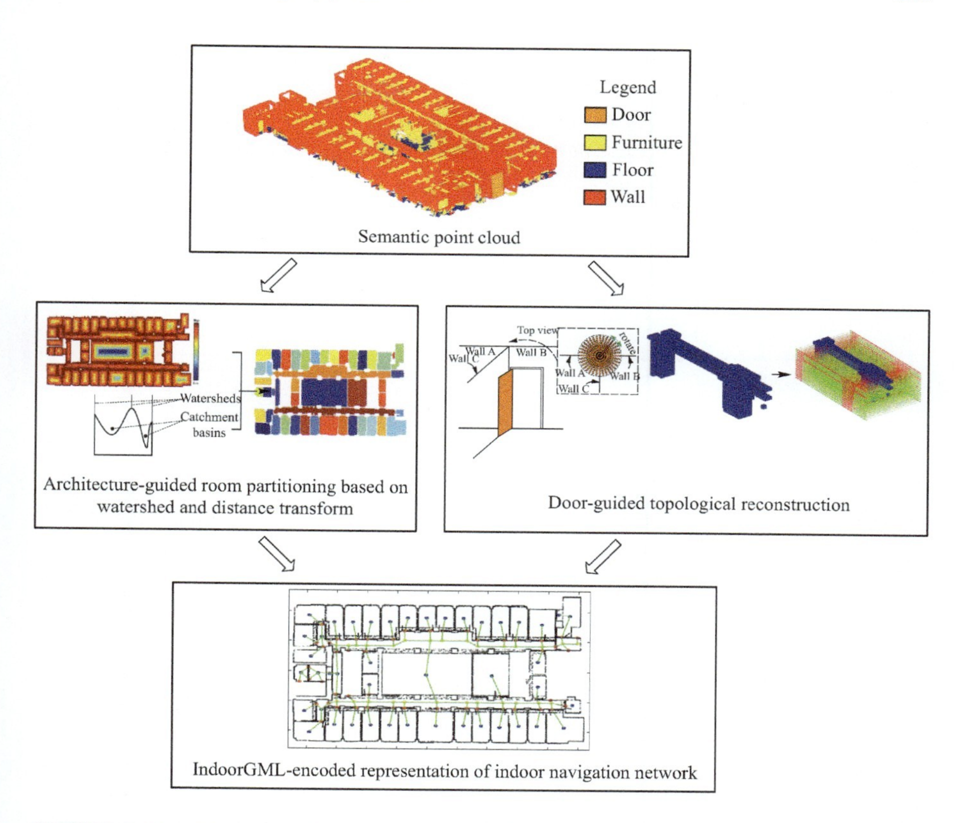

FIGURE 5.38 Flowchart of semantics-guided indoor navigation network construction and encoding.

Figure 5.39(d) exhibits the connectivity relationship after correction, where the position of the dotted door is the true position of the door for Room #2.

In terms of the virtual door-guided construction, there exist navigable common boundary surfaces through which users are allowed to move from one space to another, instead of common doors. These special boundary surfaces are called "virtual doors," which are different from the common doors but have similar functions. To detect these defined virtual doors, we establish the 3D occupancy map (Figure 5.40) of the common boundary surface between cells by using the occupancy analysis method and find the navigable "holes" on the established 3D occupancy map.

Finally, an indoor topological network is generated based on the non-overlapping subspaces and their topological relationship, as in the bottom subgraph of Figure 5.38. To support the indoor navigation system of different platforms, the generated indoor topological network is encoded according to the definition of IndoorGML in the Open Geospatial Consortium (OGC) standard. The IndoorGML based on the XML schema can easily transform and encode the generated indoor navigation network and can be applied to any platform.

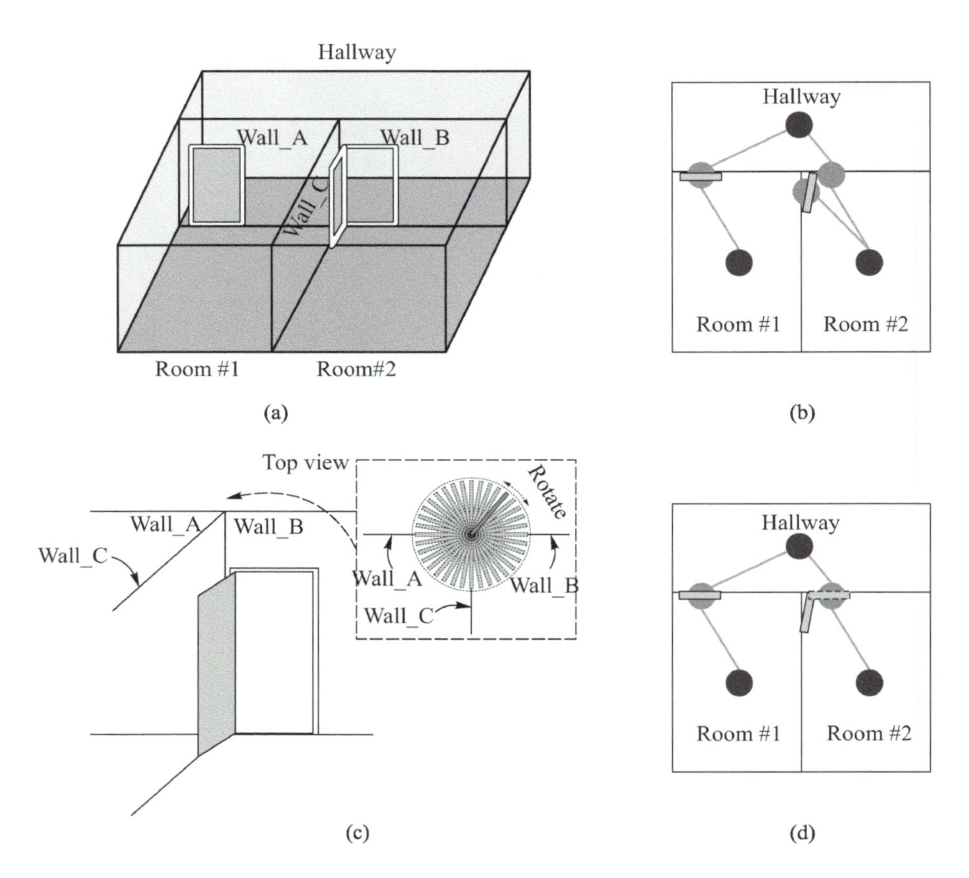

FIGURE 5.39 Conceptual door models for space transition correction. (a) Conceptual model of open door (Room #2); (b) associated connectivity relationship; (c) designed door model; (d) connectivity relationship after correction.

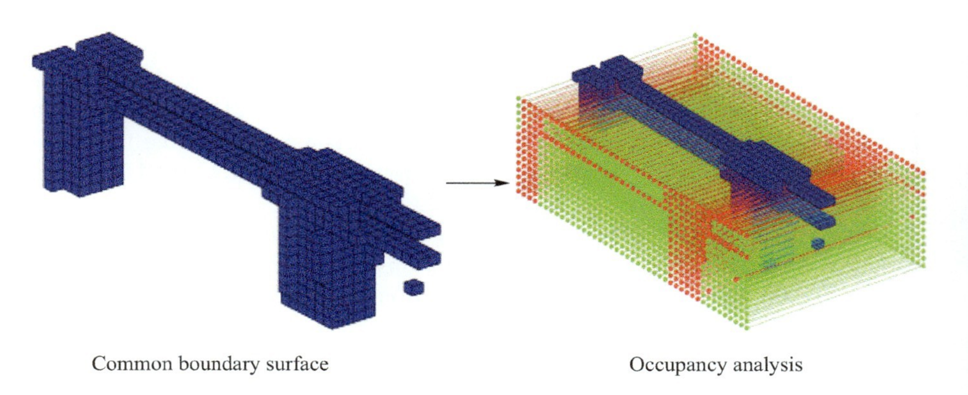

FIGURE 5.40 Occupancy analysis of common boundary surface through the simulated ray.

5.8.2 Point Cloud–Based Indoor Navigation

Indoor navigation allows users to perceive the surrounding environment through sensors and obtain their own position and orientation to reach the target position in the indoor environment with obstacles. Indoor navigation methods are mainly classified into two categories: map-based indoor navigation and SLAM-based indoor navigation, which are specifically described as follows.

5.8.2.1 Map-Based Indoor Navigation

This method first establishes a global map and saves it in the on-board database. The global map contains the geometric, semantic, topological, and other spatial information and signals (such as geomagnetic or Wi-Fi fingerprint). When implementing positioning and navigating, the user perceives the surrounding environment through sensors to construct the local map. The local map is matched in real time with the global map stored in the on-board database to produce the real-time position and orientation in the global environment.

5.8.2.2 SLAM-Based Indoor Navigation

This method uses the on-board sensors (such as LiDAR and optical cameras) to determine the position and orientation by observing the surrounding features during movement and constructs a map of the unknown environment incrementally. Compared with the cameras, LiDAR is widely used in indoor scene mapping and construction due to its advantages of insensitivity to illumination changes. In the LiDAR-SLAM framework, the point cloud frame-by-frame registration is conducted by point cloud feature extraction and the ICP algorithm to achieve the map construction. During the map construction, the sensor's position and orientation are updated to recover its trajectory. Compared with the map-based indoor navigation method, the LiDAR-SLAM framework can process the point cloud data in real time or online, and its trajectory can be used as auxiliary information for room partitioning and semantic labeling tasks, which would offer reliable support for real-time extraction of indoor elements and fast topology construction.

In summary, 3D modeling of the indoor environment strongly promotes the rapid development of indoor applications such as indoor location and navigation services. This section introduces the relevant methods for 3D fine modeling and navigation applications of indoor environments to offer significant reference for the development of relevant research. The existing geometric construction methods of indoor environments are usually based on the strong Manhattan assumption. The multi-task collaborative optimization under the deep learning hierarchical framework will become an active topic to achieve 3D indoor modeling with more complex spatial layouts. Moreover, for complex and dynamic indoor scenes, dynamic objects exhibit similar shapes and appearances in both temporal and spatial dimensions. Therefore, the investigation of spatiotemporal consistency of 3D information will also be an increasingly important issue. In addition, since indoor and outdoor environments have significant differences in physical structures and spatial relationships between entities, seamless modeling of indoor-outdoor spaces is one of the focuses of future studies.

5.9 UNDERWATER MAPPING

More than 70% of the Earth is covered by water, including rivers, lakes, and oceans. Underwater mapping, which consists mainly of water surface elevation (water level) and water depth at survey points, contributes to the development, protection, and utilization of water resources. Traditional water level monitoring mainly relies on limited hydrological stations, which consume large amounts of human and material resources. The shipborne single-beam or multi-beam sounders can acquire water depth data, but easily run aground in shallow water areas, making it difficult to meet the needs of long-term series of large-area dynamic change monitoring (Xi et al., 2022). As an active detection technology, LiDAR can quickly and efficiently obtain 3D coordinate information of target features. Part of the laser wavelengths can penetrate the water surface to reach the bottom, which has become the main means of underwater mapping (Lao et al., 2022). This chapter introduces the application of LiDAR technology in water level extraction and bathymetry.

5.9.1 Water Level Extraction

LiDAR extracts the water level by actively emitting laser pulses and receiving the pulse signal reflected from the water surface. Airborne LiDAR for water level measurement is similar to land surveying, and the technology is relatively mature. This subsection introduces the application of satellite-based LiDAR for water level extraction and the combination with optical remote sensing images to achieve large-scale, long-term, and cost-effective monitoring of lake water storage.

5.9.1.1 Water Level Extraction Based on Spaceborne Full-Waveform LiDAR

The satellite-based full-waveform LiDAR—ICESat/GLAS uses an 1064-nm near-infrared laser pulses that do not easily penetrate water, enabling water level extraction and change monitoring. The GLAS data are utilized to obtain the water level of typical lakes on the Qinghai-Tibet Plateau in China. At the same time, the water reserve change can be monitored in combination with lake areas extracted from optical remote sensing images. The flow chart is shown in Figure 5.41. The data processing steps are as follows.

1. Extract water level from the GLAS data. The GLA14 product provides the global elevation (H_{ele}) corrected for tidal, tropospheric, and atmospheric errors, which is the reference ellipsoid system used by TOPEX/Poseidon (T/P) satellites. In addition, waveform saturation correction ($H_{satCorr}$) and

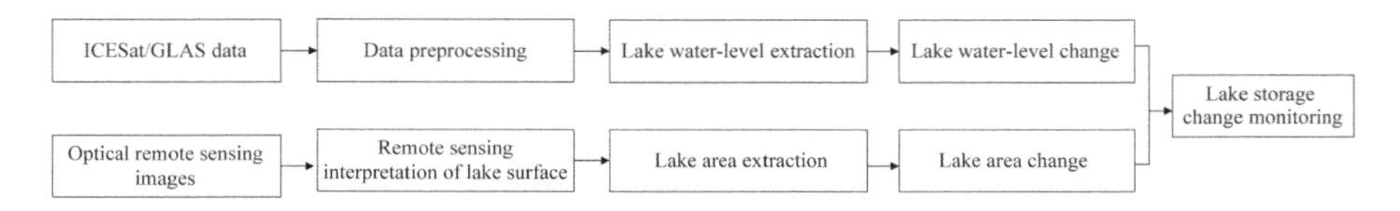

FIGURE 5.41 Flowchart of monitoring lake storage changes by combining GLAS waveform and optical remote sensing images.

elevation anomaly correction (H_{ab}) are required to correct H_{ele}. $H_{satCorr}$ is provided by GLA14 data, and H_{ab} is extracted using the Earth gravity model EGM2008. It is also necessary to convert the elevation from the T/P reference ellipsoid to the WGS84 reference ellipsoid, where the latitude and longitude of the two reference systems are almost indistinguishable, and the difference in elevation Δh is about 0.7 m. With these corrections, the final water level elevation H is calculated as in Equation (5.15).

$$H = H_{ele} - H_{satCorr} - H_{ab} - \Delta h \tag{5.15}$$

2. Extract water area from the optical image. Satellite optical images corresponding to the time phase of GLAS data are collected, and appropriate remote sensing data processing methods are used to extract the water area (A).
3. Calculate changes in water storage. The lake basin is approximated as a table structure, and the lake storage change is computed by Equation (5.16).

$$\Delta V = \frac{1}{3}(H_{i+1} - H_i)\times\left(A_i + A_{i+1} + \sqrt{A_i \times A_{i+1}}\right) \tag{5.16}$$

where ΔV is the changes of lake water reserves at time i and $(i+1)$, and H_i, H_{i+1}, A_i, and A_{i+1} represent the water level elevation and area of the lake at time i and $(i+1)$, respectively.

Figure 5.42 illustrates the changes in water level, area, and storage of Nam Co, Selin Co, and Qinghai Lake on the Qinghai-Tibet Plateau from 2003 to 2009. Among them, Selin Co is the high plateau lake with the fastest increase in water level (0.42 m/a), area (32.59 km^2/a), and storage (1.2 km^3/a), followed by Namu Co. In contrast, Qinghai Lake has shown a decrease in these parameters over the years.

5.9.1.2 Water Level Extraction Based on Spaceborne Photon-Counting LiDAR

In comparison with ICESat, ICESat-2 provides an improved temporal and spatial resolution (six beams with overlapping footprints of ~17 m in diameter every ~0.7 m in the along-track direction), thereby showing great potential for the monitoring of water levels. The following part uses Danjiangkou Reservoir as an example to introduce the extraction of water level change information using ICESat-2/ATLAS from 2018 to 2022.

Like the process of extracting water level information from GLAS data, the water level and corresponding water area are extracted using ATL13 data and optical imagery, respectively. Then the area–water level relationship model is established based on the obtained data. The time varying curve of water area and photon water level in Danjiangkou area is shown in Figure 5.43. By comparing the water area and water level variations, it is observed that the water area of the Danjiangkou Reservoir exhibits a highly consistent pattern with the water level changes. Specifically, as the water area increases, the water level synchronously rises. The correlation between the two variables is as high as 94.1%.

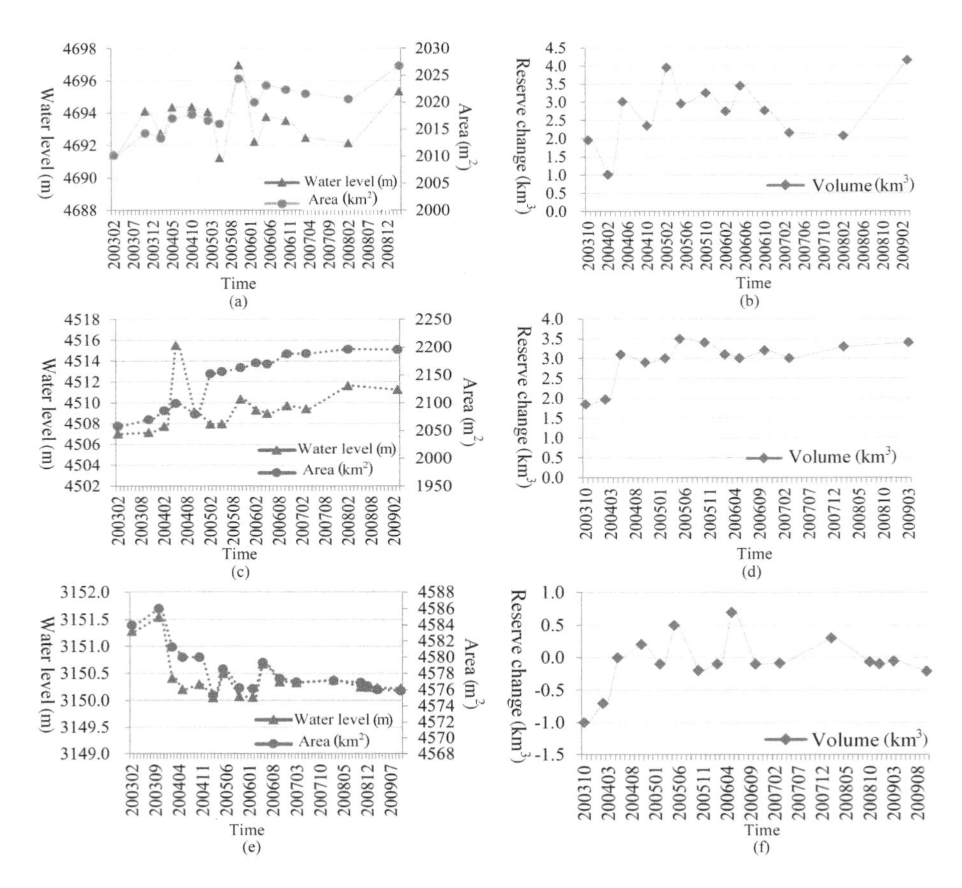

FIGURE 5.42 Major lake change in Qinghai-Tibet Plateau. (a) Water level and area changes and (b) storage changes of Nam Co from 2003 to 2009. (c) Water level and area changes and (d) storage changes of Selin Co from 2003 to 2009. (e) Water level and area changes and (f) storage changes of Qinghai Lake from 2003 to 2009.

5.9.2 Water Depth Survey

The blue-green laser with wavelengths ranging from 520 nm to 535 nm is known as the ocean optical window. It experiences relatively low energy attenuation when propagating in seawater. Light in this wavelength range can be used for LiDAR water depth measurement. This subsection introduces the airborne LiDAR bathymetry and satellite-based photon counting LiDAR bathymetry.

5.9.2.1 Airborne LiDAR Bathymetry

Airborne LiDAR bathymetry is a novel active airborne laser mapping technique that combines laser ranging, global navigation satellite system (GNSS) positioning/attitude measurement, aerial photography, and other technologies, which plays a crucial role in efficiently acquiring high-precision nearshore seabed topography. It is

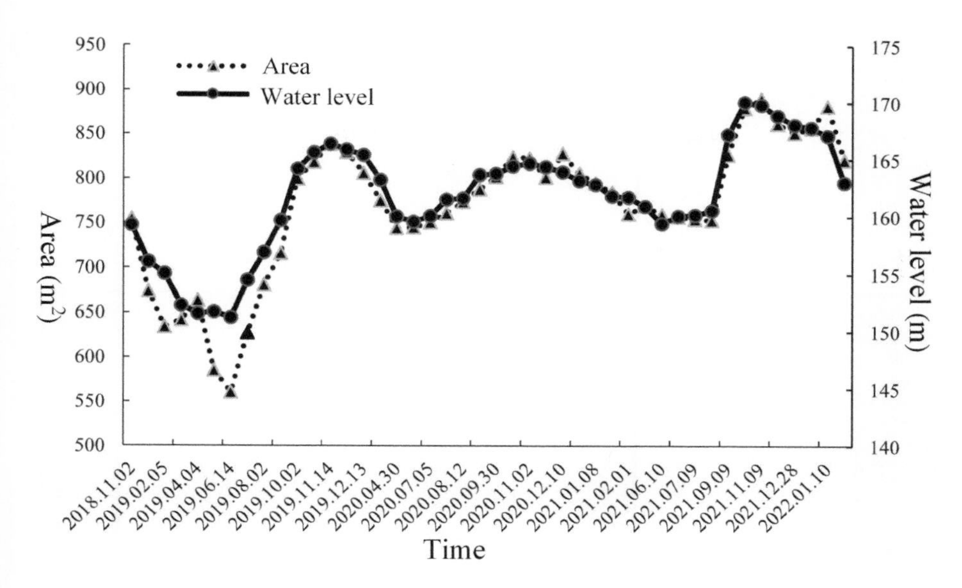

FIGURE 5.43 Changes of water area and water level in Danjiangkou area from 2018 to 2022.

widely used for surveying beaches, coastlines, shallow seas, islands, reefs, and underwater obstacles. Representative airborne LiDAR bathymetry equipment includes the Mapper5000 developed by Shanghai Institute of Optics and Fine Mechanics, the CZMIL series by Teledyne Optech, the HawkEye series by Leica, the VQ series by RIEGL, and the LADS series by Fugro.

Airborne bathymetric LiDAR uses a blue-green laser (e.g., 532 nm), which has a strong ability to penetrate to water. By transmitting and receiving the blue-green laser, it calculates the time difference between the laser echoes on the water surface and bottom and then inverts the water depth. The principle of airborne laser bathymetry is illustrated in Figure 5.44, where the first peak represents the surface echo signal, the second peak represents the bottom echo signal, and the time difference between the two peaks is the round-trip time difference between the laser on the surface and the bottom. When the laser beam is incident on the water surface at an angle θ, the measured instantaneous water depth D is roughly calculated as in Equation (5.17).

$$D = \frac{1}{2}\Delta t \times \frac{c}{n} \times \cos\theta \tag{5.17}$$

where Δt is the bidirectional transmission time of the laser beam in water, c is the speed of light in vacuum, and n is the refractive index of water with an approximate value of 1.33, which may vary slightly under different water salinity and temperature conditions.

The focus of airborne LiDAR bathymetry lies on the extraction of laser echo peaks from the water body, thereby obtaining the time corresponding to the echo peaks of the water-air interface and the bottom reflection. The echo peaks can be extracted by

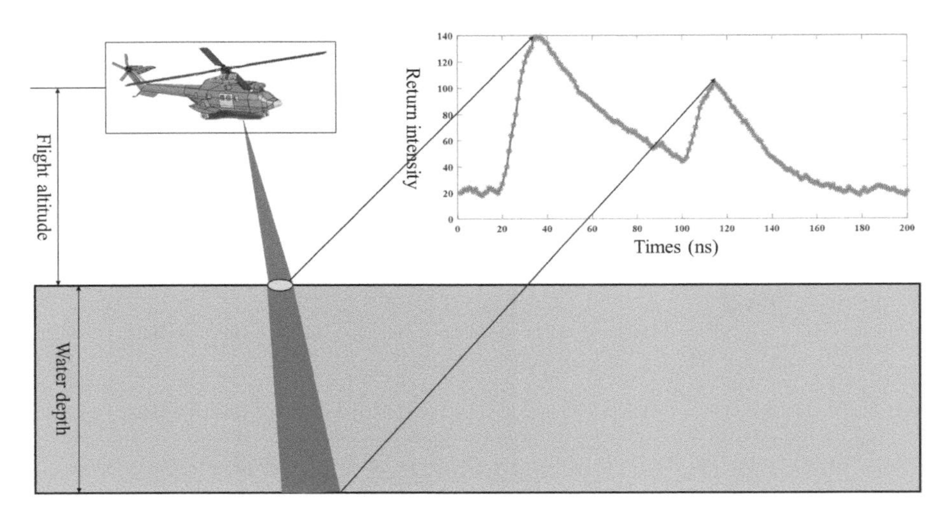

FIGURE 5.44 Schematic diagram of airborne LiDAR bathymetric survey.

using the peak detection method, deconvolution method, and mathematical simulation method. The fundamental principles of these methods are as follows:

1. Peak detection method. This method mainly relies on the shape of the echo data to search for the positions of echo peaks, which correspond to the local maxima of the echoes. It is commonly assumed that the first detected peak in the echo is the water surface reflection peak, and the last peak corresponds to the seafloor reflection peak.
2. Deconvolution method. In this approach, the contribution of the system is removed from the depth measurement echo signal to restore the original signal. Commonly used deconvolution algorithms for full waveform data from airborne LiDAR bathymetry systems include the Richardson-Lucy algorithm, Wiener filter, and least squares method.
3. Mathematical simulation method. Also known as waveform decomposition method, this approach utilizes an exponentially modified Gaussian distribution to describe the water surface reflection signal and Gaussian distribution to describe the bottom reflection signal. It constructs a mathematical model of the depth-measuring echo by simulating the echo signals. The simulated echo signals are compared with the measured echo signals to create an objective function, which is then used to calculate the unknown variables within the mathematical model. The mathematical simulation method not only extracts water depth information but also can be applied to the inversion of bottom reflectance and water quality parameters.

There is no specific algorithm that is universally suitable for all airborne bathymetric scenarios. It is necessary to choose appropriate algorithms based on specific environments and conditions. In some cases, it might even be necessary to combine multiple algorithms to enhance extraction accuracy. With the in-depth understanding and exploration of echo data, the airborne LiDAR bathymetric system can do more than just measure water depth. By radiometrically calibrating the pulses and utilizing mathematical simulation models, it can extract additional environmental parameters from the water body laser echo data, such as water attenuation coefficient, bottom reflectance, and water depth. By integrating these environmental parameters with other measurement data, it becomes possible to achieve accurate classification and information retrieval of underwater objects. This direction represents an expansion of the potential applications for future airborne LiDAR systems.

5.9.2.2 Spaceborne Photon-Counting LiDAR Bathymetry

The ATLAS system onboard the ICESat-2 employed a 532-nm-wavelength transmitter and water has the weak scattering and small attenuation coefficient in this band. The laser can penetrate a certain depth of water, enabling the detection of water depth and underwater topography information (Li et al., 2019). Taking the islands and reefs in the South China Sea as an example, this part uses the ICESat-2/ATLAS photon-counting data to extract the water depth in shallow water areas of islands and reefs. The flowchart is shown in Figure 5.45.

1. Water surface photon extraction

   Since the returning laser light will sometimes be specular over a high reflectivity surface, part of the specular echo noise close to the water surface will affect the water level acquisition, which is also the difficulty of high-precision water surface information extraction. Therefore, a new methodological framework based on statistical features to remove noise photons induced by specular reflection is proposed for a high reflectivity water surface (Wang et al., 2023). First, ICESat-2 data preprocessing is performed to improve the speed and accuracy of subsequent processing. Second, the statistical features of the outlier distance are calculated and utilized to filter out noise photons. Finally, the water surface photons are obtained according to the extreme points in the elevation distribution histogram.

2. Water bottom photon extraction

   During the depth measurement process of photon-counting LiDAR, the photon density at the water surface is higher than that at the bottom. Furthermore, as water depth increases, the density of signal photons decreases, which increases the difficulty of detecting bottom photons.

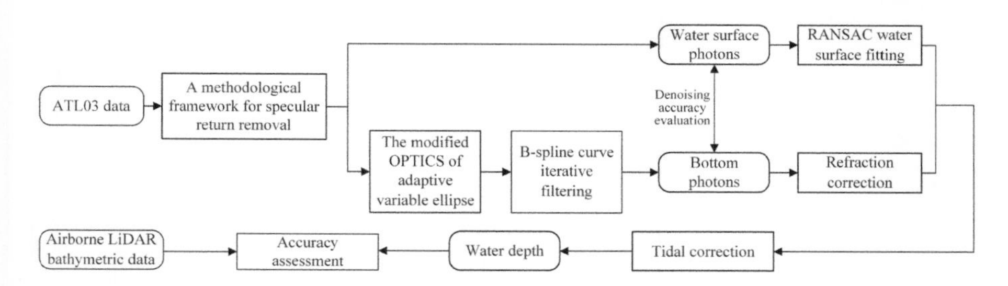

FIGURE 5.45 Bathymetry flowchart based on ICESat-2/ATLAS data.

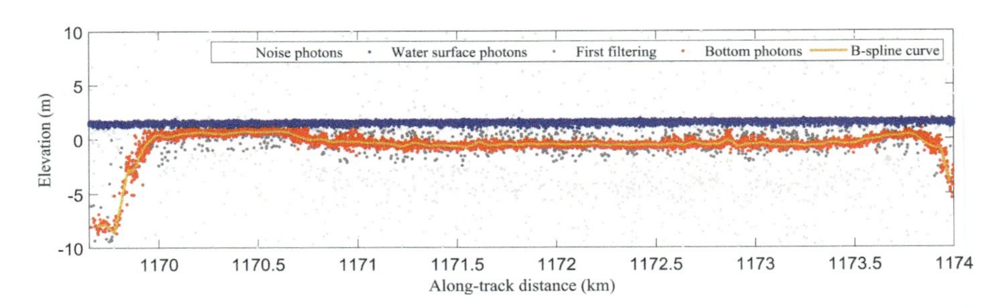

FIGURE 5.46 Result of water surface and bottom photon extraction.

Underwater photons can usually be extracted using denoising algorithms (clustering density methods, local statistics methods, raster-based photon denoising) or a combination of several denoising algorithms. This example utilizes a combination of the Ordering Points to Identify the Clustering Structure (OPTICS) algorithm of adaptive search ellipse and B-spline curve iterative filtering to extract bottom signal photons. Figure 5.46 shows the result of water surface and bottom photon extraction.

3. Bathymetry calculation

Because of the refraction of photons in water transmission and the periodic fluctuation of tidal forces on the sea surface, refraction correction and tidal correction of photons are required to improve bathymetric accuracy. Specifically, the refraction correction is performed based on Snell's law. By using the azimuth of the unit pointing vector, the offset of the photons in the along-track distance and height directions is obtained, and the latitude and longitude of photon coordinates are transformed to their actual underwater position. The result of laser bathymetry refraction correction is shown in Figure 5.47. The water depth is calculated according to the acquired surface photons and underwater photons. In addition, to obtain water depth based on the datum for sounding reduction, the bathymetric results of ATLAS and airborne LiDAR bathymetric data are subjected to ocean tide correction.

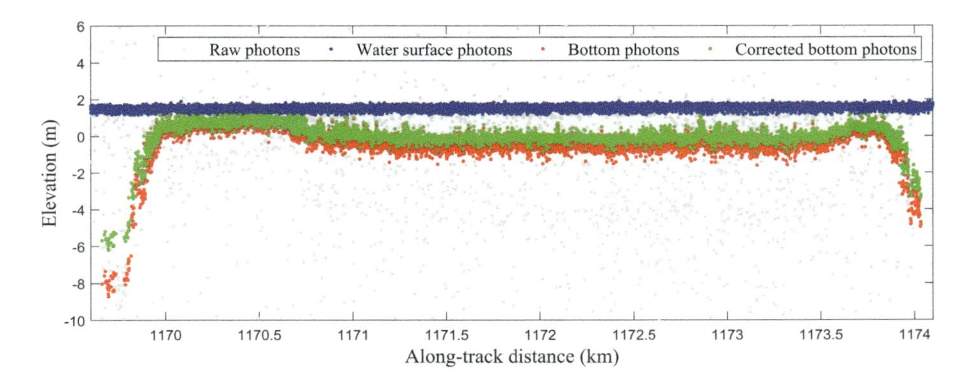

FIGURE 5.47 LiDAR bathymetry result after refraction correction.

5.10 SUMMARY

LiDAR remote sensing has many applications. Because of space constraints, this chapter introduces its application in several typical fields, such as surface model construction and contour line generation in topographic mapping; extractions of individual tree parameters and forest stand parameters, tree species classification, and forest parameter mapping at the regional scale in forest investigation; transmission corridor point cloud classification, 3D modeling, and safety analysis in power line inspection; building point cloud extraction, roof segmentation, contour extraction, and model generation in building 3D modeling; environmental perception and autonomous positioning in automatic driving; identification of crop planted areas, leaf inclination, and FPAR estimations in crop monitoring; digital archaeological and cultural heritage conservation; indoor 3D modeling and navigation; and underwater mapping.

EXERCISES

1. Briefly describe the production process of DEM, DSM, and contour line products using LiDAR technology.
2. List five vegetation structure parameters that can be extracted from LiDAR data and describe the extraction methods.
3. Introduce one individual tree segmentation algorithm based on airborne LiDAR data.
4. Briefly introduce a building 3D modeling algorithm based on airborne LiDAR point cloud data.
5. List the advantages and disadvantages of LiDAR technology in agricultural applications.
6. List the differences between a high-precision map and traditional navigation map and explain the reason why the LiDAR point cloud is used for high-precision map production.
7. Briefly describe one transmission corridor object classification method based on airborne LiDAR data.
8. Briefly describe the main steps of 3D modeling of power lines and power towers using airborne LiDAR.

REFERENCES

Cai, L., Wang, S., Yuan, G., Gu, S., & Song, D. (2018). An error model of pylons tilt monitoring using horizontal section method (in Chinese). *Bulletin of Surveying and Mapping*, (5), 71–76. https://doi.org/10.13474/j.cnki.11-2246.2018.0146.

Campbell, G. S. (1990). Derivation of an angle density function for canopies with ellipsoidal leaf angle distributions. *Agricultural and Forest Meteorology*, *49*(3), 173–176.

Cao, L. (2008). The application of multi three-dimension laser scanning in the surveying of Shanhaiguan part of the Great Wall (in Chinese). *Bulletin of Surveying and Mapping*, (3), 31–33.

Chang, Y., Zhang, F., Huang, X., & Liu, G. (2011). Dome image creation by integrating laser scanning data and high resolution images in Mogao Caves 196 and 285 (in Chinese). *Dunhuang Research*, (6), 96–100.

Chase, A. F., Chase, D. Z., & Weishampel, J. F. (2010). Lasers in the jungle: Airborne sensors reveal a vast Maya landscape. *Archaeology*, *63*(4), 27–29.

Chen, J. M., & Black, T. A. (1992). Defining leaf area index for non-flat leaves. *Agricultural and Forest Meteorology*, *57*, 1–12.

Chen, S., Wang, C., Dai, H., Zhang, H., Pan, F., Xi, X., Yan, Y., Wang, P., Yang, X., Zhu, X., & Ardana, A. (2019a). Power pylon reconstruction based on abstract template structures using airborne LiDAR data. *Remote Sensing*, *11*(13), 1579.

Chen, X., Yun, T., Xue, L., & Liu, Y. (2019b). Classification of tree species based on LiDAR point cloud data (in Chinese). *Laser and Optpelectronics Process*, *56*(12), 203–214. https://doi.org/10.3788/LOP56.122801

Doneus, M., Briese, C., Fera, N., & Janner, M. (2008). Archaeological prospection of forested areas using full-waveform airborne laser scanning. *Journal of Archaeological*, *35*(4), 882–893.

Doneus, M., Doneus, N., Briese, C., Pregesbauer, M., Mandlburger, G., & Verhoeven, G. (2013). Airborne laser bathymetry e detecting and recording submerged archaeological sites from the air. *Journal of Archaeological Science*, *40*, 2136–2151.

Edelsbrunner, H., Kirkpatrick, D., & Seidel, R. (1983). On the shape of a set of points in the plane. *IEEE Transactions on Information Theory*, *29*(4), 551–559.

Evans, D. H., Fletchera, R. J., Pottier, C., Chevance, J. B., Soutif, D., Tan, B. S., Im, S., Ea, D., Tin, T., & Kim, S. (2013). Uncovering archaeological landscapes at Angkor using LiDAR. *Proceedings of the National Academy of Sciences*, *110*(31), 12595–12600.

Feng, B., Zheng, C., Zhang, W., Wang, L., & Yue, C. (2020). Analyzing the role of spatial features when cooperating hyperspectral and LiDAR data for the tree species classification in a subtropical plantation forest area. *Journal of Applied Remote Sensing*, *14*(2), 1–26.

Green, P. J. (1995). Reversible jump Markov chain Monte Carlo computation and Bayesian model determination. *Biometria*, *82*(4), 711–732.

Humme, A., Lindenbergh, R., & Sueur, C. (2006). Revealing Celtic fields from LiDAR data using kriging based filtering. In: Proceedings of the ISPRS Commission V Symposium, Dresden, International Archives of Photogrammetry, and Remote Sensing and Spatial Information Sciences, Vol. XXXVI, part 5, 25–27.

Kang, Z., Zhong, R., Wu, A., Shi, Z., & Luo, Z. (2016). An efficient planar feature fitting method using point cloud simplification and threshold – independent BaySAC. *IEEE Geoscience and Remote Sensing Letters*, *13*(12), 1842–1846.

Kersten, T. P., Lindstaedt, M., & Vogt, B. (2009). Preserve the past for the future - Terrestrial laser scanning for the documentation and deformation analysis of Easter Island's Moai. *Photogrammetric Fernerkundung Geoinformation*, *1*, 79–90. https://doi.org/10.1127/0935-1221/2009/0008

Khan, S. H., Bennamoun, M., Sohel, F., & Togneri, R. (2014). Geometry driven semantic labeling of indoor scenes. *European Conference on Computer Vision*, *8689*, 670–694.

Koenig, K., & Höfle, B. (2016). Full-waveform airborne laser scanning in vegetation studies—A review of point cloud and waveform features for tree species classification. *Forest*, *7*(12), 198.

Korhonen, L., Korpela, I., Heiskanen, J., & Maltamo, M. (2011). Airborne discrete-return LiDAR data in the estimation of vertical canopy cover, angular canopy closure and leaf area index. *Remote Sensing of Environment*, *115*(4), 1065–1080.

Lao, J., Wang, C., Nie, S., Xi, X., & Wang, J. (2022). Monitoring and analysis of water level changes in Mekong River from ICESat-2 spaceborne laser altimetry. *Water*, *14*(10), 1613.

Li, Y., Gao, H., Jasinski, M. F., Zhang, S., & Stoll, J. D. (2019). Deriving high-resolution reservoir bathymetry from ICESat-2 prototype photon-counting LiDAR and Landsat imagery. *IEEE Transactions on Geoscience and Remote Sensing*, *57*(10), 7883–7893.

Li, Z., Liu, Q., & Pang, Y. (2016). Review on forest parameters inversion using LiDAR (in Chinese). *Journal of Remote Sensing*, *20*(5), 1138–1150. https://doi.org/10.11834/jrs.20165130

Li, Y., Ma, L., Zhong, Z., Liu, F., Chapman, M. A., Cao, D., & Li, J. (2020). Deep learning for LiDAR point clouds in autonomous driving: A review. *IEEE Transactions on Neural Networks and Learning Systems*, *38*, 1–21.

Li, W., Niu, Z., Wang, C., Huang, W., Chen, H., Gao, S., Li, D., & Muhammad, S. (2015). Combined use of airborne LiDAR and satellite GF-1 data to estimate leaf area index, height, and aboveground biomass of maize during peak growing season. *IEEE Journal of Selected Topics in Applied Earth Observations & Remote Sensing*, *8*(9), 4489–4501.

Li, X., & Strahler, A. H. (1988). Modeling the gap probability of a discontinuous vegetation canopy. *IEEE Transactions on Geoscience and Remote Sensing*, *26*(2), 161–170.

Liu, C., Wu, J., & Furukawa, Y. (2018). FloorNet: A unified framework for floorplan reconstruction from 3D scans. *European Conference on Computer Vision*, *11210*, 203–219.

Luo, S., Wang, C., Xi, X., Nie, S., Fan, X., Chen, H., Yang, X., Peng, D., Lin, Y., & Zhou, G. (2019a). Combining hyperspectral imagery and LiDAR pseudo-waveform for predicting crop LAI, canopy height and above-ground biomass. *Ecological Indicators*, *102*, 801–812.

Luo, S., Wang, C., Xi, X., Nie, S., & Zhou, G. (2019b). Estimating forest aboveground biomass using small-footprint full-waveform airborne LiDAR data. *International Journal of Applied Earth Observation and Geoinformation*, *83*, 101922.

Maas, H. G., & Vosselman, G. (1999). Two algorithms for extracting building models from raw laser altimetry data. *ISPRS Photogrammetry and Remote Sensing*, *54*(2), 153–163.

Morris, C. N. (1983). Parametric empirical Bayes inference: Theory and applications. *Journal of the American Statistical Association*, *78*(381), 47–55.

Qin, H., Wang, C., Pan, F., Xi, X., & Luo, S. (2017). Estimation of FPAR and FPAR profile for maize canopies using airborne LiDAR. *Ecological Indicators*, *83*, 53–61.

Riaño, D., Meier, E., Allgower, B., Chuvieco, E., & Ustin, S. L. (2003). Modeling airborne laser scanning data for the spatial generation of critical forest parameters in fire behavior modeling. *Remote Sensing of Environment*, *86*, 177–186.

Sittler, B. (2004). Revealing historical landscapes by using airborne laser scanning. A 3-D model of ridge and furrow in forests near Rastatt (Germany). In: M. Thies, B. Koch, H. Spiecker, & H. Weinacker (Eds.), *Proceedings of Natscan, Laser-Scanners for Forest and Landscape Assessment - Instruments, Processing Methods and Applications.* International Archives of Photogrammetry, and Remote Sensing and Spatial Information Sciences, Volume XXXVI, Part 8/W2, pp. 258–261.

Su, W., Huang, J., Liu, D., & Zhang, M. (2019). Retrieving corn canopy leaf area index from multitemporal Landsat imagery and terrestrial LiDAR data. *Remote Sensing*, *11*, 572.

Su, W., Zhu, D., Huang, J. X., & Guo, H. (2018). Estimation of the vertical leaf area profile of corn (*Zea mays*) plants using terrestrial laser scanning (TLS). *Computers and Electronics in Agriculture*, *150*, 5–13.

Tang, G. A., Zhao, M. D., Yang, X., & Zhou, Y. (2010). *Geographic information systems* (2nd ed., in Chinese). Science Press.

Wan, Y., Xi, X., Wang, C., & Wang, F. (2014). 3D reconstruction of ornamental column based on terrestrial laser scanning data (in Chinese). *Bulletin of Surveying and Mapping*, (11), 57–59. https://doi.org/10.13474/j.cnki.11-2246.2014.0363

Wang, Y., Huo, H., Wang, G., & Hu, C. (2011). Application of terrestrial LiDAR technology in the deformation analysis of multi-door chorten in Baiju Temple (in Chinese). *Journal of Beijing University of Civil Engineering and Architecture*, *27*(4), 6. https://doi.org/10.3969/j.issn.1004-6011.2011.04.003

Wang, Y., Li, G., Ding, J., Guo, Z., Tang, S., Wang, C., Huang, Q., Liu, R., & Chen, J. (2016). A combined GLAS and MODIS estimation of the global distribution of mean forest canopy height. *Remote Sensing of Environment*, *174*, 24–43.

Wang, Z., Nie, S., Xi, X., Wang, C., Lao, J., & Yang, Z. (2023). A methodological framework for specular return removal from photon-counting LiDAR data. *International Journal of Applied Earth Observation and Geoinformation*, *122*, 103387.

Wang, P., Xing, Y., Wang, C., & Xi, X. (2019). A graph cut-based approach for individual tree detection using airborne LiDAR data (in Chinese). *Journal of University of Chinese Academy of Sciences*, *36*(3), 385–391. https://doi.org/10.7523/j.issn.2095-6134.2019.03.012

Wu, H., Yue, H., Xu, Z., Yang, H., Liu, C., & Chen, L. (2021). Automatic structural mapping and semantic optimization from indoor point clouds. *Automation in Construction*, *124*, 103460.

Xi, X., Wang, Z., & Wang, C. (2022). Bathymetric extraction method of nearshore based on ICESat-2/ATLAS data (in Chinese). *Journal of Tongji University (Natural Science)*, *50*(07), 940–946.

Yang, J., Kang, Z., Zeng, L., Akwensi, P. H., & Sester, M. (2021). Semantics – Guided reconstruction of indoor navigation elements from 3D colorized points. *ISPRS Journal of Photogrammetry and Remote Sensing*, *173*, 238–261.

Yang, B., & Lee, J. (2019). Improving accuracy of automated 3-D building models for smart cities. *International Journal of Digital Earth*, *12*, 209–227.

Yang, T., Wang, C., Li, G., Luo, S., Xi, X., Gao, S., & Zeng, H. (2015). Forest canopy height mapping over China using GLAS and MODIS data (in Chinese). *Science China Earth Sciences*, *58*(1), 96–105.

Yang, X., Wang, C., Pan, F., Nie, S., Xi, X., & Luo, S. (2019). Retrieving leaf area index in discontinuous forest using ICESat/GLAS full-waveform data based on gap fraction model. *ISPRS Journal of Photogrammetry and Remote Sensing*, *148*, 54–62.

Yurtsever, E., Lambert, J., Carballo, A., & Takeda, K. (2020). A survey of autonomous driving: Common practices and emerging technologies. *IEEE Access*, *8*, 58443–58469.

Zhou, R., Jiang, W., Huang, W., Xu, B., Jiang, S. (2017). A heuristic method for power pylon reconstruction from airborne LiDAR data. *Remote Sensing*, *9*(11), 1172.

Zhou, Y., & Tuzel, O. (2018). Voxelnet: End-to-end learning for point cloud based 3D object detection. Proceedings of The IEEE Conference on Computer Vision and Pattern Recognition (CVPR), IEEE, 4490–4499.

Zhu, X., Nie, S., Wang, C., Xi, X., Lao, J., & Li, D. (2022). Consistency analysis of forest height retrievals between GEDI and ICESat-2. *Remote Sensing of Environment*, *281*, 113244.

6 LiDAR Remote Sensing Prospects

6.1 CONTINUOUSLY OPTIMIZED SENSOR PERFORMANCE

Since the beginning of the 21st century, the development of commercial LiDAR systems has entered an explosive period. Numerous scientific research institutions and commercial companies in the world have carried out extensive research and development of LiDAR systems, and the market of LiDAR systems has started to grow. With the maturity of hardware techniques, LiDAR sensors are developing in terms of high performance, low cost, and miniaturization. Various LiDAR sensors have increased battery life, ranging accuracy, scanning range, scanning frequency, and narrower beam divergence.

Low-cost, lightweight, and small-size LiDAR systems have gradually become the market mainstream. They show great advantages in terms of data acquisition and data quality. Under the premise of ensuring the high performance of the equipment, reducing the sensor price, size, and weight has become the key for businesses to increase the LiDAR market share. For example, the iPad Pro 2020 released by Apple in the United States in 2020 is equipped with a laser scanner, which made it possible for LiDAR to enter the consumer market. The LiDAR on iPad Pro 2020 has a strong ability to perceive the position of objects within a certain distance and reads the depth information of the three-dimensional (3D) space, directly enhancing the user's sense of augmented reality (AR). In addition, the iPad Pro LiDAR can achieve 3D scanning of ground objects (Figure 6.1). The collected data can be converted into a point cloud through third-party software.

6.2 A FLOOD OF NEW LiDAR SENSORS

With the increasing demand for remote sensing applications, various new LiDAR systems continue to emerge. Hyperspectral/multi-spectral, solid-state, and quantum LiDAR systems enrich the depth and breadth of LiDAR remote sensing and its applications.

6.2.1 Hyperspectral/Multi-Spectral LiDAR

As mentioned earlier in this book, most LiDAR systems currently use single-wavelength (near-infrared wavelength, such as 1064 nm or 1550 nm) or dual-wavelength (green 532 nm and near-infrared 1064 nm). Hyperspectral/multi-spectral LiDAR, a new LiDAR technique, has the capability to integrate 3D spatial and spectral information detection (Morsdorf et al., 2009). It is proven a promising technique in the field of Earth observation.

DOI: 10.1201/9781032671512-6

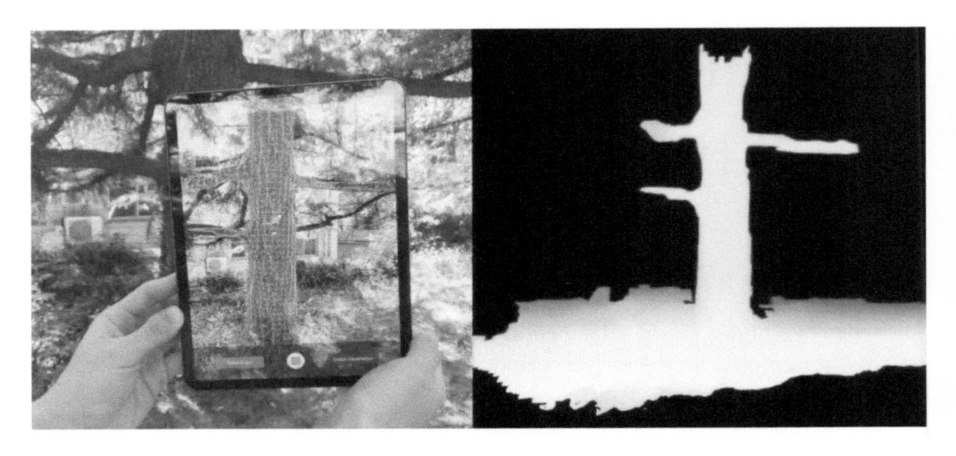

FIGURE 6.1 The iPad Pro LiDAR for object scanning.

An important application of hyperspectral/multi-spectral LiDAR systems is the retrieval of vertical distribution of vegetation physiological parameters. The traditional single-wavelength LiDAR lacks abundant spectral information. It can only detect the 3D structure of the vegetation canopy, rather than capture the information of vegetation biochemical content. Hyperspectral/multi-spectral imaging technology is capable of acquiring abundant spectral information of surface objects. It has been widely applied to retrieving the vegetation optical properties and biochemical contents, such as reflectance, transmittance, chlorophyll, water content, and nitrogen content. However, the results of passive imaging are only in two-dimensional space. Hyperspectral/multi-spectral LiDAR can acquire the LiDAR returns in different wavelengths and object spectrums at different vertical heights (Figure 6.2). These characteristics make the hyperspectral/multi-spectral LiDAR technique promising for accurate inversion of vegetation physiological parameters (e.g., vertical profile of

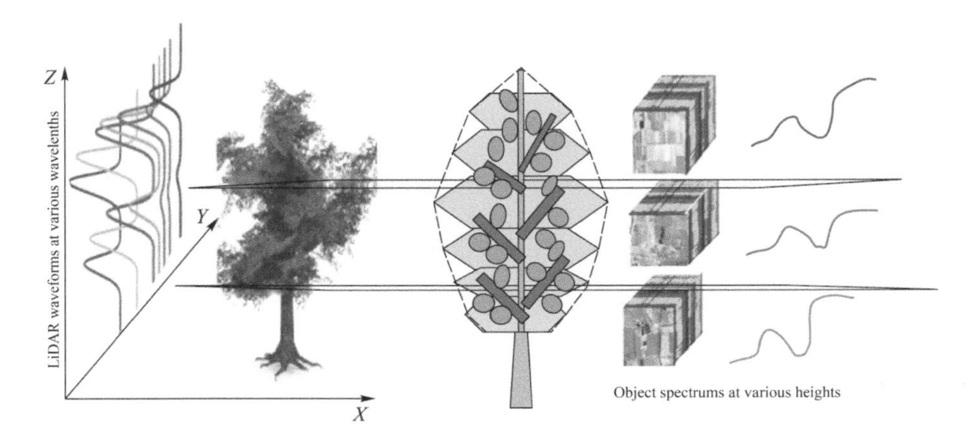

FIGURE 6.2 Schematic diagram of hyperspectral/multi-spectral LiDAR data.

vegetation biochemical parameters) in the future, greatly improving the applications of remote sensing in forest pest monitoring and crop yield estimation, among many other applications.

At present, many institutions in the world have carried out research on multi-spectral LiDAR systems. For example, Optech, Inc., launched the world's first airborne multi-spectral LiDAR system, "Titan," equipped with three independent laser bands (532 nm, 1064 nm, and 1550 nm) (Fernandez-Diaz et al., 2016). It can acquire 3D spectral information. The acquired multi-spectral LiDAR data has achieved accurate land cover classification by using only very simple classification algorithms. In China, Wuhan University and the Aerospace Information Research Institute (Chinese Academy of Sciences) have also conducted researches on hyperspectral/multi-spectral LiDAR systems and achieved preliminary results (Niu et al., 2015). Compared with multi-spectral LiDAR, hyperspectral LiDAR can obtain much more spectral information (Gong et al., 2021).

6.2.2 Solid-State LiDAR

The traditional LiDAR uses mechanical scanning, which is relatively large in size, expensive, and difficult to assemble precisely, directly limiting the practical applications. Under such circumstances, solid-state LiDAR receives extensive attention as a promising solution. This technique targets a laser beam across the environment without deploying motors or gears. At present, the realization methods of solid-state LiDAR include the micro-electro-mechanical system (MEMS), flash technique, and optical phased array (OPA) technique.

The MEMS uses a micro-scanning galvanometer to deflect the laser beam, and the scanning range is limited by the deflection range of the galvanometer. The flash technique uses a camera-like working mode. Each pixel of the photosensitive element can record the time-of-flight information of laser photon to produce a "3D" image with depth information. However, the technique has a limited field of view and a low scanning speed. The OPA is a new laser beam pointing control technique based on the microwave phased array scanning method. It has the advantages of being a non-inertial device with good accuracy, stability, and arbitrary direction control.

The working principle of OPA is described here. First, the laser power is evenly divided into a multi-channel phase modulator array; then the optical field is transmitted through the optical antenna and coherently superimposed in the spatial far field to form a beam with strong energy. Next, the optical field through specific phase modulation is transmitted at the transmitting antenna end and generates the tilt of the wavefront, further causing the deflection of the beam in the far field. Finally, through different phases, the beams are emitted toward different directions and a series of scanning beams are formed without mechanical scanning.

The compact chip of the solid-state technique not only makes LiDAR more robust but also saves structural space within the scanner, allowing the scanner to be designed with extremely small size and low cost. Due to its small size, solid-state LiDAR can be integrated into vehicles, infrastructures, and buildings. Especially in the automotive field, "those who get solid-state LiDAR will get the world of autonomous driving" has become an industry consensus. The solid-state LiDAR provides both beautiful and robust solutions for automotive sensors.

China's LiDAR companies, such as LeiShen Intelligent System Co., Ltd., Beike Tianhui Co., Ltd., RoboSense Co., Ltd., and HeSai Technology Co., Ltd., have conducted research on MEMS LiDAR but have not yet produced commercialized devices on a large scale. Other companies, such as Benewake Co., Ltd., Guangbo Intelligence, and Huake Bochuang, have also launched relevant products in terms of flash LiDAR. Quanergy Co., Ltd., brought the OPA LiDAR to the commercial horizon and is developing the full-solid-state LiDAR for in-vehicle sensing systems and driverless cars. In summary, the advanced all-solid-state and miniaturized LiDAR with high power, large scanning angle, and high resolution still requires further research.

6.2.3 Quantum LiDAR

The quantum LiDAR is a combined product of LiDAR and quantum information techniques. Compared with traditional LiDAR, it has the advantages of anti-interference, high sensitivity, high-range resolution, and high angular resolution, which has become a research hotspot in the LiDAR field. The advantages of quantum LiDAR are mainly in the following two aspects:

1. The information carrier of quantum information technique is a single quantum. The object of signal generation, modulation, reception, and detection is all based on the single quantum. The entire receiving system has extremely high sensitivity; that is, the noise floor of the quantum receiving system is extremely low. Specifically, the noise floor of quantum systems is lower by several orders of magnitude than that of the traditional LiDAR receivers. For example, ignoring some factors such as operating frequency, clutter, and dynamic range, the operating distance of quantum LiDAR can be increased several times, or even dozens of times, compared with classic LiDAR. Thus, quantum LiDAR has shown great improvements in terms of detecting weak targets and even stealth targets.
2. The modulation object of the quantum information technique is a quantum state. The quantum state has higher-order information than the classical time domain, frequency domain, polarization, etc.; that is, the modulation information has a higher dimension. From the perspective of information theory, better performance can be derived by operating on high-dimensional information. For target detection, high-order information modulation can reduce the noise floor without affecting the accumulated benefits, thereby improving the capability of detecting weak targets from noise. From the perspective of signal analysis, the signal can be processed by quantum high-order micro-modulation. This makes traditional signal analysis hard to interpret the signal and extract the modulated information, thereby improving the anti-interception capability in the electronic countermeasure environment.

In August 2016, the world's first quantum science experimental satellite, "Micius," independently developed by China, was successfully launched. The 14th Research

Institute of China Electronics Technology Group Corporation (CETC) developed a quantum radar and conducted studies on quantum detection mechanism, target scattering characteristics, and experimental validation. They also developed the first quantum LiDAR system based on a single-photon detector.

6.3 ARRIVAL OF THE LiDAR BIG DATA ERA

The rapid development of multi-platform LiDAR observation systems and intelligent computing technology provides a rare opportunity for the progress and revolution of LiDAR remote sensing science and technology. In terms of data processing, LiDAR is gradually entering a remote sensing big data era driven by data models and characterized by intelligent analysis of big data, after going through, in turn, the era of statistical mathematical models and the era of quantitative LiDAR remote sensing marked by the physical quantification of remote sensing information.

6.3.1 Multi-Platform Coexistence, Multi-Source Data Fusion

With the advancement and commercialization of LiDAR technology, the data sources continue to emerge, providing massive, multi-source (multiple platforms, various densities, point cloud, waveform, photon counting, various precisions) LiDAR data for various applications. It is essential to efficiently integrate these massive LiDAR datasets to extract reliable 3D information to support various applications.

The LiDAR data obtained from different platforms have their own advantages. Extracting surface information from multi-source LiDAR data has become an important means of practical applications. Take forestry applications as an example. Airborne LiDAR can acquire surface spatial structure at a plot scale. However, in dense forest areas, airborne LiDAR cannot accurately capture the information of the tree diameter and crown base height due to canopy occlusion. In contrast, terrestrial LiDAR can acquire fine forest structures, but is unable to capture points on top of the canopy in many cases. The fusion of airborne and terrestrial LiDAR data can complement each other and provide complete structural information for forest investigations (Goodwin et al., 2017).

In practice, accurate registration of multi-source LiDAR data is always a pivotal step in data processing and quantitative analysis. However, differences of point cloud densities and mismatch of various observation scales may bring great challenges to the registration and quantitative analysis of LiDAR data. Fusing multi-source LiDAR data and mining rich surface information are still challenging in practical applications. To date, some advanced artificial intelligence and deep learning algorithms provide potential research directions for multi-source LiDAR data fusion (Xi et al., 2020).

Various observation platforms and multi-type LiDAR sensors provide abundant data sources for surface investigation. Hyperspectral/multi-spectral imagery can provide spectral information that cannot be provided by LiDAR, and LiDAR data can provide vertical structure information that spectral images lack. Fusing the two types of data can improve the accuracy of land use classification, vegetation parameter retrievals, and crop monitoring (Luo et al., 2016; Wang et al., 2017). However, most of the existing studies are based on empirical models between LiDAR feature parameters and surface properties, without considering physical mechanisms. It is necessary to establish physical relationships between LiDAR signals and the Earth's surface characteristics and accurately mine abundant surface information contained in LiDAR data and other remotely sensed data.

6.3.2 LiDAR Big Data Era

LiDAR data in the era of big data have the characteristics of large volume, high variety, low value density, and high velocity. The arrival of the LiDAR big data era poses both challenges and opportunities for people to control data, analyze data, and utilize data. To respond to the challenges, a few thoughts are listed next.

1. Establish LiDAR data sharing platforms with a resource catalog and metadata as the core. With the rapid development of LiDAR technology, many scientists and practitioners have conducted various LiDAR data acquisition experiments and collected massive amounts of LiDAR data. However, extensive sharing of LiDAR data is still difficult for various reasons, and the benefits of the data are not fully exploited. Therefore, LiDAR data sharing platforms should be established to integrate discrete LiDAR data resources. This will help build an intelligent management and sharing service system for the whole society and promote the needs and services of regional LiDAR data sharing.
2. Develop LiDAR deep learning algorithms to intelligently extract object 3D information. The massive LiDAR point cloud data and requirements for accuracy and efficiency bring new challenges to LiDAR data processing. Deep learning algorithms spawned in the era of big data need to be introduced to the field of LiDAR data processing and analysis. With the continuous breakthrough of deep learning and the availability of 3D point clouds, 3D deep learning has achieved a series of remarkable results, such as point cloud semantic segmentation and scene understanding based on 3D deep learning, object detection, and object classification. However, at present, the theoretical and technical research of 3D deep learning is still far from meeting the needs of LiDAR practical applications. The next step is to further improve the LiDAR data deep learning algorithms and their applications in terms of network structure, learning algorithm, and performance analysis.
3. Develop free LiDAR data intelligent processing softwares with high processing performance for users. At present, some enterprises and academic institutions have developed a variety of commercial software and open-source software for point cloud processing, e.g., TerraSolid, CloudCompare, Point Cloud Magic (PCM), and LiDAR360. The main features of these software programs are management of point cloud, point cloud filtering for digital elevation model (DEM) production, building 3D reconstructions, forest structural parameter retrievals, power line inspection, etc. More features need to be added, especially the intelligence and automation of point cloud processing and engineering applications.

6.4 COMPREHENSIVE APPLICATIONS OF LiDAR

LiDAR has been widely used in many fields, such as basic surveying and mapping, forest resource investigation, power line inspection, digital cities, unmanned driving, heritage protection, polar icesheet monitoring, and mining surveys. It provides technology support for the national economy, social development, and scientific research and generates huge economic and social benefits. With the rapid development of computer technology, automatic control technology, and navigation and positioning technology, LiDAR is changing as each day passes, with a huge market potential.

6.4.1 LiDAR Satellite Missions Promote the Development of Global Surface Products toward the Direction of High Precision and High Resolution

As an important means of Earth observation, laser altimetry satellites can actively obtain 3D information of the global surface and provide services for global stereoscopic mapping. ICESat/GLAS plays an important role in the production of global surface elevation products (Lefsky, 2010). However, limited by factors such as laser sampling frequency, the standard products derived by interpolation or extrapolation have low resolution and low precision. Since 2018, the United States' ICESatat-2/ATLAS and GEDI and China's GaoFen-7 and Terrestrial Ecosystem Carbon Inventory Satellite (TECIS-1) have been successfully launched. The LiDAR Surface Topography (LIST) mission of the United States is also being deployed and implemented (Figure 6.3). The laser altimetry of these satellites would acquire the more precise surface information, which brings new opportunities for mapping surface elevation control points, vegetation structure parameters, forest biomass, terrain slope, and other features on a global scale and promotes the development of these products in the direction of high precision and high resolution.

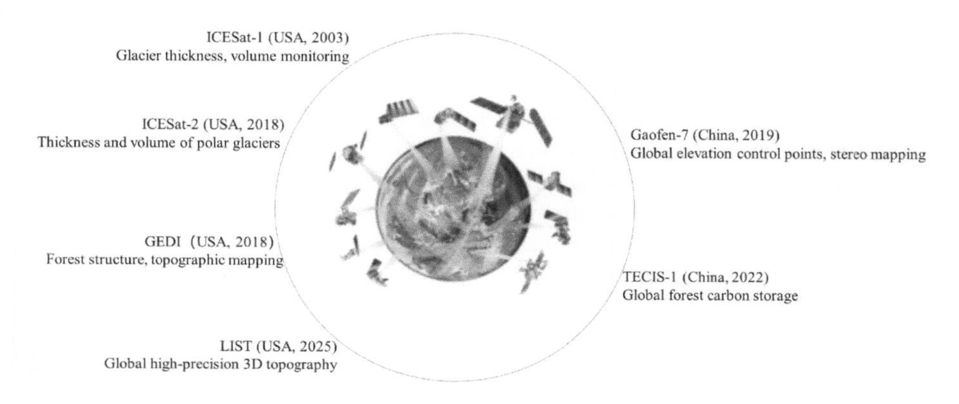

FIGURE 6.3 Laser altimetry satellite missions of the United States and China.

6.4.2 LiDAR Improves Natural Resource Monitoring from Two-Dimensional to Three-Dimensional Space

Efficient mapping and monitoring of the Earth's surface features and phenomena are crucial for achieving high-quality and sustainable economic development, as well as better understanding of the Earth system. Optical remote sensing has the advantage of a wide range, fast speed, and high precision, which has become an effective means of renewable natural resource monitoring such as forest monitoring. However, optical remote sensing is affected by day/night and weather conditions. With the emergence of LiDAR, remote sensing has entered the 3D era, and the observation platforms have become more diverse. These new LiDAR technologies bring new opportunities to the 3D dynamic monitoring of natural resources, including land, forest, ocean, mineral, grassland, water, and wetland across different spatial scales (Guo et al., 2014). Spaceborne LiDAR can obtain 3D measured data on a regional or global scale because of its high orbit and wide observation range. Airborne/unmanned aerial vehicle (UAV)–based LiDAR is widely used to extract 3D structure information and monitor the dynamic changes on a landscape or regional scale. Terrestrial LiDAR provides an important means for acquiring fine 3D information on a small scale (i.e., landscape level, ecosystem level, and plot level) or individual targets. Different LiDAR platforms show advantages and potential in the monitoring of multi-scale natural resources. Meanwhile, with the development of the photon-counting LiDAR, multi-spectral and hyperspectral LiDAR, it is believed that the collaborative applications of multi-source, multi-platform (spaceborne/airborne/UAV/vehicle/terrestrial, etc.), and multi-scale LiDAR will hopefully solve the current difficulty of obtaining full-coverage, high-precision, multi-spectral LiDAR data, as well as show enormous potential in 3D dynamic monitoring of full elements and full attributes of natural resources in the future.

6.4.3 LiDAR Promotes the Further Development of Smart City

Since the beginning of the 21st century, the concepts of "digital Earth" and "smart Earth" have been increasingly accepted by the public. The smart city has become the new trend of urban development. In the process of smart city construction, the Internet of Things (IOT) platforms are built with 3D geospatial information as the carrier, with various front-end sensors as the foundation and 5G communication as the core. The bridges between 3D information and smart humanities, smart medicine, smart transportation, smart security, and other applications are built by taking advantage of artificial intelligence (AI), which improves the informatization and intelligence of urban administration and achieves further iterative upgrading of the city. Smart city construction relies on high-efficiency and high-precision acquisition of urban 3D information. LiDAR technology plays an important role in data acquisition and technical support for reconstruction and dynamic update of 3D real scenes. Virtual reality (VR) and AR can achieve the transformation from 3D information acquisition to 3D scene reconstruction and meet the requirements of real 3D spatial cognition. In addition, the combination of the three (LiDAR, AR, and VR) can promote the rapid and efficient development of real scene 3D and meet the data demand for smart city construction in the information age.

6.4.4 LiDAR Assists with Automatic Driving

With the development of AI technology, various industries around the world have quietly changed. Among them, the motor industry, which provides convenience for people to travel, is integrating into the development of AI with a new posture. Automatic driving seems to be the future of the motor industry. Automatic positioning, environmental awareness, and high-precision mapping are inevitable topics for automatic driving. LiDAR technology plays a very important role in automatic driving, mainly in three aspects. First, high-precision positioning requires a combination of global navigation satellite system (GNSS), LiDAR, and inertial measurement unit (IMU). Second, the realization of an autopilot needs sensors such as LiDAR to sense environmental information actively in real time. Third, a high-precision map provides a priori map information for vehicle operation and plays an important role in high-precision positioning, environmental awareness, path planning, and simulation experiments. LiDAR provides a guarantee for centimeter-level high-precision maps.

6.5 SUMMARY

LiDAR has become an indispensable technical means for 3D spatial information acquisition and is evolving rapidly. Based on the systematic analysis of the current research progress of LiDAR remote sensing in the world, this chapter briefly introduced the future development trends of LiDAR from the aspects of LiDAR sensors, LiDAR data processing in the era of big data, and comprehensive LiDAR applications.

EXERCISES

1. Based on your understanding, what technical problems still need to be solved in the field of LiDAR remote sensing, and what are the future research directions?
2. What global surface products can LiDAR produce?
3. What are the advantages of hyperspectral/multi-spectral LiDAR compared with commonly used single-wavelength LiDAR?
4. Please choose an industry where LiDAR has been widely used, talk about the specific applications of LiDAR in this industry, and discuss the possible development of the application directions.

REFERENCES

Fernandez-Diaz, J. C., Carter, W. E., Glennie, C., Shrestha, R. L., Pan, Z. G., Ekhtari, N., Singhania, A., Hauser, D., & Sartori, M. (2016). Capability assessment and performance metrics for the titan multispectral mapping LiDAR. *Remote Sensing*, *8*(11), 936.

Gong, W., Shi, S., Chen, B., Song, S., Niu, Z., Wang, C., Guan, H., Li, W., Gao, S., Lin, Y., Sun, J., Yang, J., & Du, L. (2021). Development and prospect of hyperspectral LiDAR for earth observation (in Chinese). *National Remote Sensing Bulletin*, *25*(1), 501–513. https://doi.org/10.11834/jrs.20210086

Goodwin, N. R., Armston, J. D., Muir, J., & Stiller, I. (2017). Monitoring gully change: A comparison of airborne and terrestrial laser scanning using a case study from Aratula, Queensland. *Geomorphology*, *282*, 195–208.

Guo, Q., Liu, J., Tao, S., Xue, B., Li, L., Xu, G., Li, W., Wu, F., Li, Y., Chen, L., & Pang, S. (2014). Perspectives and prospects of LiDAR in forest ecosystem monitoring and modeling (in Chinese). *China Science Bulletin*, 59, 459–478. https://doi.org/10.1360/972013-592

Lefsky, M. A. (2010). A global forest canopy height map from the moderate Resolution imaging spectroradiometer and the geoscience laser altimeter system. *Geophysical Research Letters*, *37*(15), L15401.

Luo, S., Wang, C., Xi, X., Zeng, H., Li, D., Xia, S., & Wang, P. (2016). Fusion of airborne discrete-return LiDAR and hyperspectral data for land cover classification. *Remote Sensing*, *8*(1), 3.

Morsdorf, F., Nichol, C., Malthus, T., & Woodhouse, I. H. (2009). Assessing forest structural and physiological information content of multi-spectral LiDAR waveforms by radiative transfer modelling. *Remote Sensing of Environment*, *113*(10), 2152–2163.

Niu, Z., Xu, Z., Sun, G., Huang, W., Wang, L., Feng, M., Li, W., He, W., & Gao, S. (2015). Design of a new multispectral waveform LiDAR instrument to monitor vegetation. *IEEE Geoscience and Remote Sensing Letters*, *12*(7), 1506–1510.

Wang, C., Nie, S., Xi, X., Luo, S., & Sun, X. (2017). Estimating the biomass of maize with hyperspectral and LiDAR data. *Remote Sensing*, *9*(1), 11.

Xi, Z. X., Hopkinson, C., Rood, S. B., & Peddle, D. R. (2020). See the forest and the trees: Effective machine and deep learning algorithms for wood filtering and tree species classification from terrestrial laser scanning. *ISPRS Journal of Photogrammetry and Remote Sensing*, *168*, 1–16.

Appendix: LiDAR Terminology

This appendix lists commonly used LiDAR terms and introduces their concepts.[1]

A.1 Point cloud

A large collection of discrete points. In addition to 3D laser scanning, the point cloud can be obtained through image dense matching, sonar, and other means. The raw point cloud is usually stored in the standard binary file exchange format (LAS file). Each point contains not only *X*, *Y*, and *Z* coordinates but also point classification, return intensity, return number, number of returns, scanning angle, and color (red/green/blue [RGB]) information. Before using LiDAR data, the point cloud denoising, point cloud filtering, and point cloud classification are usually required. As defined by the American Society for Photogrammetry and Remote Sensing (ASPRS), laser point clouds are usually categorized into building, high vegetation, medium vegetation, low vegetation, ground, water, and unclassified point cloud.

A.2 Point cloud density

The number of points per unit area is an important index describing LiDAR data. For airborne LiDAR systems, the point cloud density is closely related to the flight altitude, speed, scanning frequency, and other factors. Repeated scanning of a strip, reducing the flight altitude, or reducing the flight speed can increase the point cloud density but increase the cost of data acquisition at the same time. The factors affecting point cloud density include laser emission frequency, angular resolution, laser target distance, angle of incidence, target material, and target surface reflectivity.

A.3 Point cloud feature

Geometric features in a point cloud, such as points, lines, and surfaces, which can express the real object surfaces or the edges of objects, are important references for feature classifications, identifications, and reconstruction. In terms of scale, point cloud features are divided into local and global features. The local features include normal, curvature, and others. The global features are the topological features of the points and others. In terms of statistical features, the point cloud features include the point cloud density, the mean and variance of the point cloud elevation in the local range (e.g., after grid transformation), and the point intensity of different object surfaces.

A.4 Point cloud filtering

The technique of separating ground points and non-ground points from the point cloud by automatic or human-computer interaction methods. It is generally believed that point cloud filtering is a preprocessing step for point cloud classification.

High-accuracy point cloud filtering is an important process in the production of digital elevation models (DEMs) based on point clouds. The human-computer interactive point cloud filtering is based on the visual discrimination of the displayed point cloud, which distinguishes whether it is a ground point or a non-ground point and marks the category accordingly. The automated point cloud filtering separates the ground point from the non-ground point through specific algorithms, such as algorithms based on the triangulated irregular network (TIN), mathematical morphological operations, slope, scanning line, point cloud leveling, and moving surface filtering. Although the various automated filtering algorithms are implemented in different ways, they mostly follow two basic assumptions: (1) the ground point must be the lowest point in the local range, which is mostly used for the selection of the initial ground point, and (2) the ground undulation is gentle, which is used for the intermediate process of filtering. Considering the existence of surface roughness, to ensure the first assumption is met, the raw point cloud must be denoised first before filtering.

A.5 Point cloud classification

The technique of separating some points of one or more object types from the point cloud and labeling them with a specific class. According to whether the sampling training is required, the point cloud classification algorithms are divided into two categories. The first one is the supervised classification. It uses the point cloud samples whose class is known as the training data, then calculates the features of the point cloud and establishes the discriminant function to classify the test point cloud. That is, its process is to select typical representative features from the training samples, establish a classification model of the sample data, and then classify the test data based on this model. The other one is unsupervised classification, which uses the statistical differences between the different classes in the point cloud feature space as the classification basis. It does not require sampling training, but uses the statistical clustering method to discover the implicit class relationships of the point cloud.

There are two main methods to assess the accuracy of point cloud classification, i.e., qualitative and quantitative methods. Qualitative methods usually use visual discrimination and combine this with a priori knowledge to determine whether the classification results are correct or not. They do not need the true value as a reference. In contrast, quantitative methods need the true value as a reference. They use some quantitative indices to assess the classification accuracy. Commonly used quantitative indices include the Kappa coefficient, F-Score, and classification accuracy (CACC).

A.6 Point cloud segmentation

The technique of clustering similar points based on certain similarity principles. The commonly used algorithms are as follows. (1) Algorithms based on boundary extraction. They include two steps, i.e., boundary extraction and clustering of points within the boundary. Their performance mainly depends on the boundary extraction algorithm. (2) Algorithms based on scanning lines. Each line of the original-depth image is treated as a single scan line, and then the objects are segmented by analyzing and clustering the scan lines. (3) Algorithms based on region growing. First, select seed points as the initial point set. Then, based on the similarity measure, continuously merge adjacent points with similar characteristics into the current point set until the termination condition is met or there are no points to be merged. Their performance is affected by the seed point selection, merging rules, and termination condition. (4) Algorithms based on clustering. These algorithms divide the point cloud into subsets according to certain criteria. Commonly used methods include K-means clustering and spectral clustering. The results of clustering algorithms are easily affected by the distance function and the designed weighted graph. (5) Primitive extraction algorithms. They extract basic shape primitives directly from the point cloud through prior knowledges. The main methods are random sample consensus (RANSAC), Hough transform, and so on. With the development of point cloud and image technology, assisting point cloud segmentation through image information is getting more and more attention.

A.7 Point cloud fitting

A technique for computing the model parameters of the object from discrete three-dimensional (3D) point coordinates. The commonly used models to fit points include regular geometric models, such as line, surface, body models, free curve, and free surface. For curves or surfaces of arbitrary shape, spline functions are used for fitting, such as Bezier surfaces and non-uniform rational B-spline (NURBS) surfaces. If fitting regular geometric models, the constructed equations of feature parameters are solved directly.

Since the models expressed in the discrete point cloud are unknown, various, and plentiful, the fitting of regular geometric models generally includes two steps. The first step is model detection. The Hough transform is commonly used. The second step is model parameter solving. Since the discrete 3D point cloud contains the noise point produced by the sensor and the interference points from other models, the calculation of model parameters requires robust methods. The commonly used methods include least squares fitting and random sample consensus (RANSAC). In addition, there are some other point cloud–fitting algorithms, such as expectation maximization and generalized principal component analysis.

A.8 Point cloud simplification

Also called "point cloud–based surface simplification." It is a technique for efficiently streamlining a large number of redundant points so that the feature points describing a three-dimensional (3D) model remain clear. For a given point cloud P (corresponding to a surface S), given a target sampling frequency $n < |P|$, find a new set of points P' (corresponding to a surface S') such that $|P'| = n$. The purpose of point cloud simplification is to reduce the amount of data and the complexity of the model representation. In practice, it is very difficult to find a globally optimal solution for point cloud simplification. Most existing point cloud simplification algorithms use methods based on local error measures, such as the simplification methods based on clustering, iteration, and particle simulation. In addition, the way of jumping point reading is often used for the simplification of LiDAR point cloud data. This is because in the collection process, the LiDAR sensors store the data in the file sequential memory, generally in chronological order. The points acquired in the adjacent time have a strong correlation in the spatial position. This method is simple and fast, but tends to lose the object features. Therefore, it is generally used for visualization of the point cloud.

A.9 Point cloud refinement

The processing techniques to improve the point cloud quality or visual effect, which is commonly used to generate three-dimensional (3D) models from the point cloud or to visualize and render point cloud data. Raw point cloud data are often affected by sensor performance, optical properties of observed targets, background environment, and occlusion of features caused the observation viewpoint, and are prone to some data quality or visualization problems, such as noise, undersampling, rough surfaces, voids, cracks, and jumping boundaries. These problems can be effectively mitigated by using specific data processing methods or auxiliary data. For example, assuming the LiDAR point cloud is the basic data source, the problems of insufficient description of edges by discrete sampling of the laser scanner and the absence of a point cloud in the weak reflection area can be handled by combining with the target 3D geometric information obtained from high-resolution images using the multi-view density matching technique. This would make the geometry and color of the measured target in the point cloud be significantly restored or enhanced. Depending on whether processed in real time or not, point cloud refinement is divided into online processing and offline processing. Online processing is mainly oriented toward the improvement of a real-time visual effect of a point cloud or 3D models based on a point cloud. Offline processing is mainly oriented toward the improvement of the point cloud quality and further improving the visual effect of point cloud or 3D models based on the point cloud.

A.10 Point cloud denoising

Techniques for reducing and removing surface noises and outliers produced in the measurement process from the point cloud. In the process of point cloud acquisition, due to the influence of the measurement environment, personnel, equipment defects, and other factors, the data acquired by measurement might contain some outliers (noise points). The existence of noise points has an obvious negative impact on the point cloud filtering, point cloud classification, surface reconstruction, etc. It would exacerbate the complexity of the algorithm and affect the efficiency of data processing. Hence, it is necessary to use suitable methods to identify and eliminate them. The commonly used algorithms include algorithms based on the local neighborhood fitting, histogram statistics, and spatial mesh division.

A.11 Point cloud registration

The mathematical computation process of converting three-dimensional (3D) spatial data point sets with two or more coordinate systems into a unified coordinate system. The key is the acquisition of common features and the robust calculation of coordinate conversion parameters. The common features are mainly categorized into point, line, and surface features, as well as multi-feature mixing that has appeared in recent years. The identification methods include manual identification, laying artificial targets, optical/depth/reflection image alignment, 3D point feature alignment, etc. The seven-parameter method is commonly used for calculating coordinate transformation parameters based on point features. The methods for calculating coordinate transformation parameters based on linear and regular surface features are generally accomplished by constructing a system of feature parameter equations containing coordinate transformation parameters. Algorithms such as least squares parity and random sample consensus (RANSAC) are used to eliminate mismatched features. Global alignment is used in multi-site ground-based 3D laser scanning to weaken the two-site cumulative alignment error. The main indicators to evaluate the accuracy of point cloud alignment are the deviation of rotation and translation, the distance between common points, and the distance from the point to the corresponding surface. The iterative closest neighbor (ICP) algorithm is often used for fine alignment after the coarse alignment based on common feature identification. Some strategies such as station optimization and introduction of external control conditions in the field data acquisition stage can also improve the alignment accuracy.

A.12 Pulse measurement

A method of determining the distance between a sensor and a target by measuring the time delay between transmitted and received laser pulses. It is a type of time-of-flight measurement. A laser transmits short laser pulses to a target in space and receives the signals backscattered from the object. The distance between the target in space and the laser is calculated by measuring the time delay between the transmitted and received pulses.

A.13 Laser scanning

A technique for active and rapid measurement of spatial coordinates. The laser pulses are emitted and then reflected by the object surfaces. The distance between each reflection point and the laser is calculated and further used to calculate the spatial coordinates of each reflection point. Laser scanning can be accomplished on ground-based, shipboard, vehicle-mounted, airborne, and satellite-mounted platforms. Laser scanning originated in the early 1960s and received widespread attention due to the Apollo 15 program. It is now widely used in mapping, geography, geology, forestry, atmosphere, unmanned vehicles, and other fields.

A.14 Laser beam return / laser echo

The interaction between a laser emitted pulse and target would produce a return signal. Sometimes, part of the laser pulse might continue to move forward with other targets in the light path to produce more return signals. These returned signals are called laser echoes. A LiDAR receiving system might record one or more laser echoes: when only one laser echo is recorded, it is called single return; when multiple echoes are recorded, it is called multiple returns. The multiple returns are usually categorized into the first return, the intermediate returns, and the last return. For a single return, if it hits the ground directly, it represents the return from the ground; if it hits the object surface, it represents the return from the object surface. Hence the single return is reflected from either the ground or the feature. In contrast, for multiple returns, the first return is usually a non-ground echo, which is closely related to the object surface elevation. The first return is commonly used to generate a digital surface model (DSM). The intermediate returns are usually non-ground echoes, which correspond to the structural information inside the object, such as the vegetation vertical structure. The last return is usually a ground echo, but it might also be an echo returned from the vegetation, buildings, and other targets. The last return is used to generate a digital elevation model (DEM) after data filtering. The LiDAR receiving system keeps a detailed record of the number of returns and the sequence of returns for each laser pulse.

A.15 Return intensity

Return intensity is the returned energy of a laser pulse that is reflected by the target surface and then received by the LiDAR receiver, which is also known as the backscattered intensity or energy. The return intensity is a quantitative description of the backscattering from the target. It is related to the energy emitted by the laser, the target reflectivity, the atmospheric conditions, and the distance between the LiDAR and the target. It can be quantitatively described by equation:

$$P_r = \rho \frac{\eta^2 A_r}{\pi R^2} \mathrm{P}_t$$

where P_r is the received energy, P_t is the transmitted energy, ρ is the target reflectivity, η is the atmospheric transmittance, R is the distance between the LiDAR and the target, and A_r is the area of received aperture.

A.16 Discrete return LiDAR

LiDAR systems that record a finite number of returns usually include the first return, the intermediate returns, and the last return. Discrete return LiDAR is usually small-footprint LiDAR, and the footprint diameter is usually less than 1 m. A discrete return LiDAR system can accurately obtain the three-dimensional coordinate information of the target. Additionally, it contains the information of return intensity, number of returns, scanning angle, etc. Discrete return LiDAR systems are widely used in topographic mapping, digital city planning, cultural heritage digitization, electric power line inspection, and so on.

A.17 Full-waveform LiDAR

LiDAR systems that continuously digitize the laser return signals at very small sampling intervals according to the time sequence. They record in detail the vertical structure information of all detected targets within the laser footprint. According to the size of the ground footprint, the full-waveform LiDAR systems are divided into large-footprint and small-footprint full-waveform LiDAR. According to different platforms, they are divided into spaceborne, airborne, and terrestrial full-waveform LIDAR.

A.18 Waveform decomposition

The process of decomposing the waveform signal of each target within the footprint from the LiDAR full waveform through some specific algorithms to extract the spatial information of targets. The waveform decomposition obtains the sub-waveform (e.g., Gaussian wave) of each target within the footprint, which can be considered the result of the interaction of the emitted pulse with a certain target. The sub-waveform that is farthest from the laser transmitter is generally considered the result of the interaction of the pulse with the ground. Through waveform decomposition, the waveform signals corresponding to the objects are determined, and the object structural information is extracted for subsequent applications.

A.19 Waveform feature extraction

The LiDAR system records the energy distribution of the transmitted pulse (emitted waveform) and the energy distribution of the returned pulse (received waveform). Based on the energy distribution of the waveform sampling points, various metric parameters that indicate the energy distribution features are calculated, i.e., waveform features. These parameters can be used to extract the corresponding object features. The theoretical basis of the waveform feature extraction is waveform decomposition. Through waveform decomposition, various waveform features are generated to determine the waveform signals corresponding to the targets. The commonly used waveform feature parameters include waveform height indices and waveform energy indices.

A.20 Height of median energy (HOME)

A waveform feature parameter used to describe full-waveform LiDAR data, referring to the distance from the height position corresponding to one half of the waveform energy to the location of the ground waveform peak. The parameter depends on the height of one half of the waveform energy and the location of ground waveform peak. The location of the ground waveform peak is determined by waveform decomposition. Specifically, waveform decomposition detects multiple sub-waveforms from the LiDAR returns, and one of the sub-waveforms is determined as the ground waveform (usually the last sub-waveform). The height of one half of the waveform energy is closely related to the vegetation vertical distribution and the canopy closure. This parameter is not sensitive to the terrain slope, which has important applications in the inversions of vegetation parameters.

A.21 LiDAR height percentile

A common LiDAR feature parameter for estimating vegetation structural parameters, which are closely related to vegetation height and biomass. The p-th percentile height of LiDAR h means that p percent of the laser return heights are lower than h and (100–p) percent of the laser return heights are higher than h. The median of LiDAR heights is just the 50th percentile height (HOME). To calculate the LiDAR height percentile, we first categorize the laser point cloud and then calculate the relative height of the non-ground points. The LiDAR height percentile is calculated based on the relative height of the non-ground points.

A.22 Laser interception index (LII)

The ratio of the number of vegetation canopy returns to the number of total returns, which is often used to indicate the vegetation coverage. The LII can also be defined as the ratio of the vegetation canopy intensity to the total return intensity. The LII is widely used in the inversion of vegetation coverage, Leaf Area Index (LAI), and forest biomass to improve the accuracy of the vegetation parameter inversions. The LII is calculated as in the following equation:

$$\mathrm{LII} = \frac{N_{\mathrm{canopy}}}{N_{\mathrm{canopy}} + N_{\mathrm{ground}}}$$

where N_{canopy} is the number of canopy returns or the canopy waveform intensity and N_{ground} is the number of ground returns or the ground waveform intensity.

NOTE

1 Translated from the LiDAR remote sensing entries in the book *Encyclopaedia of China* (third edition).

Index

Q

R

S

T

U

V

W